A Themed Issue in Honor of Professor Reinhard Schlickeiser on the Occasion of His 70th Birthday

A Themed Issue in Honor of Professor Reinhard Schlickeiser on the Occasion of His 70th Birthday

Editors

Martin Pohl
Peter H. Yoon
Horst Fichtner

Basel • Beijing • Wuhan • Barcelona • Belgrade • Novi Sad • Cluj • Manchester

Editors

Martin Pohl
Institut für Physik
und Astronomie
Universität Potsdam
Potsdam, Germany

Peter H. Yoon
Institute of Physical Science
and Technology
University of Maryland
College Park, MD, USA

Horst Fichtner
Institute of
Theoretical Physics
Ruhr University Bochum
Bochum, Germany

Editorial Office
MDPI
St. Alban-Anlage 66
4052 Basel, Switzerland

This is a reprint of articles from the Special Issue published online in the open access journal *Physics* (ISSN 2624-8174) (available at: https://www.mdpi.com/journal/physics/special_issues/ ReinhardSchlickeiser70).

For citation purposes, cite each article independently as indicated on the article page online and as indicated below:

Lastname, A.A.; Lastname, B.B. Article Title. *Journal Name* **Year**, *Volume Number*, Page Range.

ISBN 978-3-0365-9140-7 (Hbk)
ISBN 978-3-0365-9141-4 (PDF)
doi.org/10.3390/books978-3-0365-9141-4

Contents

 *physics*

Article

Landau Damping of Langmuir Waves: An Alternative Derivation

Andreas Shalchi

Department of Physics and Astronomy, University of Manitoba, Winnipeg, MB R3T 2N2, Canada; Andreas.Shalchi@umanitoba.ca

Abstract: In this paper, a discussion of the Landau damping of Langmuir waves is presented together with a simple derivation which does not require the application of methods of complex analysis. A general dispersion relation is derived systematically which corresponds to a nonlinear equation. The latter equation is solved numerically but asymptotic limits are also discussed.

Keywords: Langmuir waves; Landau damping; particle wave interactions

Citation: Shalchi, A. Landau Damping of Langmuir Waves: An Alternative Derivation. *Physics* **2021**, *3*, 940–954. https://doi.org/10.3390/physics3040059

Received: 31 August 2021
Accepted: 8 October 2021
Published: 29 October 2021

Publisher's Note: MDPI stays neutral with regard to jurisdictional claims in published maps and institutional affiliations.

1. Introduction

In the current paper, the damping of plasma waves, also known as Langmuir waves [1], is explored. Plasma waves are related to electron oscillations and, thus, they are sometimes called plasma oscillations. The standard description of those waves is based on the momentum equation (Euler equation) for the electrons, their continuity equation, and Gauss' law of electrodynamics. Furthermore, it is assumed that the electrons behave like an ideal gas. In order to find the dispersion relation of Langmuir waves, the aforementioned fluid equations are linearized leading to the Bohm–Gross dispersion relation (see [2–4]),

$$\omega^2 = \omega_p^2 + 3v_e^2 k^2. \tag{1}$$

Therein, the wave number k, the plasma frequency

$$\omega_p = \sqrt{\frac{n_0 e^2}{m_e \epsilon_0}}, \tag{2}$$

and the electron thermal speed

$$v_e = \sqrt{\frac{k_B T}{m_e}} \tag{3}$$

are used in terms of the electron particle density (particles per volume), n_0, the elementary charge, e, the electron mass, m_e, the permittivity of free space, ϵ_0, Boltzmann's constant, k_B, and the absolute temperature, T (in Kelvin). Note, the electron thermal speed comes straight from the equipartition theorem which states that each degree of freedom has the average kinetic energy $k_B T/2$. Therefore,

$$\frac{1}{2} m_e v_e^2 = \frac{1}{2} k_B T, \tag{4}$$

is a consequence of having only one degree of freedom due to the one-dimensional motion of the electrons. This relation can be straight rearranged to obtain Equation (3).

A useful parameter in the description of plasma waves is the Debye length corresponding to the distance over which significant charge separation occurs; it is given by

$$\lambda_D = \sqrt{\frac{k_B T \epsilon_0}{n_0 e^2}}. \tag{5}$$

Then, the electron thermal speed (3) reads:

$$v_e = \omega_p \lambda_D, \tag{6}$$

and the plasma wave dispersion relation (1) becomes:

$$\omega^2 = \omega_p^2 \left(1 + 3\lambda_D^2 k^2\right). \tag{7}$$

By considering linearized fluid and Maxwell equations, one can show that Langmuir waves are longitudinal in nature.

In this paper, Landau damping, describing the damping of plasma waves due to the interaction between the wave and the electrons in the plasma, is revisited. This is a special case of wave–particle interactions in a plasma. According to [5], the physical mechanism of Landau damping can be understood as follows: particles having velocities slightly less than the phase velocity of the wave are accelerated by the wave's electric field to move with the wave phase velocity. Therefore, the group of particles moving slightly slower than the phase velocity gain energy from the wave. In a collisionless plasma characterized by a Maxwellian distribution (see Section 3), the number of slower particles is greater than the number of faster particles. Therefore, energy gained from the waves by slower particles is more than the energy given to the waves by faster particles, thus leading to net damping of the waves. An example plot is given in Figure 1.

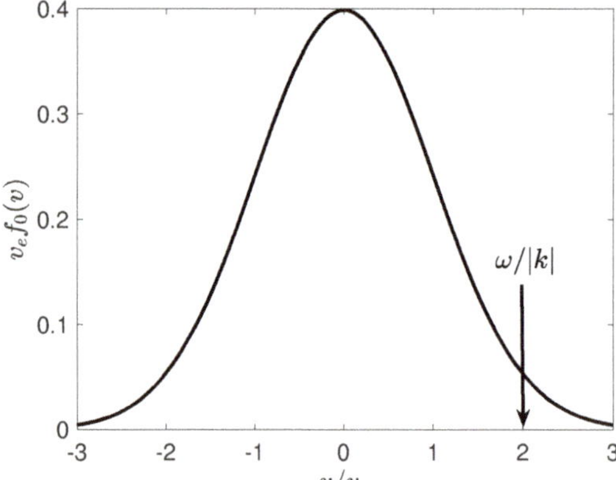

Figure 1. Maxwellian distribution. The arrow points to the location of the phase speed $\omega/|k|$ of the Langmuir wave. In the considered case, the wave speed is larger than the electron thermal speed. Particles to the left of the arrow are slower, and particles to the right of the arrow are faster than the wave. Here we have used the wave number k, the frequency ω, the electron speed v, and the electron thermal speed v_e.

The first theoretical description of Landau damping was presented in [6] and was based on the Vlasov equation [7]. Although Landau damping was originally derived for plasma waves, it can also be considered in the context of magnetohydrodynamic waves [8].

The derivation of traditional Landau damping, as provided in [6], is based on Laplace transforms and complex contour integrations. In [9], it was shown that Landau damping can also be described by using a Fourier transform and normal mode expansion. The latter modes are usually called van Kampen modes. In [10], it was demonstrated that Landau's and van Kampen's treatments of plasma oscillations are equivalent. Alternative and simpler derivations of Landau damping can be found in textbooks, see, e.g., [11]. However, all the aforementioned descriptions are based on contour integrals and other tools of complex analysis.

In order to avoid using tools of complex analysis, an alternative derivation of Landau damping is presented in the current article. This approach can be particularly useful for teaching this topic to undergraduate and graduate students in introductory plasma physics and magnetohydrodynamics courses.

The reminder of this paper is organized as follows. Section 2 lists the basic equations used to describe Landau damping. This includes the Vlasov equation, Gauss' law, and linearizing those equations. Section 3 contains the derivation of a nonlinear equation for the dispersion relation, $\omega = \omega(k)$. Numerical and analytical solutions of this nonlinear relation are discussed in Sections 4 and 5, respectively. Finally, Section 6 discusses and summarizes the findings.

2. Basic Equations

In order to describe the statistical behavior of a physical system not in a state of equilibrium the Boltzmann equation is used. Let us assume that the plasma electrons are described via the distribution function $f(\vec{x}, \vec{v}, t)$ which depends on position, $\vec{x}$, and velocity, $\vec{v}$, of the particles, as well as the time, t. Furthermore, the considered particles experience an external force, $\vec{F}$, and collisions. With the help of Liouville's theorem [12], one finds:

$$\frac{d}{dt} f(\vec{x}, \vec{v}, t) = S(\vec{x}, \vec{v}, t), \tag{8}$$

where $S(\vec{x}, \vec{v}, t)$ describes the collisions. For the total time-derivative of the distribution function,

$$\frac{d}{dt} f(\vec{x}, \vec{v}, t) = \frac{\partial f}{\partial t} + \sum_n \dot{x}_n \frac{\partial f}{\partial x_n} + \sum_n \dot{v}_n \frac{\partial f}{\partial v_n}, \tag{9}$$

can be used. Here, the dot defines time derivative.

Equations (8) and (9) can be combined to find the Boltzmann equation. The collisionless Boltzmann equation, on the other hand, is given by

$$\frac{\partial f}{\partial t} + \sum_n \dot{x}_n \frac{\partial f}{\partial x_n} + \sum_n \dot{v}_n \frac{\partial f}{\partial v_n} = 0. \tag{10}$$

In what follows, only this collisionless case is considered. Furthermore, $\dot{v}_n$ is replaced by using Newton's second law, $F_n = m\dot{v}_n$, to find:

$$\frac{\partial f}{\partial t} + \sum_n v_n \frac{\partial f}{\partial x_n} + \frac{1}{m} \sum_n F_n \frac{\partial f}{\partial v_n} = 0, \tag{11}$$

with m being the particle mass.

For the case considered in this paper, where F_n describes Coulomb interactions, Equation (11) is usually called the Vlasov equation [7]. The Vlasov equation, discussed here, is commonly used to investigate the interaction between particles and fields. In the case of Landau damping, this equation describes the interaction between Langmuir waves and the electrons of the plasma. However, the Vlasov equation is also used as a starting point to describe the interaction between magnetic turbulence and energetic particles such as cosmic rays (see, e.g., [13–15]).

For the investigations, presented here, it is sufficient to consider the one-dimensional case for which the Vlasov equation reads:

$$\frac{\partial f}{\partial t} + v \frac{\partial f}{\partial x} + \frac{1}{m} F \frac{\partial f}{\partial v} = 0. \tag{12}$$

In the case considered, the force is just the electric force and, thus,

$$F = qE = q\delta E, \tag{13}$$

can be used, where q is the electric charge of the particle. The electric field, δE, therein is the wave field of the plasma wave. Using this in Equation (12) yields:

$$\frac{\partial f}{\partial t} + v\frac{\partial f}{\partial x} + \frac{q}{m}\delta E\frac{\partial f}{\partial v} = 0. \tag{14}$$

In order to solve the Vlasov equation, a perturbation approach of the form,

$$f(x,v,t) = f_0(v) + \delta f(x,v,t), \tag{15}$$

is employed.

Here, f_0 is the unperturbed distribution often assumed to be Maxwellian (see Section 3 below). The quantity δf describes the fluctuations corresponding to the deviations from the unperturbed distributions. These fluctuations are a consequence of the interactions with the electric field of the wave and are assumed to be small so that the perturbation approach can be justified. Using Equation (15) in Equation (14) yields in first-order:

$$\frac{\partial \delta f}{\partial t} + v\frac{\partial \delta f}{\partial x} + \frac{q}{m}\delta E\frac{\partial f_0}{\partial v} = 0. \tag{16}$$

Equation (16) is called the linearized Vlasov equation. As a second equation, Gauss' law,

$$\vec{\nabla} \cdot \vec{E} = \frac{1}{\epsilon_0}\rho_e, \tag{17}$$

is employed.

Note that SI units are used here rather than Gaussian or cgs units. In the one-dimensional case considered, Equation (17) becomes:

$$\frac{\partial}{\partial x}\delta E = \frac{1}{\epsilon_0}\delta\rho_e = \frac{q}{\epsilon_0}\int_{-\infty}^{+\infty} dv\, \delta f(x,v,t), \tag{18}$$

where the electric charge density $\delta\rho_e$ is replaced by the velocity integral over the fluctuations δf. In order to solve Vlasov and Gauss equations, a Fourier ansatz,

$$\begin{aligned}
\delta f(x,v,t) &\propto \delta f(k,v,\omega)e^{i(kx-\omega t)}, \\
\delta E(x,t) &\propto \delta E(k,\omega)e^{i(kx-\omega t)},
\end{aligned} \tag{19}$$

is used.

Using Equation (19) in Equations (16) and (18) yields:

$$-i\omega\delta f + ivk\delta f + \frac{q}{m}\delta E\frac{\partial f_0}{\partial v} = 0, \tag{20}$$

and

$$ik\delta E = \frac{q}{\epsilon_0}\int_{-\infty}^{+\infty} dv\, \delta f(k,v,\omega). \tag{21}$$

The first equation can be rearranged to read:

$$\delta f = -i\frac{q}{m}\frac{\delta E}{\omega - vk}\frac{\partial f_0}{\partial v}. \tag{22}$$

Using this in Equation (21) yields:

$$ik\delta E = -i\frac{q^2}{\epsilon_0 m}\int_{-\infty}^{+\infty} dv\,\frac{\delta E}{\omega - vk}\frac{\partial f_0}{\partial v}. \tag{23}$$

As the electric field does not depend on the particle velocity v, one can cancel δE in the latter equation. Therefore, the following dispersion relation,

$$1 = -\frac{q^2}{\epsilon_0 mk}\int_{-\infty}^{+\infty} dv\,\frac{1}{\omega - vk}\frac{\partial f_0}{\partial v}, \tag{24}$$

is obtained.

The function f_0 therein corresponds to the unperturbed particle density per volume. In the following, the replacement $f_0 \to f_0 n_0$ is made and the plasma frequency, defined via Equation (2), is used. This leads Equation (24) to be rewritten as:

$$1 + \frac{\omega_p^2}{k^2}\int_{-\infty}^{+\infty} dv\,\frac{1}{\omega/k - v}\frac{\partial f_0}{\partial v} = 0. \tag{25}$$

Note, f_0 is now the number of particles per velocity meaning that due to normalization, one needs:

$$\int_{-\infty}^{+\infty} dv\, f_0(v) = 1, \tag{26}$$

which fixes constants in the distribution function f_0.

3. The Dispersion Relation

Let us now discuss the obtained dispersion relation as given by Equation (25). In order to evaluate this further, the Maxwellian distribution,

$$f_0(v) = \frac{1}{\sqrt{2\pi}}\frac{1}{v_e}e^{-v^2/(2v_e^2)}, \tag{27}$$

is employed.

The latter function is visualized via Figure 1. Note, this is a Maxwellian distribution in one dimension and, therefore, it is in coincidence with a Gaussian distribution. The used form is correctly normalized meaning that it satisfies Equation (26). Furthermore, the electron thermal speed, given by Equation (3), is used. From Equation (27), it follows that:

$$\frac{\partial f_0}{\partial v} = -\frac{1}{\sqrt{2\pi}}\frac{v}{v_e^3}e^{-v^2/(2v_e^2)}. \tag{28}$$

Using this, Equation (25) reads:

$$1 = \frac{\omega_p^2}{\sqrt{2\pi}kv_e^3}\int_{-\infty}^{+\infty} dv\,\frac{v}{\omega - vk}e^{-v^2/(2v_e^2)}. \tag{29}$$

Note that it looks like there is a singularity at $\omega = vk$. However, this is not so as long as ω is a complex number with non-vanishing imaginary part.

The aim here is to derive the dispersion relation $\omega = \omega(k)$. To this end, let us rewrite the integral occurring in Equation (29) as:

$$I(\omega, k) = \int_{-\infty}^{+\infty} dv\,\frac{v}{\omega - vk}e^{-v^2/(2v_e^2)}. \tag{30}$$

To continue,

$$\int_{-\infty}^{+\infty} dv\, e^{-v^2/(2v_e^2)} = \int_{-\infty}^{+\infty} dv\, \frac{\omega - vk}{\omega - vk} e^{-v^2/(2v_e^2)}$$

$$= \omega \int_{-\infty}^{+\infty} dv\, \frac{1}{\omega - vk} e^{-v^2/(2v_e^2)} \tag{31}$$

$$- k \int_{-\infty}^{+\infty} dv\, \frac{v}{\omega - vk} e^{-v^2/(2v_e^2)},$$

is considered.

The integral on the left-hand-side of Equation (31) is a usual Gaussian integral which can be solved via

$$\int_{-\infty}^{+\infty} dx\, e^{-cx^2} = \sqrt{\frac{\pi}{c}} \qquad \text{if} \qquad \mathrm{Re}(c) > 0. \tag{32}$$

The second integral on the right-hand-side of Equation (31) is the desired integral $I(\omega, k)$ as defined via Equation (30).

Thus, Equation (31) can be rewritten as:

$$\sqrt{2\pi}v_e = \omega \int_{-\infty}^{+\infty} dv\, \frac{1}{\omega - vk} e^{-v^2/(2v_e^2)} - kI(\omega, k). \tag{33}$$

Alternatively, this can be written as;

$$I(\omega, k) = \frac{\omega}{k} \int_{-\infty}^{+\infty} dv\, \frac{1}{\omega - vk} e^{-v^2/(2v_e^2)} - \frac{\sqrt{2\pi}v_e}{k}. \tag{34}$$

Therewith the dispersion relation (29) becomes:

$$1 = \frac{\omega_p^2}{\sqrt{2\pi}kv_e^3} \left[\frac{\omega}{k} \int_{-\infty}^{+\infty} dv\, \frac{1}{\omega - vk} e^{-v^2/(2v_e^2)} - \frac{\sqrt{2\pi}v_e}{k} \right].$$

In the remaining integral, the substitution,

$$x = \frac{v}{\sqrt{2}v_e} \tag{35}$$

is employed to derive

$$1 = \frac{\omega_p^2}{\sqrt{2\pi}kv_e^3} \left[\frac{\omega}{k} \sqrt{2}v_e \int_{-\infty}^{+\infty} dx\, \frac{e^{-x^2}}{\omega - \sqrt{2}v_e kx} - \frac{\sqrt{2\pi}v_e}{k} \right]. \tag{36}$$

To continue, the parameter,

$$\alpha := \frac{\omega}{\sqrt{2}v_e k}, \tag{37}$$

is defined.

Note that the latter parameter is a complex number in the general case. Therewith, one can write Equation (36) as:

$$1 = \frac{\omega_p^3}{\sqrt{2\pi}k^3 v_e^3} \left[\frac{\omega}{\omega_p} \int_{-\infty}^{+\infty} dx\, \frac{1}{\alpha - x} e^{-x^2} - \frac{\sqrt{2\pi}v_e k}{\omega_p} \right]. \tag{38}$$

In the result obtained, Equation (6) can be now used to finally arrive at

$$1 + \frac{1}{\sqrt{2\pi}(\lambda_D k)^3} \left[\sqrt{2\pi}(\lambda_D k) - \frac{\omega}{\omega_p} J(\alpha) \right] = 0, \tag{39}$$

where the integral,

$$J(\alpha) = \int_{-\infty}^{+\infty} dx\, \frac{1}{\alpha - x} e^{-x^2}, \tag{40}$$

is used.

Let us solve this integral. First, the integral is generalized via

$$J(\alpha, \beta) = e^{-\alpha^2} \int_{-\infty}^{+\infty} dx \, \frac{1}{\alpha - x} e^{\beta(\alpha^2 - x^2)}. \tag{41}$$

Note, we are looking for $J(\alpha, \beta = 1) \equiv J(\alpha)$. One can derive that

$$
\begin{aligned}
\frac{d}{d\beta} J(\alpha, \beta) &= e^{-\alpha^2} \int_{-\infty}^{+\infty} dx \, \frac{\alpha^2 - x^2}{\alpha - x} e^{\beta(\alpha^2 - x^2)} \\
&= e^{-\alpha^2} \int_{-\infty}^{+\infty} dx \, (\alpha + x) e^{\beta(\alpha^2 - x^2)} \\
&= \alpha e^{-\alpha^2} \int_{-\infty}^{+\infty} dx \, e^{\beta(\alpha^2 - x^2)}.
\end{aligned}
\tag{42}
$$

Here, the symmetry of the integral is used and, at the end, one would only need to solve a Gaussian integral. With the help of Equation (32), one finally derives:

$$\frac{d}{d\beta} J(\alpha, \beta) = \alpha \sqrt{\pi} e^{-\alpha^2} \beta^{-1/2} e^{\beta \alpha^2}. \tag{43}$$

Note, here and above, β is assumed to be a positive real number. To continue, the result is integrated over β:

$$J(\alpha, \beta = 1) = J(\alpha, \beta = 0) + \alpha \sqrt{\pi} e^{-\alpha^2} \int_0^1 d\beta \, \beta^{-1/2} e^{\beta \alpha^2}. \tag{44}$$

In the integral therein the substitution $\beta = \xi^2$ is used:

$$\frac{d\beta}{\sqrt{\beta}} = 2d\xi. \tag{45}$$

Therewith:

$$J(\alpha, \beta = 1) = J(\alpha, \beta = 0) + 2\alpha \sqrt{\pi} e^{-\alpha^2} \int_0^1 d\xi \, e^{\alpha^2 \xi^2}. \tag{46}$$

In Appendix A, it is demonstrated that

$$J(\alpha, \beta = 0) = -i\pi e^{-\alpha^2}. \tag{47}$$

Furthermore, $J(\alpha, \beta = 1) = J(\alpha)$ and, thus, Equation (46) reads:

$$J(\alpha) = 2\alpha \sqrt{\pi} e^{-\alpha^2} \int_0^1 d\xi \, e^{\alpha^2 \xi^2} - i\pi e^{-\alpha^2}. \tag{48}$$

Then, using the substitution $t = \alpha \xi$:

$$J(\alpha) = 2\sqrt{\pi} e^{-\alpha^2} \int_0^\alpha dt \, e^{t^2} - i\pi e^{-\alpha^2}. \tag{49}$$

The remaining integral corresponds to an imaginary error function (see, e.g., [16]):

$$\mathrm{Erfi}(\alpha) = \frac{2}{\sqrt{\pi}} \int_0^\alpha dt \, e^{t^2}. \tag{50}$$

Using this in Equation (49) finally yields:

$$J(\alpha) = \pi e^{-\alpha^2} \big[\mathrm{Erfi}(\alpha) - i \big]. \tag{51}$$

After using Equation (51) in Equation (39), one finds for the dispersion relation:

$$1 + \frac{1}{\sqrt{2\pi}\left(\lambda_D k\right)^3}\left\{\sqrt{2\pi}(\lambda_D k) - \pi\frac{\omega}{\omega_p}\left[\mathrm{Erfi}(\alpha) - i\right]e^{-\alpha^2}\right\} = 0, \tag{52}$$

where the parameter α is given by Equation (37). Alternatively, the result obtained can be written as:

$$\begin{aligned}1 + \frac{1}{\left(\lambda_D k\right)^2} &- \sqrt{\frac{\pi}{2}}\frac{\omega}{\omega_p\left(\lambda_D k\right)^3}\mathrm{Erfi}\left(\alpha\right)e^{-\alpha^2} \\ &+ i\sqrt{\frac{\pi}{2}}\frac{\omega}{\omega_p\left(\lambda_D k\right)^3}e^{-\alpha^2} = 0.\end{aligned} \tag{53}$$

Furthermore, the parameter α can be rewritten as:

$$\alpha = \frac{1}{\sqrt{2}}\frac{\omega}{\omega_p}\frac{1}{\lambda_D k}. \tag{54}$$

Equation (53) is a nonlinear equation for the dispersion relation $\omega = \omega(k)$. It depends only on two parameters: ω_p and λ_D. Below, numerical and analytical solutions of Equation (53) are considered.

4. Numerical Solution for the General Case

Instead of dealing with Equation (53), let us return and use Equation (39) as starting point for numerical investigations. Equation (48) can be written as:

$$J(\alpha) = 2\alpha\sqrt{\pi}e^{-\alpha^2}K(\alpha) - i\pi e^{-\alpha^2}, \tag{55}$$

where

$$K(\alpha) = \int_0^1 d\xi\, e^{\alpha^2 \xi^2} \tag{56}$$

is used.

Substituting these equations into Equation (39) gives:

$$\begin{aligned}1 \;+\; & \frac{1}{\sqrt{2\pi}\left(\lambda_D k\right)^3}\Big[\sqrt{2\pi}(\lambda_D k) \\ - \; & \frac{\omega}{\omega_p}2\alpha\sqrt{\pi}e^{-\alpha^2}K(\alpha) + i\pi\frac{\omega}{\omega_p}e^{-\alpha^2}\Big] = 0.\end{aligned} \tag{57}$$

For numerical investigations, it is useful to define the dimensionless quantities,

$$x := \lambda_D k \quad\text{and}\quad y := \omega/\omega_p, \tag{58}$$

so that the parameter α, defined via Equation (54), becomes:

$$\alpha = \frac{y}{\sqrt{2}x}. \tag{59}$$

Therewith, Equation (57) can be written as:

$$1 + \frac{1}{x^2} - \frac{y^2}{x^4}e^{-y^2/(2x^2)}K\left(\frac{y}{\sqrt{2}x}\right) + i\sqrt{\frac{\pi}{2}}\frac{y}{x^3}e^{-y^2/(2x^2)} = 0. \tag{60}$$

The solution of the latter equation is the complex quantity y as a function of the real variable x. It is straightforward to solve Equation (60) by using a standard Newton-solver in combination with MATLAB; the details are given in Appendix B. The numerical solutions are shown in Figure 2, where the analytical solution obtained for asymptotic limits is also given. This analytical solution is derived in Section 5 below.

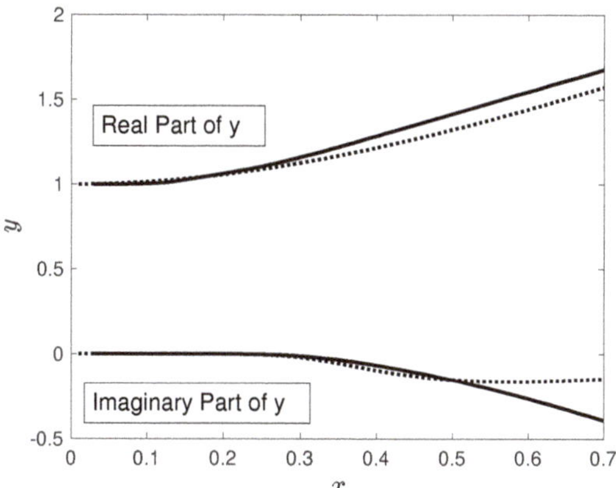

Figure 2. Visualized is the numerical solution of Equation (60). Shown are the real part of $y = \omega/\omega_p$ (top two lines) as well as its imaginary part (bottom two lines) versus the variable $x = \lambda_D k$. Compared are the numerical solution (solid lines) with the analytical limits (dotted lines) given by Equations (82) and (91), respectively. ω_p is the plasma frequency and λ_D is the Debye length.

5. Analytical Solutions in Asymptotic Limits

In Section 3, the nonlinear Equation (53) was derived and solved numerically. In order to obtain pure analytical solutions, we consider the case of

$$|\alpha| \equiv \frac{|\omega|}{\sqrt{2}v_e k} \gg 1, \tag{61}$$

meaning that the phase speed of the wave, ω/k, is assumed to be much larger than the electron thermal speed, v_e. Due to this assumption, one can employ the following asymptotic expansion (see, e.g., [16]),

$$\mathrm{Erfi}(\alpha) \approx \frac{1}{\sqrt{\pi}\alpha} e^{\alpha^2}\left[1 + \left(2\alpha^2\right)^{-1} + 3\left(2\alpha^2\right)^{-2}\right]. \tag{62}$$

Using the expansion (62) in Equation (53) yields:

$$1 + \frac{1}{\left(\lambda_D k\right)^2} - \sqrt{\frac{\pi}{2}}\frac{\omega}{\omega_p\left(\lambda_D k\right)^3}\frac{1}{\sqrt{\pi}\alpha}\left[1 + \frac{1}{2\alpha^2} + \frac{3}{4\alpha^4}\right]$$
$$+ i\sqrt{\frac{\pi}{2}}\frac{\omega}{\omega_p\left(\lambda_D k\right)^3}e^{-\alpha^2} = 0. \tag{63}$$

With the help of Equation (54) this equation can be rewritten as:

$$1 - \frac{1}{\left(\lambda_D k\right)^2}\frac{1}{2\alpha^2}\left[1 + \frac{3}{2\alpha^2}\right] + i\sqrt{\pi}\frac{\alpha}{\left(\lambda_D k\right)^2}e^{-\alpha^2} = 0. \tag{64}$$

Using definition (58) of x in Equation (64) gives:

$$x^2 - \frac{1}{2\alpha^2} - \frac{3}{4\alpha^4} + i\sqrt{\pi}\alpha e^{-\alpha^2} = 0. \tag{65}$$

To further evaluate Equation (65), let us write

$$\omega = \omega_R + i\delta\omega, \tag{66}$$

and, therefore,

$$\alpha = \alpha_R + i\delta\alpha. \tag{67}$$

In the following, it is assumed that $\delta\omega$ and $\delta\alpha$ are small. As $\delta\omega$ corresponds to the imaginary part of ω, it describes the damping of the Langmuir wave. Thus, the assumption of small $\delta\omega$ and $\delta\alpha$ corresponds to weak damping.

To continue, let us Taylor-expand and take into account only terms linear in $\delta\alpha$:

$$\begin{aligned}
\frac{1}{\alpha^2} &= \frac{1}{\left(\alpha_R + i\delta\alpha\right)^2} \approx \frac{1}{\alpha_R^2} - \frac{2i}{\alpha_R^3}\delta\alpha, \\
\frac{1}{\alpha^4} &= \frac{1}{\left(\alpha_R + i\delta\alpha\right)^4} \approx \frac{1}{\alpha_R^4} - \frac{4i}{\alpha_R^5}\delta\alpha, \\
e^{-\alpha^2} &= e^{-\left(\alpha_R + i\delta\alpha\right)^2} \approx \left(1 - 2i\alpha_R\delta\alpha\right)e^{-\alpha_R^2}.
\end{aligned} \tag{68}$$

Using the above equations in Equation (65) yields:

$$\begin{aligned}
x^2 &- \frac{1}{2\alpha_R^2} + \frac{i\delta\alpha}{\alpha_R^3} - \frac{3}{4\alpha_R^4} + \frac{3i\delta\alpha}{\alpha_R^5} \\
&+ i\sqrt{\pi}\left(\alpha_R - 2\alpha_R^2 i\delta\alpha + i\delta\alpha\right)e^{-\alpha_R^2} = 0.
\end{aligned} \tag{69}$$

Taking the real part of the latter equation gives:

$$x^2 - \frac{1}{2\alpha_R^2} - \frac{3}{4\alpha_R^4} + \left(2\sqrt{\pi}\alpha_R^2\delta\alpha - \sqrt{\pi}\delta\alpha\right)e^{-\alpha_R^2} = 0. \tag{70}$$

As soon as $\alpha_R \gg 1$ is focused on (see, e.g., Equation (61)), the term with the exponential function can be neglected. Thus, in the lowest-order:

$$x^2 - \frac{1}{2\alpha_R^2} - \frac{3}{4\alpha_R^4} = 0. \tag{71}$$

Due to Equation (37), one gets:

$$\alpha_R = \frac{\omega_R}{\sqrt{2}v_e k}. \tag{72}$$

Furthermore, one finds

$$x = \lambda_D k = \frac{v_e}{\omega_p}k \tag{73}$$

as a consequence of Equation (6).

Combining all the above findings, allows us to finally derive from Equation (71):

$$1 - \frac{\omega_p^2}{\omega_R^2}\left(1 + 3\frac{v_e^2 k^2}{\omega_R^2}\right) = 0. \tag{74}$$

This can be rewritten as:

$$\omega_R^4 - \omega_p^2\omega_R^2 - 3\omega_p^2 v_e^2 k^2 = 0, \tag{75}$$

which can be understood as a quadratic equation for ω_R^2. It has the solutions

$$\omega_R^2 = \frac{1}{2}\left[\omega_p^2 \pm \sqrt{\omega_p^4 + 12v_e^2 k^2\omega_p^2}\right]. \tag{76}$$

Therein, taking the plus-sign and employing Equation (6), one finds:

$$\omega_R^2 = \frac{1}{2}\omega_p^2\left[1 + \sqrt{1 + 12\lambda_D^2 k^2}\right]. \tag{77}$$

So far it was assumed that the restriction, given by Equation (61), is valid. Furthermore, the damping effect was assumed to be small meaning that $\delta\omega$ is small. This essentially means that $\omega_R \gg \delta\omega$. Furthermore, the condition, given by Equation (61), turns into

$$\alpha_R^2 \equiv \frac{\omega_R^2}{2v_e^2 k^2} \gg 1. \tag{78}$$

Using therein Equations (6) and (77) allows us to rewrite this condition:

$$1 + \sqrt{1 + 12\lambda_D^2 k^2} \gg 4(\lambda_D k)^2. \tag{79}$$

It is clear that this can only be valid if

$$\lambda_D k \ll 1. \tag{80}$$

Using condition (80) in Equation (77) allows us to Taylor-expand so that one obtains:

$$\omega_R^2 = \omega_p^2 \left(1 + 3\lambda_D^2 k^2\right). \tag{81}$$

One can find that the result obtained concides with the plasma wave dispersion relation of Equation (7). In dimensionless quantities (58):

$$y_R = \sqrt{1 + 3x^2}. \tag{82}$$

Considering the imaginary part of Equation (69), one finds:

$$\frac{\delta\alpha}{\alpha_R^3} + \frac{3\delta\alpha}{\alpha_R^5} + \sqrt{\pi}\alpha_R e^{-\alpha_R^2} = 0. \tag{83}$$

This equation can be rearranged:

$$\delta\alpha = -\sqrt{\pi}\frac{\alpha_R^6}{3 + \alpha_R^2}e^{-\alpha_R^2}. \tag{84}$$

As soon as $\alpha_R \gg 1$:

$$\delta\alpha = -\sqrt{\pi}\alpha_R^4 e^{-\alpha_R^2}. \tag{85}$$

From Equation (54) it follows that

$$\alpha_R = \frac{1}{\sqrt{2}}\frac{\omega_R}{\omega_p}\frac{1}{\lambda_D k} \quad \text{and} \quad \delta\alpha = \frac{1}{\sqrt{2}}\frac{\delta\omega}{\omega_p}\frac{1}{\lambda_D k}. \tag{86}$$

Therefore, one finds:

$$\delta\omega = -\sqrt{\frac{\pi}{8}}\frac{\omega_R^4}{\omega_p^3}\frac{1}{(\lambda_D k)^3}e^{-\alpha_R^2}. \tag{87}$$

Combining Equation (81) and condition (80), one gets $\omega_R \approx \omega_p$ in the lowest order. Therewith, the result obtained for $\delta\omega$ reads:

$$\delta\omega = -\sqrt{\frac{\pi}{8}}\frac{\omega_p}{(\lambda_D k)^3}e^{-\alpha_R^2}. \tag{88}$$

Therein one can use Equations (81) and (86):

$$
\begin{aligned}
\alpha_R^2 &= \frac{\omega_R^2}{2\omega_p^2}\frac{1}{(\lambda_D k)^2} \\
&= \frac{1}{2(\lambda_D k)^2}\left(1 + 3\lambda_D^2 k^2\right) \\
&= \frac{1}{2(\lambda_D k)^2} + \frac{3}{2},
\end{aligned}
\tag{89}
$$

and finds:

$$
\delta\omega = -\sqrt{\frac{\pi}{8}}\frac{\omega_p}{(\lambda_D k)^3}e^{-\frac{1}{2\lambda_D^2 k^2}-\frac{3}{2}}.
\tag{90}
$$

In dimensionless quantities (58), one gets:

$$
\delta y = -\sqrt{\frac{\pi}{8}}\frac{1}{x^3}e^{-\frac{1}{2x^2}-\frac{3}{2}}.
\tag{91}
$$

Equations (82) and (91) are visualized in Figure 2 together with the numerical solution discussed in Section 4. One can immediately see that the agreement between the numerical solution and the analytical one is much better for smaller values of x corresponding to small wave numbers. The assumption of small wave numbers is part of the analytical calculation presented above (see, e.g., Equation (80)) and, therefore, the deviation between the analytical and numerical solutions for larger wave numbers was predictable.

Note that as soon as the case $\lambda_D k \ll 1$ is considered, one can further approximate Equation (90) by

$$
\delta\omega = -\sqrt{\frac{\pi}{8}}\frac{\omega_p}{(\lambda_D k)^3}e^{-\frac{1}{2\lambda_D^2 k^2}},
\tag{92}
$$

which is in perfect agreement with Equation (17) of [6]. A discussion about the validity and possible improvements of asymptotic formulas for the Landau damping rates of electrostatic waves can be found in [17].

Using the form (66) in the electric field given by Equation (19), yields:

$$
\delta E(x,t) \propto \delta E(k,\omega)e^{i(kx-\omega t)} \propto e^{-i\omega_R t+\delta\omega t}.
\tag{93}
$$

As $\delta\omega < 0$ was derived, one finds damping of the electric field associated with the wave. This damping is the renowned Landau damping and the corresponding damping rate is given by Equations (90) or (92) depending on the desired accuracy.

6. Discussion and Conclusions

The derivations and discussions, presented in the current paper, are based on lecture notes I have developed for a course about magnetohydrodynamics. The aim was to obtain a mathematically simpler description of Landau damping compared to what Landau originally developed in his seminal paper [6]. This means, in particular, that tools of complex analysis, such as contour integrals, are avoided and one just solves standard integrals in combination with imaginary error functions.

Although using methods of complex analysis was avoided throughout the current paper, one still arrives at the correct Langmuir dispersion relation as given by Equation (81) as well as the correct Landau damping rate as given by Equation (90). These results were obtained analytically by employing Equations (61) and (80). The more general dispersion relation is given by Equation (57) corresponding to a nonlinear equation. However, it can be solved by using standard tools of computational physics. All findings are visualized in Figure 2.

Although the main intention behind the current paper was to present a simpler derivation of known results, Equation (53) is quite general since it does not require to consider asymptotic limits. Therefore, the method, proposed in this paper, could lead to an improved understanding of Landau damping in some other limits. Furthermore, it could be possible that the method developed here can be used to explore other effects such as the dispersion relation of electron-acoustic waves.

It needs to be emphasized that the exploration of Landau damping is still an active field of research pursued both in mathematics and in theoretical physics. A comprehensive overviews of Landau damping which includes some historical remarks, derivations, and mathematical details can be found in [18,19].

Funding: This research was funded by the Natural Sciences and Engineering Research Council (NSERC) of Canada.

Data Availability Statement: This study does not report any data.

Conflicts of Interest: The author declares no conflict of interest.

Appendix A. Some Mathematical Details

Equation (41) can be written as

$$J(\alpha, \beta) = e^{(\beta-1)\alpha^2} \int_{-\infty}^{+\infty} dz \, \frac{1}{\alpha - z} e^{-\beta z^2}. \tag{A1}$$

Note that α is a complex number, given by Equation (37), and β is a positive real number. Equation (A1) can be rewritten as:

$$\begin{aligned} J(\alpha, \beta) &= e^{(\beta-1)\alpha^2} \int_{-\infty}^{+\infty} dz \, \frac{\alpha+z}{\alpha^2-z^2} e^{-\beta z^2} \\ &= e^{(\beta-1)\alpha^2} \int_{-\infty}^{+\infty} dz \, \frac{\alpha}{\alpha^2-z^2} e^{-\beta z^2}, \end{aligned} \tag{A2}$$

where the symmetry properties of one of the integrals is used. In the limit $\beta \to 0$,

$$J(\alpha, \beta \to 0) = e^{-\alpha^2} \int_{-\infty}^{+\infty} dz \, \frac{\alpha}{\alpha^2 - z^2}. \tag{A3}$$

The remaining integral can be solved via substitution:

$$z = \alpha \tanh(y), \tag{A4}$$

and then,

$$dz = \alpha \left[1 - \tanh^2(y)\right] dy, \tag{A5}$$

in the integral. Furthermore, for the inverse hyperbolic tangent function, the relations,

$$\tanh^{-1}(+\infty) = -\frac{1}{2} i\pi, \tag{A6}$$

and

$$\tanh^{-1}(-\infty) = \frac{1}{2} i\pi, \tag{A7}$$

are employed. Therefore, one gets:

$$\int_{-\infty}^{+\infty} dz \, \frac{\alpha}{\alpha^2 - z^2} = -i\pi, \tag{A8}$$

and, thus:

$$J(\alpha, \beta \to 0) = -i\pi e^{-\alpha^2}. \tag{A9}$$

Appendix B. The Newton-Solver

The task is to solve the nonlinear relation given by Equation (60). For convenience, let us write this equation down again:

$$1 + \frac{1}{x^2} - \frac{y^2}{x^4} e^{-y^2/(2x^2)} K(\alpha) + i\sqrt{\frac{\pi}{2}} \frac{y}{x^3} e^{-y^2/(2x^2)} = 0. \tag{A10}$$

In the latter, the function,

$$K(\alpha) = \int_0^1 dz\, e^{\alpha^2 z^2}, \tag{A11}$$

is used, as well as:

$$x = \lambda_D k, \qquad y = \omega/\omega_p, \qquad \text{and} \qquad \alpha = \frac{y}{\sqrt{2}x}. \tag{A12}$$

Therewith, one can write:

$$K(x,y) = \int_0^1 dz\, e^{\frac{y^2}{2x^2} z^2}. \tag{A13}$$

The nonlinear relation above is solved via a standard Newton method; see, e.g., in [20]. The task is to find y for a given x; therefore, it is needed to compute y for each x. This y is obtained via the iteration method,

$$y_{n+1} = y_n - \frac{f(y_n)}{f'(y_n)}, \tag{A14}$$

where, in the case considered here,

$$f(y) = 1 + \frac{1}{x^2} - \frac{y^2}{x^4} e^{-y^2/(2x^2)} K(x,y) + i\sqrt{\frac{\pi}{2}} \frac{y}{x^3} e^{-y^2/(2x^2)}. \tag{A15}$$

Furthermore, one also needs to compute the derivative of the latter function with respect to y. One derives:

$$\begin{aligned}
f'(y) \equiv \frac{\partial f}{\partial y} &= -2\frac{y}{x^4} e^{-y^2/(2x^2)} K(x,y) + \frac{y^3}{x^6} e^{-y^2/(2x^2)} K(x,y) - \frac{y^2}{x^4} e^{-y^2/(2x^2)} K'(x,y) \\
&+ i\sqrt{\frac{\pi}{2}} \frac{1}{x^3} e^{-y^2/(2x^2)} - i\sqrt{\frac{\pi}{2}} \frac{y^2}{x^5} e^{-y^2/(2x^2)},
\end{aligned} \tag{A16}$$

where

$$K'(x,y) = \int_0^1 dz\, \frac{yz^2}{x^2} e^{\frac{y^2}{2x^2} z^2} \tag{A17}$$

is used.

Combining Equations (A14)–(A17) determines y for a given x numerically. Performing these calculations for a set of x values yields the dispersion relation $y(x)$ which was looked for.

References

1. Tonks, L.; Langmuir, I. Oscillations in ionized gases. *Phys. Rev.* **1929**, *33*, 195–210. [CrossRef]
2. Bohm, D.; Gross, E.P. Theory of plasma oscillations. A. Origin of medium-like behavior. *Phys. Rev.* **1949**, *75*, 1851–1864. [CrossRef]
3. Bohm, D.; Gross, E.P. Theory of plasma oscillations. B. Excitation and damping of oscillations. *Phys. Rev.* **1949**, *75*, 1864–1876. [CrossRef]
4. Bohm, D.; Gross, E.P. Effects of plasma boundaries in plasma oscillations. *Phys. Rev.* **1950**, *79*, 992–1001. [CrossRef]
5. Tsurutani, B.; Lakhina, G. Some basic concepts of wave-particle interactions in collisionless plasmas. *Rev. Geophys.* **1997**, *35*, 491–502. [CrossRef]
6. Landau, L. On the vibration of the electronic plasma. *J. Phys. (USSR)* **1946**, *10*, 25–34. Reprinted in *Collected Papers of L.D. Landau*; Ter Haar, D., Ed.; Pergamon Press. Ltd.: Oxford, UK, 1965; pp. 445–460. [CrossRef]

7. Vlasov, A.A. On vibration properties of electron gas. *Zh. Exp. Teor. Fiz.* **1938**, *8*, 291. (In Russian); English Translation in *Sov. Phys. Uspekhi.* **1968**, *10*, 721–733. [CrossRef]
8. Schlickeiser, R.; Fichtner, H.; Kneller, M. Revised Landau damping rates of magnetohydrodynamics waves in hot magnetized equilibrium plasmas and its consequences for cosmic ray transport in the interplanetary medium. *J. Geophys. Res.* **1997**, *102*, 4725–4739. [CrossRef]
9. van Kampen, N.G. On the theory of stationary waves in plasma. *Physica* **1955**, *21*, 949–963. [CrossRef]
10. Case, K.M. Plasma oscillations. *Ann. Phys.* **1959**, *7*, 349–364. [CrossRef]
11. Krall, N.A.; Trivelpiece, A.W. *Principles of Plasma Physics*; McGraw-Hill: New York, NY, USA, 1973.
12. Liouville, J. Sur la theorie de la variation des constantes arbitraires. *J. Math. Pures Appl.* **1838**, *3*, 342–349.
13. Schlickeiser, R. *Cosmic Ray Astrophysics*; Springer: Berlin/Heidelberg, Germany, 2002. [CrossRef]
14. Shalchi, A. *Nonlinear Cosmic Ray Diffusion Theories*; Springer: Berlin/Heidelberg, Germany, 2009. [CrossRef]
15. Zank, G.P. *Transport Processes in Space Physics and Astrophysics: Lecture Notes in Physics*; Springer: New York, NY, USA, 2014. [CrossRef]
16. Abramowitz, M.; Stegun, I.A. *Handbook of Mathematical Functions*; Dover Publications: New York, NY, USA, 1965.
17. McKinstrie, C.J.; Giacone, R.E.; Startsev, E.A. Formulas for the Landau damping rates of electrostatic waves. *Phys. Plasmas* **1999**, *6*, 463–466. [CrossRef]
18. Ryutov, D.D. Landau damping: Half a century with the great discovery. *Plasma Phys. Control. Fusion* **1999**, *41*, A1–A12. [CrossRef]
19. Mouhot, C.; Villani, C. On Landau damping. *Acta Math.* **2011**, *207*, 29–201. [CrossRef]
20. Press, W.H.; Teukolsky, S.A.; Vetterling, W.T.; Flannery, B.P. *Numerical Recipes*; Cambridge University Press: New York, NY, USA, 2007.

Brief Report

Plasma Flows in Solar Filaments as Electromagnetically Driven Vortical Flows

Yuri E. Litvinenko

Department of Mathematics, University of Waikato, Hamilton 3105, New Zealand; yuril@waikato.ac.nz

Abstract: Electromagnetic expulsion acts on a body suspended in a conducting fluid or plasma, which is subject to the influence of electric and magnetic fields. Physically, the effect is a magneto-hydrodynamic analogue of the buoyancy (Archimedean) force, which is caused by the nonequal electric conductivities inside and outside the body. It is suggested that electromagnetic expulsion can drive the observed plasma counter-streaming flows in solar filaments. Exact analytical solutions and scaling arguments for a characteristic plasma flow speed are reviewed, and their applicability in the limit of large magnetic Reynolds numbers, relevant in the solar corona, is discussed.

Keywords: space physics; plasma physics

Citation: Litvinenko, Y.E. Plasma Flows in Solar Filaments as Electromagnetically Driven Vortical Flows. *Physics* **2021**, *3*, 1046–1050. https://doi.org/10.3390/physics3040065

Received: 22 August 2021
Accepted: 3 November 2021
Published: 10 November 2021

Publisher's Note: MDPI stays neutral with regard to jurisdictional claims in published maps and institutional affiliations.

1. Introduction

Solar filaments are sheets of dense and cool plasma, surrounded by the much hotter plasma of the solar corona. The relatively low temperatures and high densities of the filament material suggest that filaments are supported against gravity by a strong magnetic field **B** in the solar atmosphere [1].

Early models of solar filaments postulated static or quasi-steady equilibria and analyzed simplified magnetohydrodynamic (MHD) equations [2,3]. More recent theoretical models employed a force-free approximation, $(\nabla \times \mathbf{B}) \times \mathbf{B} \approx 0$, to successfully describe the overall filament structure [4,5].

Realistic theoretical models should satisfy a number of observational conditions for the formation and maintenance of filaments [6]. The modeling of plasma flows in filaments appears to be of particular interest in relation to their structure and evolution. Observations clearly demonstrated that even quiescent filaments are not static formations but rather are systems of jets streaming along the filaments with speeds up to 30 km s^{-1} [7]. Thermal nonequilibrium is typically invoked to explain the observed flows [8,9]. However, one of the puzzles of the small-scale dynamics in filaments is the physical mechanism of counter-streaming—the observed simultaneous flows with speeds of 5–20 km s^{-1} in opposite directions in filament barbs (feet) [10,11].

The purpose of this paper is to advocate the electromagnetic expulsion force, whose effects are well-known in engineering and industrial applications, as a mechanism of counter-streaming in solar filaments.

2. Electromagnetically Generated Vortical Flows

Following the original argument of [12], consider an incompressible conducting fluid (plasma) with density ρ_0, temperature T_0, and electric conductivity σ_0 in a magnetic field $\mathbf{B}_0$. Provided $B_0 \gg 4\pi a j_0/c$, where c is the speed of light, both $\mathbf{B}_0$ and the electric current density $\mathbf{j}_0$ can be assumed to be locally uniform (a is a typical length scale of the problem). The resulting Lorentz force is also uniform:

$$\mathbf{f}_0 = \frac{1}{c}\,\mathbf{j}_0 \times \mathbf{B}_0. \tag{1}$$

Depending on the relative orientations of the gravity force $\rho_0\mathbf{g}$, where $\mathbf{g}$ is the body accelerations, and $\mathbf{f}_0$, the Lorentz force makes the fluid effectively heavier or lighter. In either case, the uniform volume force is potential, $\nabla \times \mathbf{f}_0 = 0$, and, just like the gravity force $\rho_0\mathbf{g}$, will be balanced by the pressure gradient:

$$\nabla p_0 = \rho_0\mathbf{g} + \mathbf{f}_0. \tag{2}$$

Now consider a body (filament) of volume V, with a temperature $T_1 \neq T_0$ and the corresponding electric conductivity $\sigma_1 \neq \sigma_0$, submerged in the plasma. The total force acting on the body is as follows:

$$\mathbf{F} = \int_V (\rho_1\mathbf{g} + \mathbf{f}_1)\, dV + \oint_S p_0\mathbf{n}\, dS. \tag{3}$$

Here,

$$\mathbf{f}_1 = \frac{1}{c}\mathbf{j}_1 \times \mathbf{B}_0 \tag{4}$$

is the Lorentz force inside the filament, ρ_1 is the density of the filament, $\mathbf{j}_1$ is the current density inside the filament, $\mathbf{n}$ is the inward normal to the surface S, p_0 is the gas pressure. If the current $\mathbf{j}_0$ remains uniform in the presence of the body and $\mathbf{B}_0$ remains approximately uniform, then $\mathbf{F}$ could be expressed as

$$\mathbf{F} = \int_V (\rho_1\mathbf{g} + \mathbf{f}_1)\, dV - \int_V \nabla p_0\, dV, \tag{5}$$

and so

$$\mathbf{F} = \int_V (\rho_1 - \rho_0)\mathbf{g}\, dV + \frac{1}{c}\int_V (\mathbf{j}_1 - \mathbf{j}_0) \times \mathbf{B}_0\, dV. \tag{6}$$

Here, the first integral is the usual buoyancy (Archimedean) force. The second integral describes the electromagnetic expulsion force. It vanishes only if the electric current density remains uniform, $\mathbf{j}_1 = \mathbf{j}_0$, which happens only if $\sigma_1 = \sigma_0$.

Although the MHD expulsion force is formally similar to the Archimedean force in hydrodynamics, it is different from the well-known magnetic buoyancy force [13]. The magnetic buoyancy force is the usual buoyancy force, associated with the density difference caused by the pressure difference in a magnetostatic equilibrium. By contrast, the expulsion force is independent of the presence of gravity.

The key point for the following is that the static description is purely illustrative. In reality, the expulsion force will almost always drive plasma flows. As the current density is not uniform in the presence of a filament with conductivity $\sigma_1 \neq \sigma_0$, the resulting Lorentz force $\mathbf{j} \times \mathbf{B}/c$ is generally not potential. Hence, it cannot be balanced by potential forces such as the gas pressure gradient ∇p_0. This is why in general the convective term $\rho_0(\mathbf{v} \cdot \nabla)\mathbf{v}$, where $\mathbf{v}$ is the plasma velocity, and the viscous term $\eta\nabla^2\mathbf{v}$, where η is the scalar viscosity, must be taken into account in the equation of motion. Physically, this means that electromagnetically generated vortical flows must appear in the vicinity of a submerged body [12]. It is these flows that may naturally explain the plasma counter-streaming in solar filaments. Assuming an incompressible steady plasma flow, the equation of motion is as follows:

$$\rho_0(\mathbf{v} \cdot \nabla)\mathbf{v} = -\nabla p + \rho_0\mathbf{g} + \eta\nabla^2\mathbf{v} + \frac{1}{c}\mathbf{j} \times \mathbf{B}. \tag{7}$$

The electromagnetic expulsion force, also known as electro-magneto-phoresis, has been extensively studied under laboratory conditions, motivated by such engineering applications as the extraction of impurities in liquid metals and the separation of mechanical mixtures and biological cells [14–16].

3. Counter-Streaming in Solar Filaments

Although the parameter regime in the solar atmosphere differs significantly from that under laboratory conditions, the expulsion force might play a role in the filament dynamics. Solar filaments consist of numerous fine and dense threads with a typical radius $a \approx 10^7$ cm and temperature $T_1 \approx 10^4$ K [1]. For the coronal temperature $T_0 \approx 10^6$ K, the ratio of conductivities inside and outside the filament is

$$\frac{\sigma_1}{\sigma_0} = \left(\frac{T_1}{T_0}\right)^{3/2} \approx 10^{-3} \ll 1. \tag{8}$$

As the electric current density $j_1 \ll j_0$, expected effects of the expulsion force in filaments should be significant.

The calculation of the general expressions for the electromagnetic expulsion force and the associated vortical flows is a complicated nonlinear problem. Several particular cases, however, can be studied in detail. Analytical progress was achieved in the limit of small ordinary Re $= \rho_0 v_0 a / \eta$ and magnetic Re$_m = 4\pi\sigma_0 v_0 a / c^2$ Reynolds numbers and a small Hartmann number $M = (aB_0/c)(4\pi\sigma_0/\eta)^{1/2}$. In this limit, the convective derivative can be ignored in the equation of motion, the electromagnetic and dynamic problems decouple, and exact analytical expressions for the plasma flows and the expulsion force can be obtained for a sphere and a cylinder with radius a [12,17]. Further analytical progress was achieved for particles of other shapes by using symmetry considerations and asymptotic methods [18,19].

As a potentially relevant example, consider a solution to the problem of electromagnetically driven flows near a cylinder with radius a and electric conductivity σ_1 surrounded by the plasma with conductivity $\sigma_0 \gg \sigma_1$ and viscosity η, given the uniform electric current $\mathbf{j}_0 = (j_\perp, 0, j_\parallel)$ and an approximately uniform magnetic field $\mathbf{B}_0 = (B_x, B_y, B_z)$ outside the cylinder. If the cylinder is co-aligned with the z-axis, the solution for the plasma velocity outside the cylinder ($r > a$) is as follows:

$$v_x = 0, \quad v_y = 0, \tag{9}$$

$$v_z(r, \phi) = \frac{a^2 j_\perp}{4c\eta}(B_x \sin 2\phi - B_y \cos 2\phi)\left(1 - \frac{a^2}{r^2}\right). \tag{10}$$

Here, $r = (x^2 + y^2)^{1/2}$ and $\phi = \tan^{-1}(y/x)$ is the polar angle (the counterclockwise angle from the x-axis in the xy-plane). The solution for a viscous fluid satisfies the standard boundary condition of vanishing velocity on the surface of the cylinder [17]. The expression for $v_z(r, \phi)$ is basically a product of two functions: $(B_x \sin 2\phi - B_y \cos 2\phi)$ describes the alternating flow directions in the neighboring sectors of the xy-plane, whereas $(1 - a^2/r^2)$ for $r > a$ describes a monotonic increase in the speed with distance from the surface of the cylinder.

Two points are worth stressing. First, the trigonometric dependence of v_z on the polar angle ϕ means that the oppositely directed flows are naturally predicted by the model. Second, the flows are generated even when the external magnetic field and electric current are co-aligned ($B_y = 0$ and $j_\perp B_z = j_\parallel B_x$), corresponding to a large-scale force-free magnetic field in the solar corona, $\mathbf{j} \times \mathbf{B} = 0$. This is a clear illustration of how the action of the expulsion force differs from that of the standard $\mathbf{j} \times \mathbf{B}$ force.

The presented velocity profile provides strong motivation for modeling counter-streaming in solar filaments as electromagnetically generated plasma flows. As far as characteristic values of the key dimensionless parameters are concerned, for typically observed flow speeds in solar filaments of order $v_0 \approx 10$ km s^{-1}, the ordinary Reynolds number is of the order 10^{-2}, and thus Re $\ll 1$ as in the solution above. By contrast, the magnetic Reynolds number is large, Re$_m \simeq 10^9 \gg 1$. This is why the solution above cannot be directly applied to describe flows along the filament threads in the solar corona. Dimensional arguments, however, suggest a simple formula for the expulsion force density:

$$\mathbf{f} = -\mathbf{f}_0 \Phi(\mathrm{Re}, \mathrm{Re}_m).\tag{11}$$

The dimensionless function Φ has the following asymptotic behavior [20]:

$$\Phi \approx \begin{cases} 1, & \mathrm{Re}_m < 1, \\ 1/\mathrm{Re}_m, & \mathrm{Re}_m > 1. \end{cases}\tag{12}$$

These expressions lead to an estimate for the typical plasma speed v_0, which is consistent with observations of counter-streaming in solar filaments in the limit $\mathrm{Re}_m \gg 1$ [21].

To put the qualitative dimensional arguments on a firmer footing, it would be necessary to derive an expression for the plasma velocity in the vicinity of a cylindrical filament thread, which would remain valid for both small and large values of Re_m. Symmetry dictates that the sought-after solution in cylindrical geometry is independent of z if the cylinder is oriented along the z-axis, in which case MHD equations reduce to two second-order partial differential equations with the only parameter being the Hartmann number M. Solutions of the resulting boundary value problem for particular orientations of the magnetic field have been constructed [22]. Further analytical progress for arbitrary orientations of the electric current and magnetic field near the filament may be achieved by deriving asymptotic analytical solutions in the limits of small and large M and by matching the solutions for $M \approx 1$.

4. Discussion

Theoretical and numerical studies of solar filaments traditionally consider idealized, nearly force-free magnetic field configurations [4,5]. The boundary conditions are controlled by the photospheric plasma flows and magnetic flux emergence. The gas pressure gradient and the gravity force are among the local forces acting on a filament in the corona, whose effects can be described by perturbing a large-scale force-free model [23].

Whereas the equilibrium in filaments is primarily determined by a balance between the Lorentz and gravity forces, the observed counter-streaming remains an unsolved problem [24]. It is worth stressing that even quiescent filaments are highly dynamic formations, characterized by mass flows and changes of shape on multiple scales [7].

As argued above, what appears to be missing in the available models of solar filaments is an evaluation of the role of the electromagnetic expulsion force in the filament dynamics. The expulsion force is an MHD analogue of the usual buoyancy (Archimedean) force [12,25], which provides a well-known method for engineering applications such as impurity extraction and the separation of bioparticles [16,26]. A nonuniform distribution of temperature in filaments leads to a nonuniform electric conductivity and hence to a nonuniform Lorentz force. The resulting electromagnetic expulsion force is generally non-potential and naturally drives vortical plasma flows that may correspond to the observed counter-streaming flows in filaments [21].

Since the model calculation of the counter-streaming flows is performed in an idealized geometry, it is difficult to speculate whether the model might predict different features for the flows in different parts of a filament. It is reasonable to expect though that, as long as the flows are generated in filament barbs, they should also be sustained in the main body of the filament (along the filament spine) whose mass is supported against gravity by a magnetic force.

The proposed mechanism may provide an explanation for the observed counter-streaming in filaments. More generally, the electromagnetic expulsion force may play a role in the dynamics of cosmic plasma with nonuniform distributions of temperature and electric conductivity.

Funding: This research was partly funded by the Deutsche Forschungsgemeinschaft (project number 434200803).

Acknowledgments: It is a pleasure to contribute to this Special Issue, dedicated to Professor Reinhard Schlickeiser, who was my host during several visits to Ruhr-Universität Bochum, supported by a Humboldt Foundation fellowship between 2007 and 2011.

Conflicts of Interest: The author declares no conflict of interest.

References

1. Tandberg-Hanssen, E. *The Nature of Solar Prominences*; Kluwer Academic Publishers: Dordrecht, The Netherlands, 1995. [CrossRef]
2. Kippenhahn, R.; Schlüter, A. Eine Theorie der solaren Filamente. *Zeit. f. Astrophys.* **1957**, *43*, 36–62. Available online: http://adsabs.harvard.edu/full/1957ZA.....43...36K (accessed on 15 August 2021).
3. Brown, A. On the stability of a hydromagnetic prominence model. *Astrophys. J.* **1958**, *128*, 646–663. [CrossRef]
4. Aulanier, G.; Démoulin, P. 3-D magnetic configurations supporting prominences. I. The natural presence of lateral feet. *Astron. Astrophys.* **1998**, *329*, 1125–1137. Available online: http://adsabs.harvard.edu/full/1998A&A...329.1125A (accessed on 15 August 2021).
5. Gibson, S.E. Solar prominences: theory and models. Fleshing out the magnetic skeleton. *Living Rev. Sol. Phys.* **2018**, *15*, 7. [CrossRef]
6. Martin, S.F. Conditions for the formation and maintenance of filaments. *Sol. Phys.* **1998**, *182*, 107–137. [CrossRef]
7. Wang, Y.-M. The jetlike nature of He II λ 304 prominences. *Astrophys. J.* **1999**, *520*, L71–L74. [CrossRef]
8. Craig, I.J.D.; McClymont, A. N. Quasi-steady mass flows in coronal loops. *Astrophys. J.* **1986**, *307*, 367–380. [CrossRef]
9. Klimchuk, J.A.; Luna, M. The role of asymmetries in thermal nonequilibrium. *Astrophys. J.* **2019**, *884*, 68. [CrossRef]
10. Diercke, A.; Kuckein, C.; Verma, M.; Denker, C. Counter-streaming flows in a giant quiet-Sun filament observed in the extreme ultraviolet. *Astron. Astrophys.* **2018**, *611*, A64. [CrossRef]
11. Zirker, J.B.; Engvold, O.; Martin, S.F. Counter-streaming gas flows in solar prominences as evidence for vertical magnetic fields. *Nature* **1998**, *396*, 440–441. [CrossRef]
12. Leenov, D.; Kolin, A. Theory of electromagnetophoresis. I. Magnetohydrodynamic forces experienced by spherical and symmetrically oriented cylindrical particles. *J. Chem. Phys.* **1954**, *22*, 683–688. [CrossRef]
13. Parker, E.N. The formation of sunspots from the solar toroidal field. *Astrophys. J.* **1955**, *121*, 491–507. [CrossRef]
14. Kolin, A. Some current and potential uses of magnetic fields in electrokinetic separations. *J. Chromatogr.* **1978**, *159*, 147–181. [CrossRef]
15. Terada, T.; Akiyama, Y.; Izumi, Y.; Nishijima, S. Separation of impurity in molten metals by using superconducting magnet. *Phys. C* **2009**, *469*, 1845–1848. [CrossRef]
16. Watarai, H.; Suwa, M.; Iiguni, Y. Magnetophoresis and electrophoresis of microparticles in liquids. *Anal. Bioanal. Chem.* **2004**, *378*, 1693–1699. [CrossRef]
17. Marty, P.; Alemany, A. Écoulement dû à des champs magnétique et électrique croisés autour d'un cylindre de conductivité quelconque. *J. Mec. Theor. Appl.* **1983**, *2*, 227–243.
18. Moffatt, H.K.; Sellier, A. Migration of an insulating particle under the action of uniform ambient electric and magnetic fields. Part 1. General theory. *J. Fluid Mech.* **2002**, *464*, 279–286. [CrossRef]
19. Yariv, E.; Miloh, T. Electro-magneto-phoresis of slender bodies. *J. Fluid Mech.* **2007**, *577*, 331–340. [CrossRef]
20. Litvinenko, Y.E.; Somov, B.V. Electromagnetic expulsion force in cosmic plasma. *Astron. Astrophys.* **1994**, *287*, L37–L40.
21. Litvinenko, Y.E.; Somov, B.V. Aspects of the global MHD equilibria and filament eruptions in the solar corona. *Space Sci. Rev.* **2001**, *95*, 67–77. [CrossRef]
22. Gerbeth, G.; Thess, A.; Marty, P. Theoretical study of the MHD flow around a cylinder in crossed electric and magnetic fields. *Eur. J. Mech. B/Fluids* **1990**, *9*, 239–257.
23. Litvinenko, Y.E.; Wheatland, M.S. A simple dynamical model for filament formation in the solar corona. *Astrophys. J.* **2005**, *630*, 587–595. [CrossRef]
24. Chen, P.-F.; Xu, A.-A.; Ding, M.-D. Some interesting topics provoked by the solar filament research in the past decade. *Res. Astron. Astrophys.* **2020**, *20*, 166. [CrossRef]
25. Ščepanskis, M.; Jakovičs, A. The magnetohydrodynamic force experienced by spherical iron particles in liquid metal. *J. Magn. Magn. Mater.* **2016**, *403*, 30–35. [CrossRef]
26. Xuan, X. Recent advances in continuous-flow particle manipulations using magnetic fluids. *Micromachines* **2019**, *10*, 744. [CrossRef]

 *physics*

Editorial

Personal Reminiscences of Reinhard Schlickeiser

Ian Lerche

Institut für Geowissenschaften, Naturwissenschaftliche Fakultät III, Martin-Luther-Universität, 06108 Halle, Germany; lercheian@yahoo.com

Citation: Lerche, I. Personal Reminiscences of Reinhard Schlickeiser. *Physics* **2021**, *3*, 1051–1053. https://doi.org/10.3390/physics3040066

Received: 5 November 2021
Accepted: 8 November 2021
Published: 12 November 2021

Publisher's Note: MDPI stays neutral with regard to jurisdictional claims in published maps and institutional affiliations.

In 1979, I arrived back at the University of Chicago from a two-year stint in Australia to find a very large German post-doc eagerly awaiting me, so we could work together on transport of cosmic ray electrons perpendicular to the galactic plane. He was Reinhard Schlickeiser, with whom I have worked on and off through to the present day. Reinhard became a very good friend indeed and has just recently retired from being Professor of Theoretical Astrophysics at the Ruhr-University, Bochum, Germany. We worked very well together for his post-doc year at Chicago and, as it turned out, also my last year at Chicago.

The basic problem of including convection, adiabatic deceleration, diffusion, and synchrotron radiation loss for cosmic ray electrons in a galactic magnetic field whose strength decreases with height above the galactic plane, and of then determining the associated synchrotron radiation spectrum with galactic height, is non-trivial. The importance of the problem lies in the fact that one measures such radiation with galactic height for a number of galaxies seen edge on. Many show breaks in the spectral index with height. If the basic problem could be solved, then the spectral index break and the height at which it occurs could be used to estimate the galactic wind speed transporting the relativistic electrons out of the plane of each galaxy. Reinhard and I solved this problem in the year he was there.

I then went to work in the oil industry for five years to get enough money to put my children through college. I then took a job at the University of South Carolina doing geology, oil risking, and sundry other matters. As an astrophysical interlude, I was invited by Reinhard to go and work with him in Bonn for a couple of months in summer 1986. He and his student Wolfgang Droege had been working for a while on the repowering problem of supernova relativistic electrons. While in Australia, Jim Caswell and I had effected a clumsy and palliative solution to the problem, but neither of us was happy with it, but equally, despite large quantities of beer, neither of us could see how to take the problem any further. Reinhard and Wolfgang had obtained the basis of a solution figured out in the time in between (Alfvén wave scattering but with the waves both ahead of and behind the shock front was the crucial ingredient in correctly solving the problem). Thus, Reinhard, Wolfgang and I worked first on cleaning up the solution. At the same time, Reinhard and I were involved in what is known as the secondary to primary ratio problem. Secondarily produced cosmic rays (by spallation of primary cosmic rays from the interstellar medium) have a different power spectrum than the primaries. Just before I left to return to Columbia, South Carolina, Reinhard and I figured out how to do that problem quantitatively and obtained a theoretical result for the difference in the spectra that precisely matched that observed.

I was heavily engaged with Jolynn Carroll (now at Tromsoe, Norway) in writing the final version of the sedimentary processes volume using radionuclides. I had also been "commuting" across the Atlantic to Germany with the environmental work. On one of the trips, I stopped off at Bochum where my old friend Reinhard Schlickeiser was now Professor of Astrophysics. Reinhard had previously asked me if I would proofread a book he was doing on cosmic ray astrophysics, and I had gone through one version with lots of recommended changes for him to handle. During my visit, he asked me if I would like to visit Bochum during the summer of 2001 to handle some astrophysics problems. I told him that it had been a long time since I had been seriously involved in quantitative astrophysics and I was not sure he would get value for his money. Reinhard grinned and said "Even you cannot be so incompetent" and bet me a case of beer that I could still do problems in astrophysics. I ended up having to pay the bet of course! Thus, summer 2001 was spent pleasantly in Bochum doing astrophysics. I was also editing a special issue of a journal at the same time so my evenings were also very busy.

I spent a second summer in 2002 at Bochum solving astrophysics problems quite happily during the week and, at weekends, doing basin analysis, environmental, and economics problems in Halle. That summer turned out to be highly productive. Reinhard and I found a solution to a twenty-year-old problem in astrophysics concerning the heating and cooling of the interstellar medium. The disparity between heating and cooling rates was many orders of magnitude no matter how one tweaked the parameters. This glaring discordance had survived many attempts to find a solution, including earlier attempts Reinhard and I had also made. Everyone had worked through the complex problem assuming the interstellar turbulence was isotropic. We allowed the turbulence to be anisotropic and then figured out what the degree of anisotropy had to be in order to get heating and cooling rates to agree. This was the first time anyone had even obtained a solution to the problem. When the work appeared, the main antagonists wrote things like the solar wind is not so anisotropic so how can the interstellar medium be? Of course, they failed to understand that one could turn their question on its head and ask: Why is the solar wind turbulence so isotropic in comparison to the interstellar medium? I figure it will be at least a decade before they choose to understand, as is not unusual in science!

In 2005, there was the chance of a stay for a year at the Ruhr University in Bochum where I would be involved again in plasma astrophysics. Approval came through for a one year Mercator Professorship for me to do astrophysics at the Ruhr University in Bochum so again I would be interacting intensively with Reinhard Schlickeiser. The first day of work I was greeted by Reinhard who had a list of astrophysics problems he thought needed to be solved and had figured I was the guy to do them. My brain was still locked into some geology and economic risk problems. I had not had time to finish (or even start) while at Leipzig so cranking my brain into the astrophysics mode once more was not easy. Indeed, I ended up doing astrophysics during the day and evenings were spent hammering away (another pun!) at the geology problems.

My time in Bochum was also spent commuting between Bochum and Halle at weekends. The nominal travel time on the train was 4 h. However, a change had to be made at Hannover and, cleverly, the German rail system had arranged a magnificent 6 minute overlap time for the connection. Of course, one train was always ten minutes late and, cunningly, the connecting train was always punctual and so had left the station before the first train arrived. This punctual lateness of the German rail system meant that the elapsed time was closer to 6 h rather than the official 4 h. One famous day, when I had to give the physics colloquium at Bochum at midday, I left Halle on the 5 a.m. train to ensure I would have plenty of time to set up the projector and such. Amazingly, the train connection was superbly delayed long enough that I arrived just 10 min before my talk—Reinhard was having kittens with the panic at the thought I was somewhere lost in the labyrinthian train system.

On one such trip, I was trying hard to solve one of Reinhard's problems on the train. The inebriated football fan next to me sort of bleary-eyed looked at what I was writing and asked "What language is that?" I answered it was mathematics. A long pause before he came back with the ultimate question "Can you speak it fluently as well as read and write it? " Ah yes, German train rides can be amazingly educational. The long train rides were actually good for solving problems because there were no interruptions from students and I cherished the time to concentrate. In this way, a majority of the problems were indeed roughed out.

By the time my stay in Bochum was over, all of the problems on Reinhard's list had been solved ranging from a variety of new plasma instabilities (mostly done with Robert Tautz and some with Anne Stockem) through to ways of balancing the heating and cooling rates of the interstellar medium (a basic problem that had defied prior solution for nearly twenty years and the general solution extended enormously the funny anisotropic solution Reinhard and I had worked out in 2002) that was completed with Reinhard and Felix Spanier, and ending with the solution of a highly nonlinear problem for radiation production and polarization from gamma ray emission objects (done mainly with Urs Schaeffer-Rolffs and Robert Tautz). The final farewell dinner was fun not the least reason being that a chess set was given to me with pictures of the whole group where the chess pieces normally were placed. The reason for this gift was due to the fact that at lunchtime, in Bochum, there was a group of us who played chess and simultaneously discussed science problems. Depending on the difficulty of the science problem, one was guaranteed to lose concentration on the chess game and thus lose the game—a ploy I suspect was often used by weaker chess players.

This ends my short overview of how it was to be involved via Reinhard in many of the astrophysical problems of the last few decades. I hope you enjoyed the short description as I had fun doing the writing thereof.

Conflicts of Interest: The authors declare no conflict of interest.

 *physics*

Article

Cylindrical and Spherical Nucleus-Acoustic Solitary and Shock Waves in Degenerate Electron-Nucleus Plasmas

A A Mamun [1,2]

1   Department of Physics, Jahangirnagar University, Savar Dhaka 1342, Bangladesh; mamun_phys@juniv.edu or director.wmsrc@juniv.edu
2   Wazed Miah Science Research Centre, Jahangirnagar University, Savar Dhaka 1342, Bangladesh

Abstract: The basic characteristics of cylindrical as well as spherical solitary and shock waves in degenerate electron-nucleus plasmas are theoretically investigated. The electron species is assumed to be cold, ultra-relativistically degenerate, negatively charged gas, whereas the nucleus species is considered a cold, non-degenerate, positively charged, viscous fluid. The reductive perturbation technique is utilized in order to reduce the basic equations (governing the degenerate electron-nucleus plasmas under consideration) to the modified Korteweg-de Vries and Burgers equations. The latter are numerically solved and analyzed to detect the basic characteristics of solitary and shock waves in such electron-nucleus plasmas. The nonlinear nucleus-acoustic waves are found to be propagated in the form of solitary as well as shock waves in such degenerate electron-nucleus plasmas. Their basic properties as well as their time evolution are significantly modified by the effects of cylindrical as well as spherical geometries. The results of this study is expected to be applicable not only to astrophysical compact objects, but also to ultra-cold dense plasmas produced in laboratory.

Keywords: nucleus-acoustic waves; nonlinear waves; nonplanar geometries

Citation: Mamun, A.A. Cylindrical and Spherical Nucleus-Acoustic Solitary and Shock Waves in Degenerate Electron-Nucleus Plasmas. *Physics* **2021**, *3*, 1088–1097. https://doi.org/10.3390/physics3040068

Received: 4 September 2021
Accepted: 5 November 2021
Published: 16 November 2021

Publisher's Note: MDPI stays neutral with regard to jurisdictional claims in published maps and institutional affiliations.

1. Introduction

Recently, Mamun, Amina and Schlickeiser [1,2] have first used the name "nucleus-acoustic (NA) waves" for the study of NA shock waves in strongly coupled degenerate plasmas [1] as well as NA solitary waves in self-gravitating degenerate plasmas [2]. The NA waves (NAWs) are degenerate pressure driven acoustic type of waves. They are completely new since they exist in degenerate plasmas at absolute zero temperature, but they do not exist in non-degenerate plasmas at either absolute zero or finite temperature. The degenerate plasmas [3–5] containing negatively charged degenerate electron gas and positively charged nucleus or ion fluid have important applications not only in astrophysical compact objects [3–7], but also in ultra-cold dense plasmas produced in laboratory devices [8–12]. The degenerate electron gas is formed by increasing its pressure more and more so that it cannot be compressed anymore due to Pauli's exclusion principle. This implies that there is no extra space in electron gas for more electrons to exist, and as a result, the space among electrons is infinitesimally small. This corresponds to an extremely high density electron gas with $\Delta x \to 0$ and $\Delta p \to \infty$, where Δx and Δp are the uncertainties in position and momentum, respectively. This generates an extremely high pressure because of Heisenberg's uncertainty principle, $\Delta x \Delta p \geq \hbar/2$, where $\hbar$ is the reduced Planck constant. This pressure is called the electron degenerate pressure [3–6]. The latter is the function of only number density of the degenerate electron gas.

The electron degenerate pressure $\mathcal{P}_e$ can be expressed as [3–6,13]

$$\mathcal{P}_e = \mathcal{P}_{e0} \left(\frac{\mathcal{N}_e}{\mathcal{N}_{e0}} \right)^{\gamma}, \tag{1}$$

where $\mathcal{N}_e$ is the number density of the degenerate electron gas, $\mathcal{N}_{e0}$ represents $\mathcal{N}_e$ and $\mathcal{P}_{e0}$ represents $\mathcal{P}_e$ at equilibrium, $\gamma = 5/3$ and $\gamma = 4/3$ are, respectively, for non-relativistically and cold ultra-relativistically degenerate electron (CUDE) gas, according to Chandrasekhar [3–5]. However, the CUDE pressure ($\gamma = 4/3$) is of the present interest. The CUDE pressures [5,6,13], $\mathcal{P}_e$ and $\mathcal{P}_{e0}$ can, respectively, be expressed as

$$\mathcal{P}_e = \mathcal{P}_{e0} n_e^{\frac{4}{3}}, \tag{2}$$

$$\mathcal{P}_{e0} \simeq \frac{3}{4} \hbar c \mathcal{N}_{e0}^{\frac{4}{3}}, \tag{3}$$

where $n_e = \mathcal{N}_e / \mathcal{N}_{e0}$, and c is the speed of light in vacuum. It is clear that the CUDE pressure $\mathcal{P}_e$ depends only on $\mathcal{N}_e$ and on its equilibrium value, $\mathcal{N}_{e0}$.

On the basis of Mamun, Amina and Schlickeiser [1,2], a number of theoretical investigations [14–22] on nonlinear NAWs in degenerate quantum plasmas under different situations have been made during the last five years. However, in these studies the length scale, phase speed and dispersion properties of the NAWs are not defined. The dependent as well as independent variables are also not properly normalized. There are also some studies [13,14,22], where "ion-acoustic (IA) waves (IAWs)" are used instead of the NAWs. This is not correct, since at absolute zero temperature the degenerate electron gas does not allow the IAWs to exist, but does allow the NAWs to exist. This fact along with the concept of the IAWs [23,24] have lead Mamun [25] to introduce proper length scale as well as time scale of the NAWs for the study of the linear propagation of the latter. The linear dispersion relation for the NAWs [25], propagating in cold degenerate electron-nucleus plasmas (CDENPs), is given by

$$\omega = \frac{k C_q}{\sqrt{1 + k^2 \lambda_{Dq}^2}}, \tag{4}$$

where ω is the angular frequency and k is the propagation constant of the NAWs; $\lambda_{Dq} = (\mathcal{Z}\hbar c \mathcal{N}_{e0}^{1/3}/4\pi \mathcal{N}_0 \mathcal{Z}^2 e^2)^{1/2}$ and $\tau_p = \omega_p^{-1} = (m/4\pi \mathcal{N}_0 \mathcal{Z}^2 e^2)^{1/2}$ are, respectively, the length scale and the time scale (inverse of the nucleus plasma frequency) of the NAWs; $C_q = \lambda_{Dq}/\tau_p = (\gamma \mathcal{P}_{e0}/\rho_n)^{1/2} = (\mathcal{Z}\hbar c \mathcal{N}_{e0}^{1/3}/m)^{1/2}$ is the speed of the NAWs, in which $\rho_n = m\mathcal{N}_0$ is the nucleus mass density, $\mathcal{N}_0 = \mathcal{N}_{e0}/\mathcal{Z}$ is the equilibrium nucleus number density, and m ($\mathcal{Z}$) is the mass (charge state) of the nucleus species, and e is the charge of the proton.

The dispersion relation defined by Equation (4) for the long wavelength NAWs ($k\lambda_{Dq} \ll 1$) becomes $\omega \simeq k C_q$. There is an important issue on the basic differences between IAWs and NAWs since the form of their dispersion relations are identical. Their basic differences can be pinpointed as follows:

- The IAWs are driven by the electron thermal pressure depending on the electron temperature and number density, whereas the NAWs are driven by the electron degenerate pressure depending only on the electron number density.
- The non-degenerate plasmas at finite temperature allow the IAWs to exist, but do not allow the NAWs to exist.
- The degenerate plasmas at absolute zero temperature do not allow the IAWs to exist, but do allow the NAWs to exist.
- The NAWs and IAWs are completely different from the view of their length scale and phase speed.

The present paper is attempted to study the basic characteristics of cylindrical as well as spherical solitary and shock waves associated with the NAWs (defined by Equation (4)) in the CDENPs under consideration. The paper is structured as follows. The normalized basic equations describing the nonlinear dynamics of the NAWs in the CDENPs under consideration are provided in Section 2. To study cylindrical and spherical solitary waves, a modified Korteweg-de Vries (MK-dV) equation is obtained and properly examined in

Section 3. To identify the basic features of the cylindrical and spherical shock waves, a modified Burgers (MBurgers) equation is also obtained and critically examined in Section 4. A brief discussion is given in Section 5.

2. Basic Equations

The CDENPs containing the CUDE gas [3–6,26,27] and the cold viscous fluid of any nucleus like $^{1}_{1}H$ or [3–5] or $^{4}_{2}He$ or $^{12}_{6}C$ or $^{16}_{8}O$ [6,26,27] are considered. The macroscopic state of such CDENPs is described in nonplanar geometry as

$$\frac{\partial \Phi}{\partial R} = \frac{1}{eN_e}\frac{\partial P_e}{\partial R}, \tag{5}$$

$$\frac{\partial \mathcal{N}}{\partial T} + \frac{1}{R^\nu}\frac{\partial}{\partial R}(R^\nu \mathcal{N}\mathcal{U}) = 0, \tag{6}$$

$$\frac{\partial \mathcal{U}}{\partial T} + \mathcal{U}\frac{\partial \mathcal{U}}{\partial R} = -\frac{Ze}{m}\frac{\partial \Phi}{\partial R} - \eta_n\frac{\partial^2 \mathcal{U}}{\partial R^2}, \tag{7}$$

$$\frac{1}{R^\nu}\frac{\partial}{\partial R}\left(R^\nu\frac{\partial \Phi}{\partial R}\right) = 4\pi e(\mathcal{N}_e - Z\mathcal{N}), \tag{8}$$

where $\nu = 1$ and $\nu = 2$ represent the cylindrical and spherical geometries, respectivel, $\mathcal{N}$ is the nucleus fluid number density; $\mathcal{U}$ is the nucleus fluid speed, Φ is the electrostatic potential, m and Ze are, respectively, the mass and charge of the nucleus species, T and R are the time and space variables, respectively, and η_n is the coefficient of dynamic viscosity for the cold nucleus fluid. To note is that in Equation (5), the inertia of the CUDE gas is negligible compared to that of the viscous nucleus fluid, and that in Equation (7) the effects of the self-gravitational field and nucleus degeneracy are negligible in comparison with those of the electrostatic field and electron degeneracy, respectively.

To describe the equilibrium state of the CDENPs under consideration, it is reasonably assumed that $\mathcal{N} = \mathcal{N}_0, \mathcal{U} = 0$, and $\Phi = 0$ at equilibrium. Thus, the equilibrium state of the CDENPs under consideration is described by

$$\mathcal{N}_{e0} = Z\mathcal{N}_0, \tag{9}$$

$$\mathcal{P}_{e0} = \mathcal{K}, \tag{10}$$

where Equation (9) represents the equilibrium charge neutrality condition, and in Equation (10), $\mathcal{K}$ is the integration constant, given by Equation (3).

To find the expression for the normalized ultra-relativistically degenerate electron number density n_e in terms of the normalized electrostatic potential, $\phi = 3e\mathcal{N}_{e0}\Phi/4P_{e0}$, first, substitute Equation (2) and $R = r\lambda_{Dq}$ (where r is the normalized space variable) into Equation (5). Thus, Equation (5) reduces to

$$\frac{\partial \phi}{\partial r} = n_e^{-\frac{2}{3}}\frac{\partial n_e}{\partial r} = 3\frac{\partial n_e^{\frac{1}{3}}}{\partial r}. \tag{11}$$

Next, integrating Equation (11), and obtaining the integration constant as 3 (since $\phi = 0$ and $n_e = 1$ at equilibrium), n_e can finally be expressed as

$$n_e = \left(1 + \frac{\phi}{3}\right)^3. \tag{12}$$

To normalize Equations (6)–(8), $\mathcal{N} = n\mathcal{N}_0$, $\mathcal{U} = uC_q$, $\Phi = 4\phi P_{e0}/3\mathcal{N}_{e0}e$, $T = t\tau_p$, $R = r\lambda_{Dq}$, $\eta_n = \eta\lambda_{Dq}C_q$, and Equation (12) are substituted into Equations (6)–(8). The nonlinear propagation of the NAWs in CDENPs is, therefore, governed by the following normalized equations:

$$\frac{\partial n}{\partial t} + \frac{1}{r^{\nu}}\frac{\partial}{\partial r}(r^{\nu} n u) = 0, \tag{13}$$

$$\frac{\partial u}{\partial t} + u\frac{\partial u}{\partial r} = -\frac{\partial \phi}{\partial r} - \eta\frac{\partial^2 u}{\partial r^2}, \tag{14}$$

$$\frac{1}{r^{\nu}}\frac{\partial}{\partial r}\left(r^{\nu}\frac{\partial \phi}{\partial r}\right) = 1 + \phi + \frac{\phi^2}{3} + \frac{\phi^3}{27} - n. \tag{15}$$

Let us note that the length scale, (λ_{Dq}), and the NA speed, (C_q), depend on the CUDE pressure, which is given by Equation (3), and that the simple form of the normalized basic equations given by Equations (13)–(15) are obtained by the special choice of the normalization used.

3. MK-dV Equation

The MK-dV equation for the nonlinear propagation of the NAWs in the CDENPs is derived by the reductive perturbation technique (RPT) which requires first the stretching of the independent variables, r and t as [28]

$$\xi = \epsilon^{\frac{1}{2}}(r - V_p t), \tag{16}$$

$$\tau = \epsilon^{\frac{3}{2}}t, \tag{17}$$

and next the expansion of the dependent variables, n, u and ϕ as [28]

$$n = 1 + \epsilon n^{(1)} + \epsilon^2 n^{(2)} + \cdots, \tag{18}$$

$$u = \epsilon u^{(1)} + \epsilon^2 u^{(2)} + \cdots, \tag{19}$$

$$\phi = \epsilon \phi^{(1)} + \epsilon^2 \phi^{(2)} + \cdots, \tag{20}$$

where $V_p = \omega/kC_q$ is the normalized NAW phase speed, ξ is normalized by λ_{Dq}, τ is normalized by τ_p, and ϵ is a smallness parameter satisfying $0 < \epsilon < 1$.

Using Equations (16)–(20) in the system (13)–(15), taking the coefficients of $\epsilon^{3/2}$ from Equation (13) as well as from Equation (14), and the coefficients of ϵ from Equation (15), one obtains:

$$n^{(1)} = \frac{u^{(1)}}{V_p}, \tag{21}$$

$$u^{(1)} = \frac{\phi^{(1)}}{V_p}, \tag{22}$$

$$V_p = 1. \tag{23}$$

The relation (23), representing $\omega = kC_q$, is the linear dispersion relation for the long wavelength NAWs, which can also be obtained from Equation (4) for a long wavelength limit, $k\lambda_{Dq} \ll 1$. This means that the RPT, utilized here, is valid for the long wavelength NAWs, and that the phase speed of the long wavelength NAWs is directly proportional to the square root of the degenerate pressure of the CUDE gas, while inversely proportional to the square root of the mass density of cold nucleus fluid. Thus, in the NAWs, the pressure of the CUDE gas gives rise to the restoring force, and the mass density of the nucleus fluid gives rise to the inertia.

Again, using Equations (16)–(20) in Equations (13)–(15), keeping the coefficients of $\epsilon^{5/2}$ from Equation (13) as well as from Equation (14), and keeping the coefficients of ϵ^2 from Equation (15), one obtains:

$$\frac{\partial n^{(1)}}{\partial \tau} + \frac{\partial}{\partial \xi}\left[u^{(2)} + n^{(1)}u^{(1)} - \mathcal{V}_p n^{(2)}\right] + \frac{\nu}{\mathcal{V}_p \tau}u^{(1)} = 0, \tag{24}$$

$$\frac{\partial u^{(1)}}{\partial \tau} + \frac{\partial}{\partial \xi}\left[\phi^{(2)} + \frac{1}{2}[u^{(1)}]^2 - \mathcal{V}_p u^{(2)}\right] = 0, \tag{25}$$

$$\frac{\partial^2 \phi^{(1)}}{\partial \xi^2} - \phi^{(2)} - \frac{1}{3}\left[\phi^{(1)}\right]^2 + n^{(2)} = 0. \tag{26}$$

Now, using Equations (21)–(26), $\phi^{(2)}$, $u^{(2)}$ and $n^{(2)}$ can be eliminated to obtain the MK-dV equation in the form:

$$\frac{\partial \phi^{(1)}}{\partial \tau} + \frac{\nu}{2\tau}\phi^{(1)} + \mathcal{A}\phi^{(1)}\frac{\partial \phi^{(1)}}{\partial \xi} + \mathcal{B}\frac{\partial^3 \phi^{(1)}}{\partial \xi^3} = 0, \tag{27}$$

where $\mathcal{A} = 7/6$ and $\mathcal{B} = 1/2$ are the nonlinear and dispersion coefficients, respectively.

Let us note that the second term of the MK-dV Equation (27) is due to the effect of cylindrical or spherical geometry, which disappears for a large value of τ. To examine the effects of cylindrical and spherical geometries on the NA solitary waves in the CDENPs under consideration, one has to solve the MK-dV Equation (27) numerically by using the stationary solitary wave solution [29] of Equation (27) with $\nu = 0$ as an initial profile,

$$\phi = \phi_0 \operatorname{sech}^2\left(\frac{\zeta}{\Delta}\right), \tag{28}$$

where $\phi = \phi^{(1)}$, $\zeta = \xi - \mathcal{U}_0\tau$ with $\mathcal{U}_0$ and ζ being normalized by $\mathcal{C}_q$ and λ_{Dq}, respectively, and $\phi_0 = 3\mathcal{U}_0/\mathcal{A}$ and $\Delta = 2\sqrt{\mathcal{B}/\mathcal{U}_0}$ are the normalized amplitude and width of the initial pulse, respectively.

The positive values of $\mathcal{A}$ and $\mathcal{B}$ along with Equation (28) (with $\phi_0 = 3\mathcal{U}_0/\mathcal{A}$, $\Delta = 2\sqrt{\mathcal{B}/\mathcal{U}_0}$ and $\mathcal{U}_0 > 0$) indicate that the CDENPs under consideration support cylindrical as well as spherical solitary waves with $\phi > 0$. The MK-dV Equation (27) is numerically solved and analyzed for nonplanar ($\nu = 1$ and $\nu = 2$) geometries. Let us notice that $\tau < 0$ means that the solitary waves propagate inward the direction of the cylinder or sphere [30]. It is also used to converse the numerical solution of the MK-dV equation given by Equation (27). The results are displayed in Figure 1.

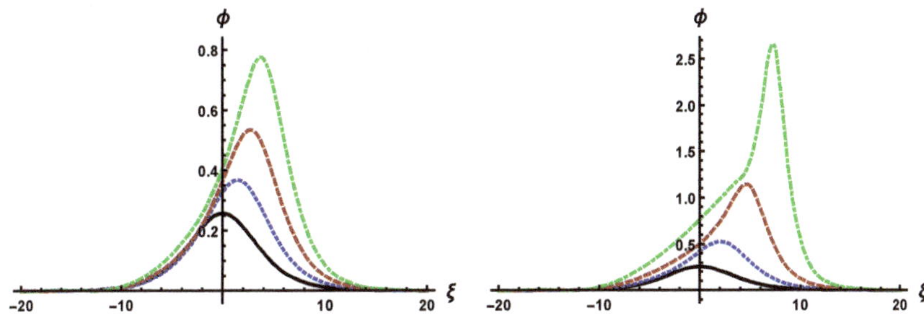

Figure 1. Time evolution of (**left panel**) cylindrical ($\nu = 1$) and (**right panel**) spherical ($\nu = 2.0$) nucleus-acoustic (NA) solitary waves in the cold degenerate electron-nucleus plasmas (CDENPs) under consideration for $\mathcal{U}_0 = 0.1$, $\tau = -20$ (solid line), -10 (dotted line), -5 (dashed line), and -2.5 (dashed-dotted line). See text for details.

It shows that the time evolution of the solitary waves in the CDENPs under consideration are significantly modified by the effects of cylindrical and spherical geometries. It is observed from Figure 1 that the amplitude of the spherical solitary waves is approximately two times higher than that of the cylindrical ones, and that the time evolution of the spherical solitary waves is faster than that of the cylindrical ones.

4. MBurgers Equation

To derive the MBurgers equation for the nonlinear propagation of the NAWs, one can again employ the RPT [28], but exploit different stretching of the independent variables r and t as [31,32]

$$\xi = \epsilon(r - V_p t), \tag{29}$$

$$\tau = \epsilon^2 t. \tag{30}$$

Now, using Equations (29), (30) and (18)–(20) in the system (13)–(15), and taking the coefficients of ϵ^2 from Equations (13) and (14), and the coefficients of ϵ from Equation (15), a set of Equations (21)–(23) is obtained. However, using Equations (29), (30), (18)–(20) in Equations (13)–(15), and again taking the coefficients of ϵ^3 from Equations (13) and (14), and the coefficients of ϵ^2 from Equation (15), one obtains:

$$\frac{\partial n^{(1)}}{\partial \tau} + \frac{\partial}{\partial \xi}\left[u^{(2)} + n^{(1)}u^{(1)} - V_p n^{(2)}\right] + \frac{\nu}{V_p \tau}u^{(1)} = 0, \tag{31}$$

$$\frac{\partial u^{(1)}}{\partial \tau} + \frac{\partial}{\partial \xi}\left[\phi^{(2)} + \frac{1}{2}[u^{(1)}]^2 - V_p u^{(2)}\right] = \eta\frac{\partial^2 u^{(1)}}{\partial r^2}, \tag{32}$$

$$\phi^{(2)} + \frac{1}{3}\left[\phi^{(1)}\right]^2 - n^{(2)} = 0. \tag{33}$$

Using Equations (21)–(23) and (31)–(33), $\phi^{(2)}$, $u^{(2)}$ and $n^{(2)}$ can be eliminated to obtain the MBurgers Equation (34) in the form:

$$\frac{\partial \phi^{(1)}}{\partial \tau} + \frac{\nu}{2\tau}\phi^{(1)} + A\phi^{(1)}\frac{\partial \phi^{(1)}}{\partial \xi} = C\frac{\partial^2 \phi^{(1)}}{\partial \xi^2}, \tag{34}$$

where $C = \eta/2$ is the dissipation coefficient. One can also see that the second term of the MBurgers Equation (34) is due to the effect of cylindrical or spherical geometry, which disappears for a large value of τ.

To define shock wave solution clearly, first, consider $\nu = 0$ in the MBurgers Equation (34). The latter (for $\nu = 0$) can be expressed as:

$$\frac{\partial \phi^{(1)}}{\partial \tau} + A\phi^{(1)}\frac{\partial \phi^{(1)}}{\partial \xi} = C\frac{\partial^2 \phi^{(1)}}{\partial \xi^2}, \tag{35}$$

which is the standard Burgers equation. To obtain the stationary shock wave solution of this standard Burgers equation, a frame moving ($\zeta = \xi - \mathcal{U}_0\tau;\ \tau' = \tau$) with the constant speed $\mathcal{U}_0$, the steady state condition ($\partial\phi^{(1)}/\partial\tau' = 0$) and $\phi^{(1)} = \phi$ are assumed. These assumptions reduce Equation (35) to

$$\frac{d\phi}{d\zeta} = -\frac{U_0}{C}\phi + \frac{A}{2C}\phi^2, \tag{36}$$

where the integration constant is found to be zero, since $\phi \to 0$ and $d\phi/d\zeta \to 0$ at $\zeta \to \infty$. Now, skipping few steps of mathematics of undergraduate level, the shock wave solution of Equation (6) is given by [31]

$$\phi = \frac{1}{2}\phi_m\left[1 - \tanh\left(\frac{\zeta}{\delta}\right)\right], \tag{37}$$

where $\phi_m = 2\mathcal{U}_0/\mathcal{A}$ and $\delta = 2\mathcal{C}/\mathcal{U}_0$ are, respectively, the amplitude and thickness of the shock waves. Equation (37) represents the stationary shock wave solution of the MBurgers Equation (35).

The positive values of $\mathcal{A}$ and $\mathcal{C}$ along with Equation (37) imply that the CDENPs under consideration support cylindrical as well as spherical shock waves with $\phi > 0$.

To examine the effects of cylindrical and spherical geometries on the NA shock waves in the CDENPs under consideration, one has to solve the MBurgers Equation (34) numerically by using the stationary shock wave solution of Equation (34) with $\nu = 0$ as the initial profile given by Equation (37). The MBurgers equation is numerically solved for cylindrical and spherical geometries. Let us again notice that $\tau < 0$ means that the shock waves propagate inward direction of the cylinder and sphere [30], and that $\tau < 0$ is used to converse its numerical solution. The results are shown in Figure 2.

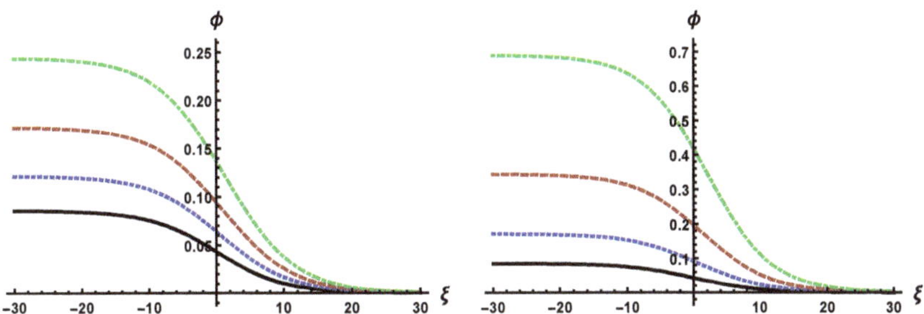

Figure 2. Time evolution of (**left panel**) cylindrical ($\nu = 1$) and (**right panel**) spherical ($\nu = 2$) NA shock waves in the CDENPs under consideration for $\mathcal{U}_0 = 0.1$, $\eta = 1$, $\tau = -20$ (solid line), -10 (dotted line), -5.0 (dashed line), and -2.5 (dashed-dotted line). See text for details.

The numerical results, shown in Figure 2, point out that the time evolution of the NA shock waves in the CDENPs under consideration are significantly modified by the effects of the cylindrical and spherical geometries. The numerical results observed in Figure 2 indicate that the amplitude of the spherical shock waves is approximately two times higher than that of the cylindrical ones, and that the time evolution of the spherical shock waves is faster than that of the cylindrical ones.

The profiles represented by the analytic solution of the standard Burgers Equation (37) or those obtained from the numerical solutions of the MBurgers Equation (34) are known as shock waves. The latter are formed when the effect of this dissipation represented by the term containing $\mathcal{C}$ is balanced by that of the nonlinearity represented by the term containing $\mathcal{A}$.

5. Discussion

The nonlinear propagation of the NAWs in the CDENPs composed of CUDE gas [3–5] and the viscous fluid of nucleus of any element like ${}^{1}_{1}\text{H}$ [3–5] or ${}^{4}_{2}\text{He}$ or ${}^{12}_{6}\text{C}$ or ${}^{16}_{8}\text{O}$ [6,26,27] has been considered to identify the characteristics of the nonlinear waves formed in the CDENPs under consideration. The results obtained from current study study are as follows:

- The phase speed of the NAWs is given by

$$C_q = \sqrt{\frac{\gamma P_{e0}}{\rho_n}} = \sqrt{\frac{\mathcal{Z}\hbar c}{m\Lambda_e}}, \tag{38}$$

where Equation (3) is used, and $\Lambda_e = \mathcal{N}_{e0}^{-1/3}$ is the inter-electron distance. This expression indicates that C_q^2 is inversely proportional to Λ_e and the mass, m, of a nucleus species, but is directly proportional to the number of protons, $\mathcal{Z}$, in the nucleus species. The phase speed does not depend on the temperature of the electron or nucleus species. This is an unique feature of the NAWs by which the NAWs appeared as new waves, and are completely different from the IAWs [23,24] which do not exist at absolute zero-temperature.

- The dimensional amplitudes of both types of nonlinear waves are determined by using Equations (3) and (38), and are expressed as:

$$\frac{9}{7}\sqrt{\frac{\hbar c}{\Lambda_e}\frac{m\mathcal{V}_0^2}{\mathcal{Z}e^2}} \quad \text{and} \quad \frac{27}{28}\sqrt{\frac{\hbar c}{\Lambda_e}\frac{m\mathcal{V}_0^2}{\mathcal{Z}e^2}}, \tag{39}$$

where $\mathcal{V}_0$ is the dimensional speed of the frame of reference. These expressions imply that the amplitudes of the both types of nonlinear waves are directly proportional to $\mathcal{V}_0$, and to the square root of the mass of the nucleus species, $\sqrt{m}$, but inversely proportional to the square root of the inter-electron distance, $\sqrt{\Lambda_e}$, and the number of the proton, $\sqrt{\mathcal{Z}}$, in a nucleus species.

- The dimensional widths of both types of nonlinear waves are given by

$$\sqrt{2}\lambda_{Dq}\sqrt{\frac{C_q}{\mathcal{V}_0}} \quad \text{and} \quad \frac{\eta_n}{\mathcal{V}_0}. \tag{40}$$

These expressions imply that the width of the solitary waves is the order of a fraction of the length scale, λ_{Dq}, of the waves, since C_q is a fraction of $\mathcal{V}_0$ for the formation of the NA solitary waves. The width of the NA shock waves increases with the dynamical viscosity coefficient, η_n, of the nucleus fluid, but decreases with the speed, $\mathcal{V}_0$.

- The amplitude (width) of the cylindrical NA solitary and shock waves is smaller (larger) than that of the spherical NA solitary and shock waves. The time evolution of the spherical solitary and shock waves is faster than that of the NA cylindrical solitary and shock waves.
- The amplitude (width) of the NA solitary waves is minimum (maximum) for a very large value of τ, which causes to neglect the effect of cylindrical and spherical geometries, and gives rise to one dimensional (1D) planar NA solitary and shock waves. Thus, for a large value of τ, 1D planar, cylindrical and spherical solitary and shock waves are found to be identical.
- The length scale as well as the phase speed, height, and thickness of the NA solitary and shock waves are completely independent of temperature. These are completely new linear and nonlinear features of the NAWs under consideration.

The exact analytical solutions of Equations (27) and (34) are difficult to be obtained because of the nonplanar term (containing ν), where a singularity arises at $\tau = 0$. A class of analytical solutions of Equation (27) was obtained from the solution of the standard K-dV equation [33,34]. However, we are interested to find a solitary wave solution of Equation (27) with the standard boundary condition, viz., $\lim_{\xi\to-\infty}\phi(\xi,\tau) = \lim_{\xi\to\infty}\phi(\xi,\tau)$. Therefore, Equation (27) was solved numerically in order to find the spatiotemporal evolution of an initially imposed solitary profile at $\tau = \tau_{\min} < 0$ with the standard boundary conditions in (ξ,τ) domain. It was also assumed that the solution $\phi(\xi,\tau)$ along with its derivative tends to zero as $\xi \to \pm\infty$. Further, the solutions of Equations (27) and (34) with $\nu = 0$ as an initial profiles (i.e., $\phi(\xi,\tau_{\min} < 0) = \phi_0 \operatorname{sech}^2(\xi/\Delta)$ for Equation (27) and $\phi(\xi,\tau_{min} < 0) = (\phi_m/2)(1 - \tanh[\xi/\delta])$ for Equation (34)) were used. The finite difference method was used for numerical solutions. On the other hand, the traveling wave solutions [35] of combined K-dV-modified K-dV equations as well as complexly coupled-K-dV equation are obtained by utilizing the technique of the Bäcklund transformation.

Recently, the trace of nuclei of massive elements, such as $^{56}_{26}\mathrm{Fe}$, $^{85}_{37}\mathrm{Rb}$, $^{96}_{42}\mathrm{Mo}$, etc. in white dwarf and neutron stars has also been predicted [36,37]. The densities of the stars are small to neglect their roles in the formation of the NA solitary and shock waves in the CDENPs [3–6,26,27] under consideration.

Let us add here that the roles of magnetic field and rotation of neutron stars in the formation of the NA solitary and shock waves are also important problems, but those are beyond. the scope of the present study. However, the theory, presented here, is valid for the long wavelength electrostatic NAWs propagating along the magnetic lines of force of white dwarfs and non-rotating neutron stars.

Funding: This research received no external funding.

Acknowledgments: The work is dedicated to an eminent space and astrophysicist, Reinhard Schlickeiser, who supervised the author during his research stay at Ruhr Universität Bochum (Germany) as a Friedrich Wilhelm Bessel Research Awardee, on the occasion of his 70th Birth Anniversary. The author wishes him a very long active and cherished life on the same occasion.

Conflicts of Interest: The author declares no conflict of interest.

References

1. Mamun, A.A.; Amina, M.; Schlickeiser, R. Nucleus-acoustic shock (waves in a strongly coupled self-gravitating degenerate quantum plasma. *Phys. Plasmas* **2016**, *23*, 094503. [CrossRef]
2. Mamun, A.A.; Amina, M.; Schlickeiser, R. Heavy NA spherical solitons in self-gravitating super-dense plasmas. *Phys. Plasmas* **2017**, *24*, 042307. [CrossRef]
3. Chandrasekhar, S. The highly collapsed configurations of a stellar mass. *Mon. Not. R. Astron. Soc.* **1931**, *91*, 456–4566. [CrossRef]
4. Chandrasekhar, S. The maximum mass of ideal white dwarfs. *Astrophys. J.* **1931**, *74*, 81–82. [CrossRef]
5. Chandrasekhar, S. The pressure in the interior of a star. *Mon. Not. R. Astron. Soc.* **1936**, *96*, 644–646. [CrossRef]
6. Shapiro, S.L.; Teukolsky, S.A. *Black Holes, White Dwarfs, and Neutron Stars: The Physics of Compact Objects*, 1st ed.; Wiley-VCH Verlag: Weinheim, Germany, 1983; pp. 15–375.
7. Potekhin, A.Y.; Chabrier, G. Thermodynamic functions of dense plasmas: Analytic approximations for astrophysical applications. *Contrib. Plasma Phys.* **2010**, *50*, 82–87. [CrossRef]
8. Fletcher, R.S.; Zhang, X.L.; Rolston, S.L. Observation of collective modes of ultracold plasmas. *Phys. Rev. Lett.* **2006**, *96*, 105003. [CrossRef]
9. Glenzer, S.H.; Redmer, R. X-ray Thomson scattering in high energy density plasmas. *Rev. Mod. Phys.* **2009**, *81*, 1625. [CrossRef]
10. Drake, R.P. Perspectives on high-energy-density physics. *Phys. Plasmas* **2009**, *16*, 055501. [CrossRef]
11. Drake, R.P. High-energy-density physics. *Phys. Today* **2010**, *63*, 18. [CrossRef]
12. Hu, S.X.; Collins, L.A.; Boehly, T.R.; Kress, J.D.; Goncharov, V.N.; Skupsky, S. First-principles thermal conductivity of warm-dense deuterium plasmas for inertial confinement fusion applications. *Phys. Rev. E* **2014**, *89*, 04105. [CrossRef]
13. Shukla, P.K.; Mamun, A.A.; Mendis, D.A. Nonlinear ion modes in a dense plasma with strongly coupled ions and degenerate electron fluids. *Phys. Rev. E* **2011**, *84*, 026405. [CrossRef]
14. Sultana, S.; Schlickeiser, R. Fully nonlinear heavy ion-acoustic solitary waves in astrophysical degenerate relativistic quantum plasmas. *Astrophys. Space Sci.* **2018**, *363*, 1–9. [CrossRef]
15. Sultana S.; Schlickeiser, R. Arbitrary amplitude nucleus-acoustic solitons in multi-ion quantum plasmas with relativistically degenerate electrons. *Phys. Plasmas* **2018**, *25*, 022110. [CrossRef]
16. Sultana, S.; Islam S.; Mamun, A.A.; Schlickeiser, R. Modulated heavy nucleus-acoustic waves and associated rogue waves in a degenerate relativistic quantum plasma system. *Phys. Plasmas* **2018**, *25*, 012113. [CrossRef]
17. Chowdhury, N.A.; Hasan, M.M.; Mannan, A.; Mamun, A.A. Nucleus-acoustic envelope solitons and their modulational instability in a degenerate quantum plasma system. *Vacuum* **2018**, *147*, 31–37. [CrossRef]
18. Karmakar, P.K.; Das, P. Nucleus-acoustic waves: Excitation, propagation, and stability. *Phys. Plasmas* **2018**, *25*, 082902. [CrossRef]
19. Das, P.; Karmakar, P.K. Nonlinear nucleus-acoustic waves in strongly coupled degenerate quantum plasmas. *Europhys. Lett.* **2019**, *126*, 10001. [CrossRef]
20. Mannan, A.; Sultana, S.; Mamun, A.A. Arbitrary amplitude heavy nucleus-acoustic solitary waves in thermally degenerate plasmas. *IEEE Trans. Plasma Sci.* **2020**, *48*, 4093–4102. [CrossRef]
21. Kaur, R.; Singh, K.; Saini, N.S. Heavy-and light-nuclei acoustic dressed shock waves in white dwarfs. *Chin. J. Phys.* **2021**, *72*, 286–298. [CrossRef]
22. Saini, N.S.; Kaur, R. Ion-acoustic solitary, breathers, and freak waves in a degenerate quantum plasma. *Waves Random Complex Media* **2021**, 1–22. [CrossRef]
23. Tonks, L.; Langmuir, I. Oscillations in ionized gases. *Phys. Rev.* **1929**, *33*, 95–210. [CrossRef]
24. Revans, R.W. The transmission of waves through an ionized gas. *Phys. Rev.* **1933**, *44*, 798–902. [CrossRef]
25. Mamun, A.A. Degenerate pressure driven modified nucleus-acoustic waves in degenerate plasmas. *Phys. Plasmas* **2018**, *25*, 024502. [CrossRef]
26. Koester, D.; Chanmugam, G. Physics of white dwarf stars. *Rep. Prog. Phys.* **1990**, *53*, 837–915. [CrossRef]
27. Koester, D. White dwarfs: Recent developments. *Astron. Astrophys. Rev.* **2002**, *11*, 33–66. [CrossRef]

28. Washimi, H.; Taniuti, T. Propagation of ion-acoustic solitary waves of small amplitude. *Phys. Rev. Lett.* **1966**, *17*, 996–997. [CrossRef]
29. Mamun, A.A.; Shukla, P.K. Effects of nonthermal distribution of electrons and polarity of net dust-charge number density on nonplanar dust-ion-acoustic solitary waves. *Phys. Rev. E* **2009**, *80*, 037401. [CrossRef] [PubMed]
30. Maxon, S.; Viecelli, J. Spherical solitons. *Phys. Rev. Lett.* **1974**, *32*, 4–6. [CrossRef]
31. Mamun, A.A.; Cairns, R.A. Dust-acoustic shock waves due to strong correlation among arbitrarily charged dust. *Phys. Rev. E* **2009**, *79*, 055401. [CrossRef]
32. Mamun, A.A. On stretching of plasma parameters and related open issues for the study of dust-ion-acoustic and dust-acoustic shock waves in dusty plasmas. *Phys. Plasmas* **2019**, *26*, 084501. [CrossRef]
33. Hirota, R. Exact solutions to the equation describing "cylindrical solitons". *Phys. Lett. A* **1979**, *71*, 393–394. [CrossRef]
34. Mannan, A.; Fedele, R.; Onorato, M.; De Nicola, S.; Jovanović, D. Ring-type multi-soliton dynamics in shallow water. *Phys. Rev. E* **2015**, *91*, 012921. [CrossRef] [PubMed]
35. Yokus, A.; Kaya, D. Comparison exact and numerical simulation of the traveling wave solution in nonlinear dynamics. *Int. J. Modern Phys. B* **2020**, *34*, 2050282. [CrossRef]
36. Witze, A. Space-station science ramps up. *Nature* **2014**, *510*, 196–197. [CrossRef] [PubMed]
37. Vanderburg, A.; Johnson, J.A.; Rappaport, S.; Bieryla, A.; Irwin, J.; Lewis, J.A.; Kipping, D.; Brown, W.R.; Dufour, P.; Ciardi, D.R.; et al. A disintegrating minor planet transiting a white dwarf. *Nature* **2015**, *526*, 546–549. [CrossRef]

Article

Studying the Influence of External Photon Fields on Blazar Spectra Using a One-Zone Hadro-Leptonic Time-Dependent Model

Michael Zacharias [1,2]

[1] Laboratoire Univers et Théorie, Observatoire de Paris, Université PSL, Université de Paris, CNRS, F-92190 Meudon, France; michael.zacharias@obspm.fr or mzacharias.phys@gmail.com

[2] Centre for Space Science, North-West University, Potchefstroom 2520, South Africa

Abstract: The recent associations of neutrinos with blazars require the efficient interaction of relativistic protons with ambient soft photon fields. However, along side the neutrinos, γ-ray photons are produced, which interact with the same soft photon fields producing electron-positron pairs. The strength of this cascade has significant consequences on the photon spectrum in various energy bands and puts severe constraints on the pion and neutrino production. In this study, we discuss the influence of the external thermal photon fields (accretion disk, broad-line region, and dusty torus) on the proton-photon interactions, employing a newly developed time-dependent one-zone hadro-leptonic code *OneHaLe*. We present steady-state cases, as well as a time-dependent case, where the emission region moves through the jet. Within the limits of this toy study, the external fields can disrupt the "usual" double-humped blazar spectrum. Similarly, a moving region would cross significant portions of the jet without reaching the previously-found steady states.

Keywords: non-thermal radiation mechanisms; relativistic jets; relativistic processes; BL Lacertae objects

Citation: Zacharias, M. Studying the Influence of External Photon Fields on Blazar Spectra Using a One-Zone Hadro-Leptonic Time-Dependent Model. *Physics* **2021**, *3*, 1098–1111. https://doi.org/10.3390/physics3040069

Received: 22 September 2021
Accepted: 8 November 2021
Published: 18 November 2021

Publisher's Note: MDPI stays neutral with regard to jurisdictional claims in published maps and institutional affiliations.

1. Introduction

The theory of the blazar emission was transformed in the early 1990s by the introduction of the so-called *external-Compton* scenario. The scenario explains the high-energy component of the spectral energy distribution (SED) through relativistic electrons inverse-Compton (IC) scattering soft, thermal photon fields that originate outside the jet. This transformation of blazar research was significantly driven by the works of Reinhard Schlickeiser and collaborators employing the accretion disk (AD) as a source for soft external photons [1–4].

Blazars, a sub-class of active galaxies, are indeed peculiar objects with—in the words of Reinhard Schlickeiser [5]—

"properties [that] include high optical polarization, extreme optical variability, flat-spectrum radio emission associated with a compact core, and apparent superluminal motion. Such properties are thought to be produced by those few, rare extragalactic radio galaxies and quasars that are favorably aligned to permit us to look almost directly down a relativistically outflowing jet of matter expelled from a supermassive black hole."

Despite the decades of research that have passed since the advent of the external-Compton model, a clear consensus on the source of the γ-ray emission in blazars has not yet been reached.

The low-energy component of the double-humped SED is the least controversial part, as synchrotron emission of relativistic electrons fits all the required properties (including the aforementioned polarization). However, the nature of the high-energy component is subject of intensive discussions. Within leptonic models, it is explained through IC emission–either with the self-made synchrotron photons (synchrotron-self Compton, SSC) or with external photons, such as the AD, photons from the broad-line region (BLR) [6], the dusty torus (DT) [7,8], or even the cosmic microwave background (CMB) [9,10]. However, if relativistic jets are also capable of accelerating protons, the γ-rays could also originate from such interactions, through direct proton-synchrotron emission, or via proton-photon interactions, causing a cascade of pairs [11–15]. Especially the production of pions would have the capability to discriminate between the leptonic and the hadronic scenario, as it would also produce neutrinos. While neutrinos have been associated with blazars recently [16,17], the significances are not yet sufficient to claim a real detection. Nonetheless, the discussion is ongoing [18], and the upcoming neutrino observatories KM3NET and IceCube-Gen2 may provide the definitive answer.

In this study, the hadro-leptonic model is combined with the external soft photons, to study their influence on the resulting pair cascade and the jet emission. A newly developed time-dependent, one-zone hadro-leptonic code—*OneHaLe*—is introduced in Section 2. It is used in Section 3 to study the influence of the external photon fields by first calculating steady-state spectra at various locations within the jet, as the region of influence of the soft photon fields on the jet is strongly distance-dependent. Subsequently, we present the case of an emission region moving outward passing through the various external photon fields. We note that the study conducted is a toy model: In order to properly identify the influence of the external fields, all other parameters of the emission region remain the same, irrespective of the location. This may have significant consequences for the emerging spectra. Section 4 provides the discussion of the results and the conclusions.

2. Code Description

The code is based on the recently developed extended hadro-leptonic steady-state code *ExHaLe-jet* [19]. In fact, the fundamental equations governing the particle and radiation processes are the same, and we only provide a brief overview here describing the free parameters. In the following, quantities in the host galaxy frame are marked with a hat, while quantities in the observer's frame are marked by the superscript "obs". Unmarked quanitites are either in the co-moving frame of the emission region or invariant.

A spherical emission region is assumed with radius R located a distance z_0 from the black hole within the jet, pervaded by a tangled magnetic field of strength B. The emission region moves with bulk Lorentz factor Γ under a viewing angle θ^{obs} with respect to the observer's line-of-sight implying a Doppler factor, $\delta = [\Gamma(1 - \beta_\Gamma \cos \theta^{\mathrm{obs}})]^{-1}$, where $\beta_\Gamma = \sqrt{1 - \Gamma^{-2}}$.

The Fokker-Planck equation governing the time-dependent evolution of a given particle species i (protons, charged pions, muons, or electrons) with spectral density $n_i(\chi)$ is given as

$$\frac{\partial n_i(\chi,t)}{\partial t} = \frac{\partial}{\partial \chi}\left[\frac{\chi^2}{(a+2)t_{\mathrm{acc}}}\frac{\partial n_i(\chi,t)}{\partial \chi}\right]$$
$$- \frac{\partial}{\partial \chi}(\dot{\chi}_i n_i(\chi,t)) + Q_i(\chi,t) - \frac{n_i(\chi,t)}{t_{\mathrm{esc}}} - \frac{n_i(\chi,t)}{\gamma t^*_{i,\mathrm{decay}}}. \tag{1}$$

For numerical reasons, we use the normalized particle momentum, $\chi = p_i/(m_i c) = \gamma\beta$, where $p_i = \gamma m_i \beta c$ is the particle momentum, m_i is the particle mass, c the speed of light, γ the particle's Lorentz factor, and $\beta = \sqrt{1 - \gamma^{-2}}$. The first term on the right-hand side of Equation (1) describes Fermi-II acceleration through scattering of particles on magnetohydrodynamic waves. The parametrization of [20] is used with $a = 9v_s^2/4v_A^2$, v_s and v_A the shock speed and Alfvèn speed, respectively, and the energy-independent acceleration time scale, t_{acc}. This parametrization approximates the momentum diffusion through hard-sphere scattering.

The second term on the right-hand side of Equation (1) provides momentum changes $\dot{\chi}_i$ through gains (Fermi-I acceleration $\dot{\chi}_{\mathrm{FI}} = \chi/t_{\mathrm{acc}}$) and continuous losses. All charged particles lose energy through synchrotron radiation and adiabatic expansion of the emission region. Protons also lose energy through pion production and Bethe-Heitler pair production, while electrons suffer additional losses through IC scattering of ambient photon fields. These ambient fields consist of all intrinsically produced radiation fields—such as synchrotron—as well as the external photon fields, namely the AD, the BLR, the DT, and the CMB. Naturally, the pion production can turn a proton into a neutron. As we do not explicitly consider neutrons at this point, this effect is approximated here by a continuous loss process instead of a catastrophic loss. This channel is marked as "neutron" losses in Figures 2 and 6, while the nominal pion production cooling term is marked as "pion".

The remaining three terms on the right-hand side of Equation (1) mark the injection of particles, the escape of particles from the emission region, and the decay of unstable particles, respectively. $t^*_{i,\text{decay}}$ is the proper decay time scale, which is $2.6 \times 10^{-8}\,\text{s}$ for charged pions, and $2.2 \times 10^{-6}\,\text{s}$ for muons, respectively. As neutral pions decay after $2.8 \times 10^{-17}\,\text{s}$ into γ rays, Equation (1) is not solved for neutral pions, while their radiation output is directly calculated from their injection spectrum.

While Fermi-I and II acceleration terms are considered here, we treat them merely as re-acceleration processes characterized by the acceleration time scale $t_{\text{acc}} = \eta_{\text{acc}} t_{\text{esc}}$; namely, a multiple η_{acc} of the escape time scale [21]. We do not consider the primary acceleration of protons and electrons, which may take place in small sub-regions of the larger emission region [20,22], but approximate it through the injection term $Q(\chi, t)$. Here, a simple power-law injection is used with spectral index s_i between a minimum and maximum Lorentz factor, $\gamma_{\text{min},i}$ and $\gamma_{\text{max},i}$, respectively. The injection normalization, $Q_{0,i}(t)$, is given by

$$
Q_{0,i}(t) = \frac{L_{\text{inj},i}(t)}{V m_i c^2}
\begin{cases}
\dfrac{2 - s_i(t)}{\gamma_{\text{max},i}^{2-s_i(t)} - \gamma_{\text{min},i}^{2-s_i(t)}} & \text{if } s_i(t) \neq 2 \\[2ex]
\left(\ln \dfrac{\gamma_{\text{max},i}}{\gamma_{\text{min},i}} \right)^{-1} & \text{if } s_i(t) = 2
\end{cases}
\tag{2}
$$

with the injection luminosity, $L_{\text{inj},i}$, and the volume, V, of the spherical emission region. The injection functions for pions and muons are calculated directly from the photo-hadron interactions [23] and decays. Let us again emphasize that Equation (1) is explicitly solved for (charged) pions and muons.

The escape of particles is described by $t_{\text{esc}} = \eta_{\text{esc}} R / c$, a multiple η_{esc} of the light travel time. As $\eta_{\text{esc}} > 1$, this mimics the advective flow of particles through the emission region.

Equation (1) is solved with a Chang and Cooper routine [24]; for a detailed description, see [19,25].

The interaction of protons with photons can result in the creation of pions. Charged pions decay into muons, which in turn decay into electrons. During both decay processes, neutrinos are produced. The neutrino spectra are calculated following [19,26,27]. The secondary electrons produced in this decay chain are injected into the electron-Fokker-Planck equation along with the primary electrons. Additionally, secondary electrons are also produced from Bethe-Heitler pair production and γ-γ pair production.

We do not consider explicitly neutrons in this code. Their number density is low compared to the proton density [13]; so their effect is small. Nonetheless, we plan to rectify this issue in a future update of the code.

The photon density n_{ph} within the emission region is governed by the radiation transport equation:

$$
\frac{\partial n_{\text{ph}}(\nu, t)}{\partial t} = \frac{4\pi}{h\nu} j_\nu(t) - n_{\text{ph}}(\nu, t) \left(\frac{1}{t_{\text{esc,ph}}} + \frac{1}{t_{\text{abs}}} \right).
\tag{3}
$$

with the frequency ν, the Planck constant h, the emissivity j_ν, the photon escape time scale $t_{\text{esc,ph}} = 4R/3c$, and the absorption time scale t_{abs} due to synchrotron-self absorption and γ-γ pair production. For the latter, all internal and external photon fields are considered.

From the photon distribution, n_{ph}, one can calculate the spectral luminosity in the observer's frame:

$$\nu^{\mathrm{obs}} L_{\nu^{\mathrm{obs}}}^{\mathrm{obs}} = \delta^4 \frac{h\nu^2 V}{t_{\mathrm{esc,ph}}} n_{\mathrm{ph}}(\nu,t). \tag{4}$$

Equations (3) and (4) hold for all radiation processes within the emission region.

On their way from the source to the observer, γ-ray photons are subject to further absorption processes. A routine to calculate the important cases of absorption in the BLR and DT following the prescription in [28] is implemented in the code.

The AD is described with a standard Shakura-Sunyaev disk [29] implying that the disk is fully described through the mass of the supermassive black hole M_{BH} and its accretion efficiency η_{SS} (or Eddington ratio). The proper transformation of the angles into the comoving frame is considered. The BLR and DT are approximated as isotropic photon fields in the host galaxy frame within an distance $\hat{R}_{\mathrm{BLR}}$ and $\hat{R}_{\mathrm{DT}}$ from the black hole, and their energy distribution is given through a grey-body spectrum of temperature $\hat{T}_{\mathrm{BLR}}$ and $\hat{T}_{\mathrm{DT}}$ normalized to a luminosity of $\hat{L}_{\mathrm{BLR}}$ and $\hat{L}_{\mathrm{DT}}$, respectively.

The above description holds for both steady-state and time-dependent cases. The steady state is achieved if the proton and electron densities derived from Equation (1), do not vary by more than 10^{-4} compared to the respective values of the previous two time steps. Time-dependency can be achieved by varying any of the free parameters, in which case steady states may not be achieved from time step to time step.

3. Influence of the External Fields

Table 1 provides an overview of the free parameters that have been described in the previous section. The given parameter values are a toy model, which we use to perform a small parameter study. The parameters are based upon the flat spectrum radio quasar 3C 279 [30], however a direct data comparison is beyond the scope of this paper.

Table 1. Free parameters of the code along with symbols, units, and toy model values. Quantities in the host galaxy frame are marked with a hat.

Parameter		Unit	Value
Redshift	z_{red}		0.536
Location in jet	z_0	cm	1.0×10^{16}, 1.0×10^{17}, 1.0×10^{18}, 1.0×10^{19}
Magnetic field	B	G	50
Radius	R	cm	4.5×10^{15}
Bulk Lorentz factor	Γ		50
Observation angle	θ^{obs}	deg	1.3
Proton injection luminosity	$L_{\mathrm{inj},p}$	erg/s	3.0×10^{43}
Proton spectral index	s_p		2.1
Proton min Lorentz factor	$\gamma_{\mathrm{min},p}$		4.0×10^5
Proton max Lorentz factor	$\gamma_{\mathrm{max},p}$		2.5×10^8
Electron injection luminosity	$L_{\mathrm{inj},e}$	erg/s	2.0×10^{41}
Electron spectral index	s_e		3.0
Electron min Lorentz factor	$\gamma_{\mathrm{min},e}$		5.0×10^1
Electron max Lorentz factor	$\gamma_{\mathrm{max},e}$		2.0×10^3
Multiple escape time	η_{esc}		5
Multiple acceleration time	η_{acc}		30
Black hole mass	M_{BH}	$M_\odot$	3.0×10^8
AD efficiency	η_{SS}		0.08
BLR luminosity	$\hat{L}_{\mathrm{BLR}}$	erg/s	2.3×10^{44}
BLR temperature	$\hat{T}_{\mathrm{BLR}}$	K	1.0×10^4
BLR radius	$\hat{R}_{\mathrm{BLR}}$	cm	7.6×10^{16}
DT luminosity	$\hat{L}_{\mathrm{DT}}$	erg/s	3.0×10^{44}
DT temperature	$\hat{T}_{\mathrm{DT}}$	K	5.0×10^2
DT radius	$\hat{R}_{\mathrm{DT}}$	cm	4.2×10^{18}

Instead, we wish to analyze the influence of the external fields on the SED and the particle distributions. We chose four locations: close to the AD ($z_0 = 1 \times 10^{16}$ cm), within the BLR ($z_0 = 1 \times 10^{17}$ cm), within the DT ($z_0 = 1 \times 10^{18}$ cm), and outside the external fields (referred to as "jet", $z_0 = 1 \times 10^{19}$ cm). All other parameters remain unchanged, including the radius and the magnetic field of the emission region. This highlights that these are indeed toy models meant to study the influence of the external fields without any degeneracies introduced by varying other parameters.

The result is shown in Figures 1–3. While the SEDs are transformed into the observer's frame, the photon spectra are shown as they leave the emission region in the jet. The internal γ-γ absorption processes are fully considered (the corresponding optical depth $\tau_{\gamma\gamma}$ is shown in Figure 4 left), however no additional absorption of γ rays while traveling through the photon field of the host galaxy (namely, BLR and DT) or through the cosmological photon fields (extragalactic background light and CMB) are shown. Any of these photon fields could additionally (and severely) attenuate the photon flux above 10 GeV. These absorption processes are, however, not important for the conclusions of this study.

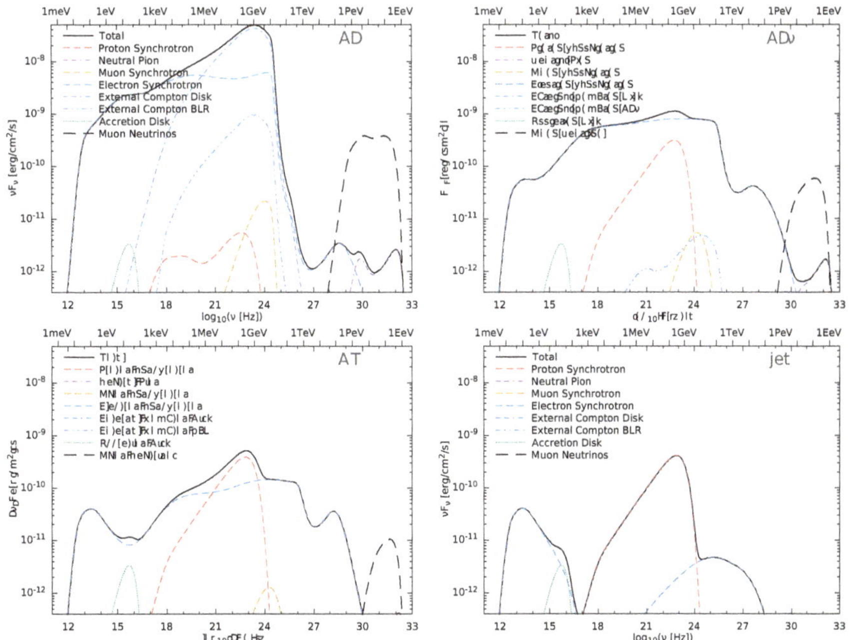

Figure 1. SEDs in the observer's frame for the four locations: close to the AD (**top left**), within the BLR (**top right**), within the DT (**bottom left**), and outside the external fields (jet, **bottom right**). The black solid line marks the total photon spectrum, while the colored lines mark individual photon components as labeled. Only those processes are labeled, which are visible in at least one panel. No external absorption is applied, implying that photon spectra are shown as they leave the jet. The black dashed line marks the neutrino spectrum.

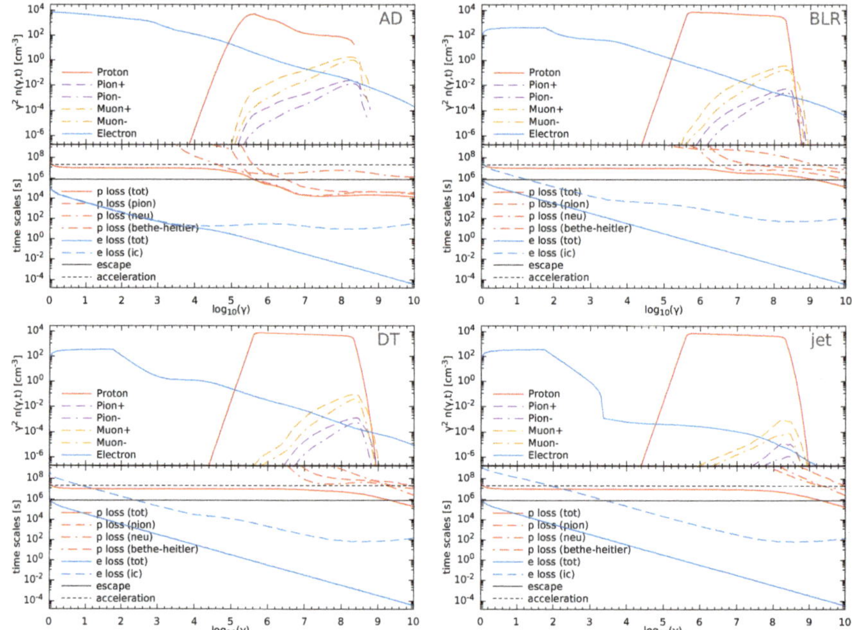

Figure 2. Steady-state particle distributions (times Lorentz factor squared), and relevant time scales as labeled as a function of Lorentz factor γ for the same locations as in Figure 1. For proton losses, the total, photo-pion, "neutron", and Bethe-Heitler loss time scales are shown. Adiabatic losses dominate at lower proton energies, where the loss time scale is constant, while synchrotron losses may contribute at the highest proton energies. For electron losses, the total, and IC loss time scales are shown. Synchrotron losses dominate, where IC losses are negligible, while adiabatic losses are irrelevant.

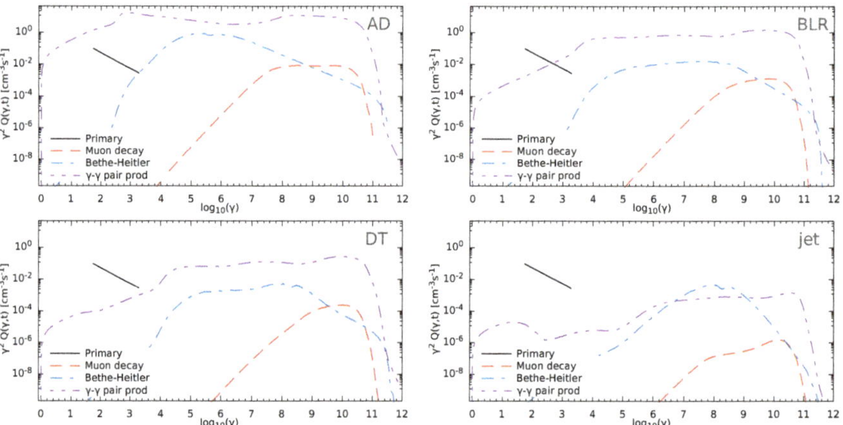

Figure 3. Steady-state electron injection rates Q (times Lorentz factor squared) as a function of the Lorentz factor γ as labeled for the same locations as in Figure 1.

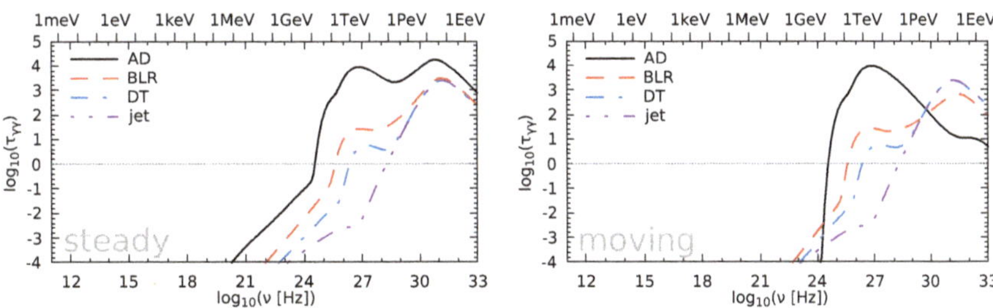

Figure 4. Optical depth $\tau_{\gamma\gamma}$ due to γ-γ pair production as a function of frequency (observer's frame) for the steady-state cases (**left**) and the moving-blob case (**right**) at the different positions within the jet, as labeled. The thin horizontal line marks $\tau_{\gamma\gamma} = 1$.

Close to the AD, the external fields are very intense, and are further enhanced through the large chosen bulk Lorentz factor of 50. In turn, the cooling of protons through proton-photon interactions is very strong (Figure 2), as indicated by the cooling time scales being dominated by pion production (indicated by the "pion" and "neutron" loss channels) at Lorentz factors $\gamma > 10^5$. This severely influences the proton distribution function and results in negligible proton synchrotron emission. The strong pion production, which can also be seen in the SED (Figure 1) through the neutral pion bump at PeV energies, results in a significant production of muons and highly relativistic electrons (Figure 3) with Lorentz factors $\gamma > 10^{10}$. Similarly, highly energetic electrons are also injected through Bethe-Heitler pair production. These electrons produce γ rays through synchrotron emission, as well as through IC emission for lower-energetic electrons. The γ rays are absorbed through γ-γ pair production with all photon fields that permeate the emission region. The strength of the γ-γ absorption is shown in the left panel of Figure 4, and manifests itself in Figure 1 by the significant flux suppression at energies above 10 GeV. In turn, a strong electron-positron cascade is initiated. This results in an electron distribution, which is dominated by secondaries (Figure 3). The resulting electron synchrotron flux (Figure 1) extends through almost the entire frequency range, destroying the familiar double-hump shape in the SED. The peak of the flux at γ rays stems from IC scattering of AD photons.

Within the BLR, the proton cooling is drastically reduced at high Lorentz factors with cooling time scales being longer than the escape time scale of particles at all (relevant) energies (Figure 2). Unlike in the AD case, where the proton distribution cuts off sharply at $\gamma_{\mathrm{max,p}}$, in this case (and the following cases) the proton distribution extends beyond the injection cut-off because of the (re-)acceleration terms present in Equation (1). The change in the spectral shape between the AD and BLR cases allows for an enhanced proton synchrotron emission in the BLR case, influencing the SED at GeV energies (Figure 1). While pion and Bethe-Heitler pair production are reduced compared to the AD case, the pair cascade is still very significant (Figure 3) because of γ-γ pair production (Figure 4 left). While the process is less severe than in the AD case, the secondaries still dominate the electron distribution (Figure 2), and produce synchrotron emission beyond PeV energies. In the BLR case, IC emission is negligible.

This trend continues in the DT case, as the cascade weakens (Figures 3 and 4 left) and the more familiar double-humped SED emerges (Figure 1). At ultra-violet (UV) energies in the SED, a minor contribution from the AD itself is visible. The γ-ray peak is dominated by proton synchrotron emission, even though the secondary electron synchrotron emission still dominates at X-ray and TeV energies. The neutral pion bump is below the shown flux scale, indicating the reduced interaction of protons with photons. In fact, the protons are completely in a slow-cooling regime (Figure 2).

Lastly, the emission region is located outside the external photon fields in the "jet" case. While secondary pairs are still being produced (Figure 3), their number is low (Figure 2) due to low absorption (Figure 4 left), and their flux contribution only shows around photon energies on the TeV-scale, but at relatively low flux values (Figure 1). Apart from that, the SED is dominated by synchrotron emission of protons at X- and γ rays, and primary electrons in the optical domain. Both peaks are cleanly separated. The AD itself is clearly visible in the UV range as a big blue bump.

The changes in the cooling strength can also be seen in the energy densities of the particles, which are given in Table 2. The particle energy densities are always dominated by protons (by several orders of magnitude compared to the electrons in most cases). The strong cooling in the AD case results in a low particle energy density, while the reduced cooling in the other cases results in increased and comparable energy densities. Given the constant value of the magnetic field in all cases, the ratio of magnetic to particle energy density decreases from case to case but is always larger than unity.

Table 2. Energy densities in particles u_{par} (in erg/cm^3) and the ratio u_B/u_{par} of magnetic to particle energy density. The magnetic energy density in all cases is $u_B = 100\,\mathrm{erg/cm}^3$. The horizontal line separates the steady-state (top) from the moving (bottom) cases.

Position	u_{par}	u_B/u_{par}
AD	13.4	7.5
BLR	55.6	1.8
DT	59.1	1.7
jet	59.7	1.7
AD	0.38	263
BLR	4.48	22.3
DT	33.3	3.0
jet	59.7	1.7

The different cases are also manifested in the emerging neutrino spectra. With the weakening production of pions and muons from case to case, the flux of neutrinos also decreases and drops below the scale of the plots in the "jet" case. The AD case produces not just the highest neutrino flux, but also a different neutrino spectral shape than the other cases with a flat maximum (or mildly double-humped structure) over almost three orders of magnitude in energy. In the BLR and DT case, the neutrino spectra show a single peak at about 100 PeV. Interestingly, all three cases would be detectable with the future IceCube-Gen2 instrument [31]. However, the unrealistic SEDs—especially in the AD and BLR cases—make it seem unlikely that neutrinos could be observed from a blazar–at least, under this simple set-up.

For the examples discussed above, we have used a bulk Lorentz factor of 50. Hence, if the emission region were moving, it would cover a lot of space in a relatively short amount of time because of the Lorentz contraction: $\hat{z} = z_0 + \Gamma\beta_\Gamma ct$, where t is the time since launch in the comoving frame, and $\hat{z}$ is the location of the emission region in the host galaxy frame. In turn, the external fields, and thus the conditions within the emission region may change quickly. We try to analyze this, by letting the emission region flow from the base (placed at six times the Schwarzschild radius (innermost stable circular orbit) of the black hole) downstream through the jet.

As before, none of the other parameters change, implying that also the primary injection of protons and electrons continues with the same rate Q and spectral shape throughout the simulation. This assumes a quasi-instantaneous acceleration of particles [32], as well as a continuous supply. This is not realistic, as the acceleration of particles also takes time [20]. Additionally, neither the magnetic field B nor the radius R vary. While the radius of the emission region may not expand as rapidly as the larger jet structure that surrounds it, it expands nonetheless while it travels through the jet [33] given the high energy densities in the emission region. While recent observational results [30,34] indicate compact emission regions beyond the BLR, and maybe even at tens of parsecs from the black hole, it is not clear whether these are indeed moving emission regions originating close to the black hole or turbulent cells within a larger flaring region. Similarly, while a high magnetic field can be expected close to the black hole, the expansion of the emission region causes a drop of the magnetic field with increasing distance. These considerations highlight once more the toy character of this study. Applying such parameter changes are interesting avenues for future studies beyond the scope of this paper.

Having obtained the full journey of the emission region through the jet, we extract the SEDs and particle distributions at roughly the same distances as in the steady-state cases. In fact, given the finite time resolution in the simulation, we extract the SEDs and particle distribution at the time step closest to the respective distances of the steady-state models (AD: 9.83×10^{15} cm, BLR: 9.75×10^{16} cm, DT: 9.64×10^{17} cm, jet: 9.75×10^{18} cm). In order to save computation time, while also properly resolving the initial steps within the BLR, an adaptive time step of $\Delta t_j = 1 \times 10^{3+j/20}$ is used, where j is the step number. This ensures reasonable accuracy and resolution, and also explains why the time values given in Figures 5 and 6 are not simple increases by a factor 10, as one would expect. This is a reasonable trade-off. The results are shown in Figures 5–7, while the optical depth due to γ-γ pair production is shown in the right panel of Figure 4. Any times and time scales discussed below are in the comoving frame.

The changes to the SEDs and the particle spectra are profound. The emission region has passed the AD position after merely 5 ks. The bright external photon fields cause proton-photon interactions, producing a significant amount of pions (Figure 6), which decay into photons or muons and pairs. In fact, Figure 7 shows that the injection of pairs from muon decay is almost at the level as in the steady state (Figure 3), but Bethe-Heitler produced pairs are about two orders of magnitude below. Similarly, γ-γ pair production is below the steady-state level, because the internal photon fields (Figure 5) have not yet been fully developed. In turn, the optical depth due to γ-γ pair production (Figure 4 right) is not at the steady-state level—merely the absorption caused by external fields is fully present. One consequence is the reduced absorption of PeV photon energies, allowing for a very strong flux in the neutral pion bump (Figure 5). The "under-development" of the internal photon fields is a consequence of the low electron and proton densities (Figure 6) compared to the steady-state values. The consequence is the absence of the "nominal" electron synchrotron bump in the infrared domain. The γ rays are dominated by IC scattering of AD photons—although orders of magnitude below the steady-state case.

The situation only changes mildly until the BLR position is reached after 5.7×10^4 s. This is still less than the escape times of photons (2×10^5 s) and particles (7.5×10^5 s). Therefore, particle and photon densities continue to increase. The spectral shape of the SED shown in Figure 5 is somewhat similar to the steady-state case (Figure 1), but at a factor of a few reduced in flux. There are a few more details where SEDs differ. In the γ-ray domain, Figure 5 shows contributions from IC scattering of both the AD and the BLR. Comparing the IC/AD spectra of the top panels in Figure 5, one notices the similarity between them. Given that not even one light-crossing time scale has passed since the launch, the photons produced below have not yet vanished from the emission region, and therefore continue to contribute to the SED even though the IC/AD production has reduced a lot at this distance. The IC/BLR spectrum shows a different spectral shape and a higher flux than in the steady-state case, which can be attributed to the slightly different shapes in the electron distributions (Figure 6 vs. Figure 2). These are a consequence of the reduced γ-γ pair production at lower energies (Figure 7). As the protons have also not reached the steady-state density, their synchrotron flux is reduced compared to the steady state, while pion and muon production are similarly reduced. Most notably, the neutral pion decay flux is barely visible at PeV to EeV energies in Figure 5—a reduction of about an order of magnitude compared to the steady state.

After 5.7×10^5 s, the emission region has reached the DT position. The time the emission region has traveled is now comparable to the escape time scales of light and particles. In turn, the SED in Figure 5 is almost equal to the steady-state case (Figure 1), except for a reduced peak γ-ray flux by a factor of a few. This can be attributed to the still lower number of protons compared to the steady state, resulting in an equally reduced proton synchrotron flux. This is also coupled to the low efficiency of proton-synchrotron emission, implying that the flux needs more time to build compared to the electron synchrotron flux, which is basically instantaneous–cf., the cooling time scales in Figure 6, where electron synchrotron cooling is faster than basically any other time scale (including the travel time), while the proton synchrotron cooling only dominates at energies beyond the cut-off of the proton distribution. The IC/AD and IC/BLR components visible in the SED (Figure 5) exhibit a flux about an order of magnitude below the flux at the BLR position. This corresponds very well to an exponential decay, as the photons leave the emission region without being replenished.

The following, relatively long cruise towards the "jet" position (reached after 5.8×10^6 s) allows for the near-complete relaxation of the emission region towards the steady state that was obtained above. At this position, SED and particle distributions are practically equal to the steady-state case.

The particle energy densities change considerably from position to position because of the accumulation of relativistic particles in the emission region. This is the reason why the particle energy density at the AD position is about a factor 35 lower than in the steady state case. This accumulation of particles continues through the other position, increasing the particle energy density along the way until the jet position, where the previous steady-state value is obtained. Similarly, the ratio of energy densities is initially very large and decreases on the way out.

The neutrino spectra shown in Figure 5 indicate as well that the interactions and distributions require time to unfold. While at the AD position lots of neutrinos are produced, their flux is a factor of a few below the steady-state flux. At the BLR position the flux reduction is almost an order of magnitude (similar to the pion flux), while it is closer to the steady-state flux at the DT position. At the "jet" position, the neutrino flux is reduced a lot, as in the steady-state case.

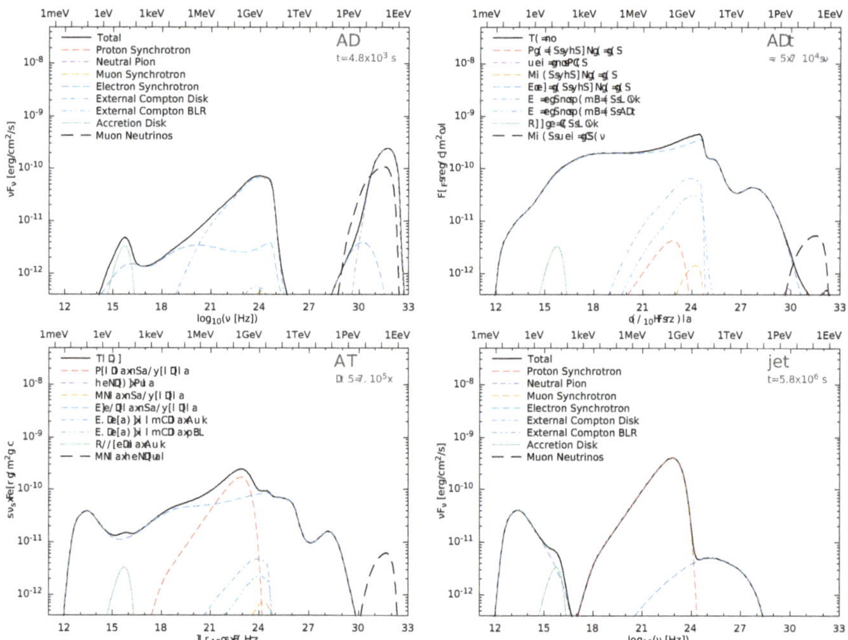

Figure 5. Same as Figure 1 but for a moving blob. In each panel, the displayed time is the time that has passed in the comoving frame since the launch.

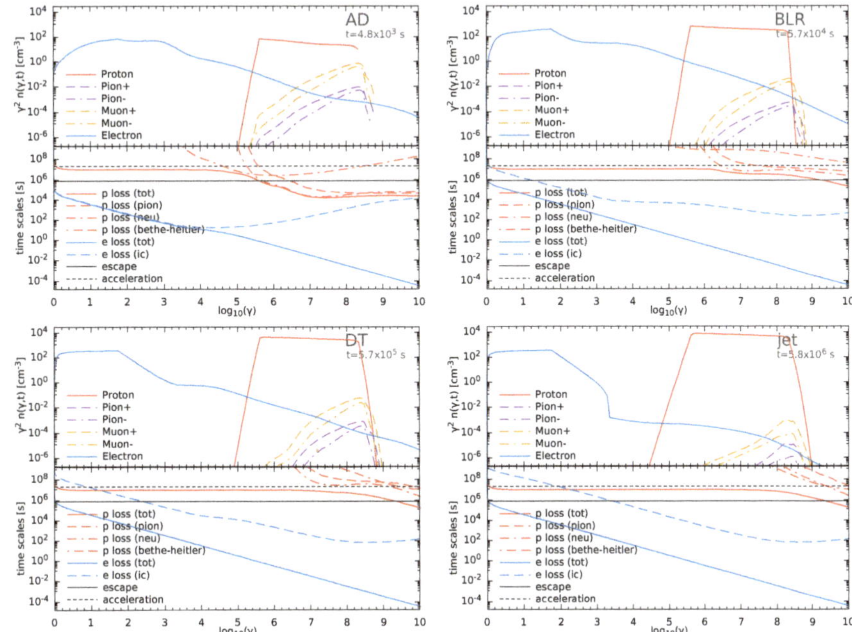

Figure 6. Same as Figure 2 but for a moving blob as in Figure 5. In each panel, the displayed time is the time that has passed in the comoving frame since the launch.

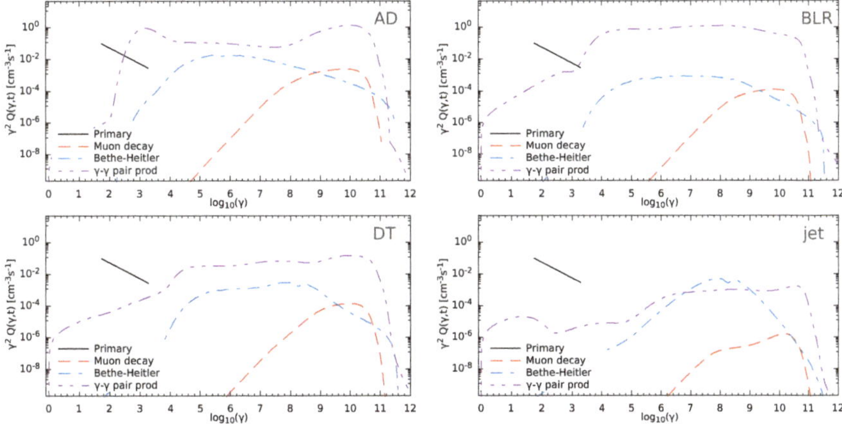

Figure 7. Same as Figure 3 but for a moving blob as in Figure 5.

4. Discussion and Conclusions

The results of the toy study presented in this paper clearly show the importance of the external fields in case of the presence of relativistic protons in the jet. Their influence on the particle evolution is significant resulting in very different steady-state SEDs at different positions in the jet. Especially at locations within the BLR, the familiar double-humped SED structure is destroyed. At the DT position, the spectrum is already comparable to "standard" blazar SEDs, while the "jet" position outside the external fields provides the cleanest separation between the low-energy and the high-energy bump.

The situation changes entirely when the motion of the emission region is taken into account. The relatively long source time scales (particle and photon accumulation, interactions, escape) compared to the fast speed imply that the external conditions change too fast for the emission region to adapt even until the edge of the DT. Only on "jet" scales is the previous steady state fully recovered. This, of course, is a consequence of the choice of $\Gamma = 50$, which is a rather extreme value. Lower values on the order of $\Gamma \sim 10$ could change the situation—especially as it would also significantly reduce the energy density of the external fields within the emission region. Steady-state solutions might be achieved at positions much closer to the black hole. Testing this, and the other potential changes to the model parameters as described above, is however beyond the scope of this paper.

Within the model parameters used in this toy study, the production of neutrinos depends strongly on the external fields with practically none produced at the "jet" position. While different parameter sets of the emission region might produce better SED shapes at positions within the external photon fields, it corroborates the results obtained by other authors [18,35,36], which makes it difficult to reconcile the neutrino and photon observations within a one-zone model.

To conclude, the production of neutrinos in a blazar jet in reasonable quantities remains a challenge, as the requirement for a reasonably dense soft photon field—in order to produce the required pions—also supports the pair cascade through γ-γ absorption and Bethe-Heitler pair production. The intrusion of a gas cloud or a star into the jet [37,38] might provide sufficient numbers of cold protons for direct proton-proton interactions [39], but the consequences (efficiency of the process, developing pair cascade, etc.) would also need further studies.

Funding: The author acknowledges postdoctoral financial support from LUTH, Observatoire de Paris.

Data Availability Statement: The *OneHaLe* code is still under development and is therefore not yet meant for public use. However, the code can be shared upon reasonable request to the author.

Acknowledgments: I am eternally grateful to Reinhard Schlickeiser for his supervision and guidance from the Bachelor thesis up to the Ph.D. thesis and beyond. Thank you very much for all the guidance along the way—and for all the travel destinations that you have made possible. Because of you, I realized quickly that this is indeed what I want to do—being a researcher, I mean. I am also grateful for stimulating discussions with Anita Reimer, Andreas Zech, Catherine Boisson, Markus Böttcher, Anton Dmytriiev and Chris Diltz, which helped in creating and improving the code. I would like to thank the referees for constructive reports that helped a lot to improve the paper. Simulations for this paper have been performed on the TAU-cluster of the Centre for Space Research at North-West University, Potchesftroom, South Africa.

Conflicts of Interest: The author declares no conflict of interest.

References

1. Dermer, C.D.; Schlickeiser, R.; Mastichiadis, A. High-energy gamma radiation from extragalactic radio sources. *Astron. Astrophys.* **1992**, *256*, L27–L30. Available online: http://articles.adsabs.harvard.edu/pdf/1992A%26A...256L..27D (accessed on 1 November 2021).
2. Dermer, C.D.; Schlickeiser, R. Model for the high-energy emission from blazars. *Astrophys. J.* **1993**, *416*, 458. [CrossRef]
3. Dermer, C.D.; Sturner, S.J.; Schlickeiser, R. Nonthermal compton and synchrotron processes in the jets of active galactic nuclei. *Astrophys. J. Suppl.* **1997**, *109*, 103–137. [CrossRef]
4. Dermer, C.D.; Schlickeiser, R. Transformation properties of external radiation fields, energy-loss rates and scattered spectra, and a model for blazar variability. *Astrophys. J.* **2002**, *575*, 667–686. [CrossRef]
5. Dermer, C.D.; Schlickeiser, R. Quasars, blazars, and gamma rays. *Science* **1992**, *257*, 1642–1647. [CrossRef]
6. Sikora, M.; Begelman, M.C.; Rees, M.J. Comptonization of diffuse ambient radiation by a relativistic jet: The source of gamma rays from blazars? *Astrophys. J.* **1994**, *421*, 153. [CrossRef]
7. Błażejowski, M.; Sikora, M.; Moderski, R.; Madejski, G.M. Comptonization of infrared radiation from hot dust by relativistic jets in quasars. *Astrophys. J.* **2000**, *545*, 107–116. [CrossRef]
8. Arbeiter, C.; Pohl, M.; Schlickeiser, R. The influence of dust on the inverse Compton emission from jets in Active Galactic Nuclei. *Astron. Astrophys.* **2002**, *386*, 415–426. [CrossRef]
9. Böttcher, M.; Dermer, C.D.; Finke, J.D. The Hard VHE γ-ray emission in high-redshift TeV blazars: Comptonization of cosmic microwave background radiation in an extended jet? *Astrophys. J.* **2008**, *679*, L9. [CrossRef]
10. Zacharias, M.; Wagner, S.J. The extended jet of AP Librae: Origin of the very high-energy γ-ray emission? *Astron. Astrophys.* **2016**, *588*, A110. [CrossRef]

11. Mannheim, K.; Biermann, P.L. Gamma-ray flaring of 3C 279: A proton-initiated cascade in the jet? *Astron. Astrophys.* **1992**, *253*, L21–L24. Available online: http://articles.adsabs.harvard.edu/pdf/1992A%26A...253L..21M (accessed on 1 November 2021).

12. Mannheim, K. The proton blazar. *Astron. Astrophys.* **1993**, *269*, 67–76. Available online: http://articles.adsabs.harvard.edu/pdf/1993A%26A...269...67M (accessed on 1 November 2021).

13. Mücke, A.; Engel, R.; Rachen, J.P.; Protheroe, R.J.; Stanev, T. Monte Carlo simulations of photohadronic processes in astrophysics. *Comput. Phys. Commun.* **2000**, *124*, 290–314. [CrossRef]

14. Dimitrakoudis, S.; Mastichiadis, A.; Protheroe, R.J.; Reimer, A. The time-dependent one-zone hadronic model. First principles. *Astron. Astrophys.* **2012**, *546*, A120. [CrossRef]

15. Petropoulou, M.; Mastichiadis, A. Bethe-Heitler emission in BL Lacs: Filling the gap between X-rays and γ-rays. *Mon. Not. R. Astron. Soc.* **2015**, *447*, 36–48. [CrossRef]

16. IceCube Collaboration; Aartsen, M.G.; Ackermann, M.; Adams, J.; Aguilar, J.A.; Ahlers, M.; Ahrens, M.; Al Samarai, I.; Altmann, D.; Andeen, K.; et al. Multimessenger observations of a flaring blazar coincident with high-energy neutrino IceCube-170922A. *Science* **2018**, *361*, eaat1378. [CrossRef]

17. Hovatta, T.; Lindfors, E.; Kiehlmann, S.; Max-Moerbeck, W.; Hodges, M.; Liodakis, I.; Lähteemäki, A.; Pearson, T.J.; Readhead, A.C.S.; Reeves, R.A.; et al. Association of IceCube neutrinos with radio sources observed at Owens Valley and Metsähovi Radio Observatories. *Astron. Astrophys.* **2021**, *650*, A83. [CrossRef]

18. Reimer, A.; Böttcher, M.; Buson, S. Cascading constraints from neutrino-emitting blazars: The case of TXS 0506+056. *Astrophys. J.* **2019**, *881*, 46. [CrossRef]

19. Zacharias, M.; Reimer, A.; Boisson, C.; Zech, A. ExHaLe-jet: An extended hadro-leptonic jet model for blazars. I. Code description and initial results. *Mon. Not. R. Astron. Soc.* **2021**, submitted.

20. Weidinger, M.; Spanier, F. A self-consistent and time-dependent hybrid blazar emission model. Properties and application. *Astron. Astrophys.* **2015**, *573*, A7. [CrossRef]

21. Diltz, C.; Böttcher, M.; Fossati, G. Time dependent hadronic modeling of flat spectrum radio quasars. *Astrophys. J.* **2015**, *802*, 133. [CrossRef]

22. Chen, X.; Pohl, M.; Böttcher, M. Particle diffusion and localized acceleration in inhomogeneous AGN jets—I. Steady-state spectra. *Mon. Not. R. Astron. Soc.* **2015**, *447*, 530–544. [CrossRef]

23. Hümmer, S.; Rüger, M.; Spanier, F.; Winter, W. Simplified models for photohadronic interactions in cosmic accelerators. *Astrophys. J.* **2010**, *721*, 630–652. [CrossRef]

24. Chang, J.; Cooper, G. A practical difference scheme for Fokker-Planck equations. *J. Comput. Phys.* **1970**, *6*, 1–16. [CrossRef]

25. Dmytriiev, A.; Sol, H.; Zech, A. Connecting steady emission and very high energy flaring states in blazars: The case of Mrk 421. *Mon. Not. R. Astron. Soc.* **2021**, *505*, 2712–2730. [CrossRef]

26. Barr, S.; Gaisser, T.K.; Lipari, P.; Tilav, S. Ratio of ν_e/ν_μ in atmospheric neutrinos. *Phys. Lett. B* **1988**, *214*, 147–150. [CrossRef]

27. Gaisser, T.K. *Cosmic Rays and Particle Physics*; Cambridge University Press: New York, NY, USA, 1990. Available online: https://ui.adsabs.harvard.edu/abs/1990cup..book.....G/abstract (accessed on 1 November 2021).

28. Böttcher, M.; Els, P. Gamma-gamma absorption in the broad line region radiation fields of gamma-ray blazars. *Astrophys. J.* **2016**, *821*, 102. [CrossRef]

29. Shakura, N.I.; Sunyaev, R.A. Black holes in binary systems. Observational appearance. *Astron. Astrophys.* **1973**, *500*, 33–51. Available online: http://adsabs.harvard.edu/full/1973A%26A....24..337S (accessed on 1 November 2021).

30. HESS Collaboration; Abdalla, H.; Adam, R.; Aharonian, F.; Ait Benkhali, F.; Angüner, E.O.; Arakawa, M.; Arcaro, C.; Armand, C.; Ashkar, H.; et al. Constraints on the emission region of 3C 279 during strong flares in 2014 and 2015 through VHE γ-ray observations with H.E.S.S. *Astron. Astrophys.* **2019**, *627*, A159. [CrossRef]

31. The IceCube-Gen2 Collaboration; Aartsen, M.G.; Abbasi, R.; Ackermann, M.; Adams, J.; Aguilar, J.A.; Ahlers, M.; Ahrens, M.; Alispach, C.; Allison, P.; et al. IceCube-Gen2: The window to the extreme universe. *J. Phys. G Nucl. Part. Phys.* **2020**, *48*, 060501. [CrossRef]

32. Böttcher, M.; Baring, M.G. Multi-wavelength variability signatures of relativistic shocks in blazar jets. *Astrophys. J.* **2019**, *887*, 133. [CrossRef]

33. Boula, S.; Mastichiadis, A. An expanding one-zone model for studying blazars emission. *Astron. Astrophys.* **2021**. Available online: https://ui.adsabs.harvard.edu/abs/2021arXiv211005325B (accessed on 1 November 2021). [CrossRef]

34. HESS Collaboration; Abdalla, H.; Adam, R.; Aharonian, F.; Ait Benkhali, F.; Angüner, E.O.; Arcaro, C.; Armand, C.; Armstrong, T.; Ashkar, H.; et al. H.E.S.S. and MAGIC observations of a sudden cessation of a very-high-energy γ-ray flare in PKS 1510–089 in May 2016. *Astron. Astrophys.* **2021**, *648*, A23. [CrossRef]

35. Gao, S.; Fedynitch, A.; Winter, W.; Pohl, M. Modelling the coincident observation of a high-energy neutrino and a bright blazar flare. *Nat. Astron.* **2019**, *3*, 88–92. [CrossRef]

36. Cerruti, M.; Zech, A.; Boisson, C.; Emery, G.; Inoue, S.; Lenain, J.P. Leptohadronic single-zone models for the electromagnetic and neutrino emission of TXS 0506+056. *Mon. Not. R. Astron. Soc.* **2019**, *483*, L12–L16. [CrossRef]

37. Bosch-Ramon, V.; Perucho, M.; Barkov, M.V. Clouds and red giants interacting with the base of AGN jets. *Astron. Astrophys.* **2012**, *539*, A69. [CrossRef]

38. Zacharias, M.; Böttcher, M.; Jankowsky, F.; Lenain, J.P.; Wagner, S.J.; Wierzcholska, A. Cloud Ablation by a Relativistic Jet and the Extended Flare in CTA 102 in 2016 and 2017. *Astrophys. J.* **2017**, *851*, 72. [CrossRef]

39. Hoerbe, M.R.; Morris, P.J.; Cotter, G.; Becker Tjus, J. On the relative importance of hadronic emission processes along the jet axis of active galactic nuclei. *Mon. Not. R. Astron. Soc.* **2020**, *496*, 2885–2901. [CrossRef]

Article

A Shock-in-Jet Synchrotron Mirror Model for Blazars

Markus Böttcher

Centre for Space Research, North-West University, Potchefstroom 2520, South Africa; Markus.Bottcher@nwu.ac.za

Abstract: Reinhard Schlickeiser has made groundbreaking contributions to various aspects of blazar physics, including diffusive shock acceleration, the theory of synchrotron radiation, the production of gamma-rays through Compton scattering in various astrophysical sources, etc. This paper, describing the development of a self-consistent shock-in-jet model for blazars with a synchrotron mirror feature, is therefore an appropriate contribution to a Special Issue in honor of Reinhard Schlickeiser's 70th birthday. The model is based on our previous development of a self-consistent shock-in-jet model with relativistic thermal and non-thermal particle distributions evaluated via Monte-Carlo simulations of diffusive shock acceleration, and time-dependent radiative transport. This model has been very successful in modeling spectral variability patterns of several blazars, but has difficulties describing orphan flares, i.e., high-energy flares without a significant counterpart in the low-frequency (synchrotron) radiation component. As a solution, this paper investigates the possibility of a synchrotron mirror component within the shock-in-jet model. It is demonstrated that orphan flares result naturally in this scenario. The model's applicability to a recently observed orphan gamma-ray flare in the blazar 3C279 is discussed and it is found that only orphan flares with mild ($\lesssim$ a factor of 2–3) enhancements of the Compton dominance can be reproduced in a synchrotron-mirror scenario, if no additional parameter changes are invoked.

Keywords: active galaxies; blazars; diffusive shock acceleration; radiative transport; gamma-rays

Citation: Böttcher, M. A Shock-in-Jet Synchrotron Mirror Model for Blazars. *Physics* **2021**, *3*, 1112–1122. https://doi.org/10.3390/physics3040070

Received: 16 September 2021
Accepted: 5 November 2021
Published: 22 November 2021

Publisher's Note: MDPI stays neutral with regard to jurisdictional claims in published maps and institutional affiliations.

1. Introduction

Blazars are a class of jet-dominated active galactic nuclei. As most convincingly argued by Reinhard Schlickeiser (RS) in 1996 [1], their broad-band non-thermal emission, ranging from radio to gamma-rays, must be strongly Doppler boosted due to relativistic motion of an emission region along the jet, oriented close to our line of sight. The spectral energy distributions (SEDs) of blazars are dominated by two broad, non-thermal radiation components. The low-frequency component, from radio to optical/UV/X-ray frequencies, is generally attributed to synchrotron radiation by relativistic electrons. Most notably, Crusius and Schlickeiser [2,3] have evaluated the angle-averaged synchrotron emission from isotropically distributed electrons in random magnetic fields, including plasma effects, which are now frequently used as the standard expressions for the low-frequency emission from blazars. However, note also an alternative suggestion by RS in 2003 [4] that the low-frequency emission may be produced as electrostatic bremsstrahlung, i.e., the scattering of electrostatic Langmuir waves excited by two-stream instabilities, as expected in the jet-inter-stellar-medium interaction scenario of Schlickeiser et al. (2002) [5].

Motivated by early γ-ray observations by the SAS-2 and COS-B satellites, already in 1979–1980, RS had considered inverse-Compton scattering as the dominant mechanism to produce high-energy γ-rays in astrophysical sources, pointing out the importance of Klein-Nishina effects in the calculation of γ-ray spectra [6–8]. Also in leptonic models for blazars, inverse-Compton scattering by relativistic electrons in the jet is considered the dominant high-energy emission mechanism. Target photons for Compton scattering can be the co-spatially produced synchrotron (or electrostatic bremsstrahlung) radiation, in which case it is termed synchrotron self-Compton (SSC) emission (e.g., [9,10]). The first suggestion of target photon fields from outside the jet involved RS in two seminal papers suggesting the photon field of the accretion disk as the dominant target photon field [11,12]. Alternative sources of external target photons may be the broad-line region (BLR) (e.g., [13]), a dusty, infra-red emitting torus (e.g., [14]), or other regions of the jet (e.g., [15,16]). The relativistic motion of the high-energy emission region in a blazar jet through these generally anisotropic external radiation fields leads to complicated transformation properties from the active galactic nucleus (AGN) rest frame into the emission-region frame, which were studied in detail by Dermer and Schlickeiser in 2002 [17]. Which of these potential radiation fields might dominate, depends critically on the location of the emission region, which can be constrained by the absence of obvious signatures of $\gamma\gamma$ absorption of high-energy and very-high-energy γ-rays by the nuclear radiation fields of the central AGN, with one of the first detailed discussions of such constraints published by Dermer and Schlickeiser in 1994 [18].

The generation of the non-thermal broadband emission from blazars requires the efficient acceleration of electrons to ultra-relativistic energies. One of the plausible mechanisms of particle acceleration acting in the relativistic jets of blazars is diffusive shock acceleration (DSA), which was studied in the context of a general derivation of the kinetic equation of test particles in turbulent plasmas by RS in two seminal papers in 1989 [19,20] for non-relativistic shock speeds, while particle acceleration by magnetic turbulence, specifically in relativistic jets was studied by Schlickeiser and Dermer in 2000 [21]. Particle acceleration at relativistic shocks has been considered by several authors, using both analytical methods (e.g., [22–24]) and Monte-Carlo techniques (e.g., [25–29]). The simulations by Niemiec and Ostrowski [28] and Summerlin and Baring [29] indicate that diffusive shock acceleration at oblique, mildly relativistic shocks is able to produce relativistic, non-thermal particle spectra with a wide range of spectral indices, including as hard as $n(p) \propto p^{-1}$, where p is the particle's momentum.

In two recent papers [30,31], we had coupled Monte-Carlo simulations of diffusive shock acceleration (DSA), using the code of Summerlin and Baring [29], with time-dependent radiation transfer, based on radiation modules originally developed by Böttcher, Mause and Schlickeiser in 1997 [32] and further developed as detailed in [33,34]. In those studies, we found that the particles' mean free path for pitch-angle scattering, λ_{pas}, which mediates the first-order Fermi process in DSA, must have a strong dependence on particle momentum, with an index $\alpha > 1$ for a parameterization of $\lambda_{\mathrm{pas}}(p) \propto p^{\alpha}$. This likely indicates a decaying level of magneto-hydrodynamic turbulence with increasing distance from the shock front. Higher-energy particles, with their larger gyro radii, then probe more distant regions from the shock front, experiencing less efficient pitch-angle scattering. Time-dependent simulations of DSA plus radiation transfer were used to fit the multi-wavelength variability of the blazars 3C279 and Mrk 501 in [31] and the X-ray variability of 1ES 1959 + 650 in [35]. Multi-wavelength flares with approximately equal flare amplitude in the low-frequency (synchrotron) and high-frequency (Compton) components of the SED were naturally produced by an increase of the power injected into shock-accelerated particles, without the need for significant changes of the plasma parameters determining $\lambda_{\mathrm{pas}}(p)$.

However, an orphan γ-ray flare on 20 December 2013, with no significant counterpart in the synchrotron emission component, reported as Flare B in [36], presented a severe challenge to this as well as any other single-zone emission model for blazars. A fit to the observed γ-ray flare was possible with a significant hardening of the DSA-generated particle spectrum as the result of a reduction of the pitch-angle-scattering mean-free path, both in overall normalization $\lambda_{\mathrm{pas}}(0)$ and index α. However, keeping the optical (synchrotron) flux approximately constant, as observed, required a reduction of the magnetic field by a factor of 8.7, followed by a gradual recovery to the quiescent-state value with a fine-tuned time dependence. While the authors argue that such magnetic-field reductions and subsequent gradual recoveries after the passage of a shock have indeed been observed in interplanetary shocks (e.g., [37]), it is worth exploring alternative ways to explain orphan γ-ray flares in blazars within the framework of the shock-in-jet model developed in [30,31].

One plausible way of producing orphan γ-ray flares in the framework of a leptonic single-zone blazar model is the temporary enhancement of an external radiation field that serves as target for inverse-Compton scattering. This is the basis of a class of models termed *synchrotron mirror models*, where the synchrotron radiation of the high-energy emission region traveling along the jet, is reflected by a cloud to re-enter the emission region at a later time. Such models were first considered by Ghisellini and Madau [38], however without proper consideration of light-travel time effects, and by Böttcher and Dermer [39] and Bednarek [40], properly treating light-travel time effects, but considering primarily the time-dependence of the target-photon energy density without detailed calculations of the emerging γ-ray spectra. The synchrotron mirror model was more recently re-visited by Vittorini et al. [41], with a fully time-dependent leptonic synchrotron mirror model applied to the spectral variability of 3C454.3 in 2010 November, and Tavani et al. [42], considering also moving mirrors and applying the model to the light curve of the same flare B of 3C279 considered by [31]. Note a similar model termed the "ring of fire" model by MacDonald et al. [43,44], where the emission region passes a static synchrotron-emitting region of an outer sheath of the jet (the "ring of fire"), which produces very similar variability features as the synchrotron mirror model.

In the present paper, the time-dependent shock-in-jet model of Böttcher and Baring [31] is extended to include self-consistently a synchrotron-mirror component. Section 2 describes the additions to the model. Section 3 presents the resulting spectral variability features from an attempt to apply this model to the orphan γ-ray flare B of 3C279. Section 4 summarizes and discusses the results.

2. Model Description

The model developed here is a further development of the time-dependent shock-in-jet model of Böttcher and Baring [31]. In addition to the radiation components already included in [31], we now introduce synchrotron emission reflected by a spherical cloud of radius R_{cl} at a distance z_{cl} from the central engine, assumed for simplicity to be located close to the path of the jet, however, not hydrodynamically interacting with it, as considered, e.g., by the jet-star interaction model [45,46] or the cloud ablation model [47,48]. A mildly relativistic, oblique shock is propagating along the jet, thus accelerating particles in the local environment of the shock which constitutes our moving emission region of radius R_b. The emission region is starting out at time $t_e = 0$ (in the AGN rest frame) at a height z_0 above the black-hole—accretion-disk system powering the jet, and is propagating with a bulk Lorentz factor Γ, corresponding to a speed of $\beta_\Gamma c$. Thus, at any given time t_e, the emission region is located at $z_e = z_0 + \beta_\Gamma c\, t_e$.

Synchrotron radiation emitted by the emission region at z_e is reflected back by the cloud to re-enter the emission region at a distance z_r from the central engine, given by

$$z_r = \frac{2\beta_\Gamma z_{cl} + z_e(1 - \beta_\Gamma)}{1 + \beta_\Gamma},\tag{1}$$

at a time (in the AGN rest frame) t_r given by

$$t_r = t_e + 2\frac{z_{cl} - z_e}{(1 + \beta_\Gamma)c}.\tag{2}$$

Equation (2) may be inverted to find the time at which reflected synchrotron radiation received at time t_r has been emitted:

$$t_e = \frac{1 + \beta_\Gamma}{1 - \beta_\Gamma}t_r - \frac{2}{1 - \beta_\Gamma}\frac{z_{cl} - z_0}{c}\tag{3}$$

implying that reflected synchrotron emission will be received starting at a time t_0 (corresponding to $t_e = 0$) given by

$$t_0 = \frac{2}{1+\beta_\Gamma}\frac{z_{cl}-z_0}{c}.$$ (4)

Reflected synchrotron radiation will be received by the emission region until it passes the cloud at $t_{pass} \approx (z_{cl}-z_0)/(\beta_\Gamma c)$.

The code writes out the observed synchrotron emission spectra, $\nu F_\nu^{sy}(t_e)$ for every time step (with times in the observer's frame) as the shock propagates along the jet. Therefore, at any time $t_{AGN} > t_0$, one can use Equation (3) to find the time (in the AGN frame) at which synchrotron radiation reflected back into the emission region, has been emitted. The synchrotorn flux irradiating the cloud, νF_ν^{cl}, is then found as

$$\nu F_\nu^{cl} = \nu F_\nu^{sy}(t_e)\frac{d_L^2}{(z_{cl}-z_e)^2}$$ (5)

where d_L is the luminosity distance to the source.

Assuming, for simplicity, that the cloud re-radiates a fraction τ_{cl} of the impinging synchroton radiation isotropically, it will emit a spectral luminosity of $\nu L_{nu}^{cl} = \pi R_{cl}^2 \tau_{cl} \nu F_\nu^{cl}$. The emission region will thus receive a flux of Reflected Synchrotron (RS—happy coincidence) radiation, in the comoving frame, of

$$\nu' F_{\nu'}^{RS}(t_r) \approx \frac{R_{cl}^2 \tau_{cl} \nu F_\nu^{sy}(t_e)\Gamma^4 d_L^2}{4\,(z_{cl}-z_e)^2\,(z_{cl}-z_r)^2}$$ (6)

where $\nu' \approx \Gamma\nu$ is the photon frequency in the co-moving frame.

The code evaluates a time-evolving reflected-synchrotron photon field in the emission region, $n_{RS}'(\epsilon',t_r')$, where $\epsilon' = h\nu'/(m_e c^2)$ is the dimensionless photon energy in the emission-region frame by the interplay of RS emission entering the emission region at a rate $dn_{RS,inj}'(\epsilon',t_r')/dt_r' = \pi R_b^2 \nu' F_{\nu'}^{RS}(t_r)/(V_b\,\epsilon'^2\,m_e c^2)$, where V_b is the volume of the emission region, and escape on an escape time scale $t_{esc}' = 3R_b/(4\,c)$, over a simulation time step $\Delta t'$ as

$$\Delta n_{RS}'(\epsilon',t_r') = \left(\frac{\pi R_b^2}{V_b}\frac{\nu' F_{\nu'}^{RS}(t_r)}{\epsilon'^2\,m_e c^2} - \frac{n_{RS}'(\epsilon',t_r')}{t_{esc}'}\right)\Delta t'.$$ (7)

In the above expressions, h is the Planck constant, m_e the electron mass, and $t_r' = t_r/\Gamma$.

The time-dependent RS photon field resulting from Equation (7) acts target for inverse-Compton scattering to produce the synchrotron-mirror Compton emission, and synchrotron-mirror Compton cooling is included self-consistently. For the evaluation of the synchrotron-mirror Compton emission, it is assumed, for simplicity, that the target photons enter the jet directly from the front, and the head-on approximation (e.g., [49]) for the Compton cross section is used. Hence, photons scattered along the viewing direction, making an angle θ_{obs}' with respect to the jet axis in the co-moving frame of the emission region, with $\mu_{obs}' \equiv \cos\theta_{obs}'$, have been scattered by a scattering angle $\mu' = -\mu_{obs}'$. The rate density at which Reflected Synchrotron Compton (RSC) emission is produced, is calculated as

$$\dot{n}_{RSC}'(\epsilon_s',\Omega_s') = c\int_0^\infty d\epsilon'\,n_{RS}'(\epsilon')\int_1^\infty d\gamma\,(1+\beta\mu_{obs}')\,n_e(\gamma)\frac{d\sigma_C}{d\epsilon_s'}(\epsilon',\epsilon_s',\mu')$$ (8)

where

$$\frac{d\sigma_C}{d\epsilon_s'} = \frac{\pi r_e^2}{\gamma\epsilon_e}\left\{y+\frac{1}{y}-\frac{2\epsilon_s'}{\gamma\epsilon_e y}+\left(\frac{\epsilon_s'}{\gamma\epsilon_e y}\right)^2\right\}$$ (9)

with $\epsilon_e = \epsilon'\gamma(1+\beta\mu_{obs}')$, $y = 1-\epsilon_s'/\gamma$ and $\beta = \sqrt{1-1/\gamma^2}$ [49].

3. Results: Spectral Variability Features

As detailed in the introduction, the study of the synchrotron mirror model developed here was motivated by the difficulties in modeling the orphan γ-ray flare B of 3C279 in December 2013 reported by Hayashida et al. [36]. We therefore start with the quiescent-state parameters of the shock-in-jet model for 3C279 used in [31]. The emission region is set to start out at $z_0 = 0.1$ pc, and the cloud acting as the mirror is assumed to be located at $z_{cl} = 1$ pc. The bulk Lorentz factor is $\Gamma = 15$, as used in [31]. The cloud radius is assumed to be $R_{cl} = 3 \times 10^{17}$ cm and its reflective fraction is $\tau_{cl} = 0.1$. The complete list of model parameters can be found in Table 1.

Table 1. Relevant model parameters for the case study motivated by the December 2013 flare of 3C279.

Parameter	Symbol	Value
Electron injection luminosity	L_{inj}	1.0×10^{43} erg s^{-1}
Pitch-angle mean free-path (m.f.p.) scaling normalization	η_0	100
Pitch-angle m.f.p. scaling index	α	3.0
Magnetic field	B	0.8 G
Electron escape time scale factor	η_{esc}	3.0
Emission region radius	R_b	2.0×10^{16} cm
Bulk Lorentz factor	Γ	15
Viewing angle	θ_{obs}	3.82°
Initial distance of the shock from the black hole (BH) along the jet	z_0	0.1 pc
Distance of the cloud from the BH	z_{cl}	1 pc
Radius of the cloud	R_{cl}	3×10^{17} cm
Reflective fraction of the cloud	τ_{cl}	0.1
Mass of the BH	M_{BH}	$5 \times 10^8\, M_\odot$
Luminosity of the accretion disk	L_d	6×10^{45} erg s^{-1}
Black-body temperature of external radiation field	T_{ext}	300 K
Energy density of external radiation field	u_{ext}	4×10^{-4} erg cm^{-3}

The resulting sequence of snap-shot SEDs (starting right before the onset of the synchrotron-mirror Compton emission) is illustrated in Figure 1. It is clear that the model does produce a significant orphan γ-ray flare, accompanied by a slight reduction of the synchrotron emission due to the increased Compton cooling of relativistic electrons. The latter is consistent with the observed evolution of the SED. However, the amplitude of the orphan γ-ray flare amounts only to an enhancement of the γ-ray flux by a factor of $\sim$2, in contrast to the observed dramatic increase by a factor of $\gtrsim$10. Even increasing the synchrotron-mirror efficiency (e.g., by increasing τ_{cl} or R_{cl}; see Equation (6)), does not increase the γ-ray flare amplitude substantially. The reason for this is that the γ-ray flux is limited by the available power injected into shock-accelerated electrons, which are already in the fast-cooling regime, thus radiating very efficiently. Any further enhancement of the reflected-synchrotron energy density will only suppress the synchrotron emission further, but not lead to a significant increase of the γ-ray flare amplitude. We therefore conclude that a pure shock-in-jet synchrotron mirror scenario is not able to produce the observed large-amplitude orphan γ-ray flare in 3C279 in December 2013. In order to achieve this, additional power would need to be injected into shock-accelerated electrons, leaving us with the same difficulties encountered in [31], i.e., requiring a fine-tuned reduction and gradual recovery of the magnetic field.

Nevertheless, in spite of its inapplicability to this particular orphan flare, it is worthwhile considering this simulation for a generic study of the expected spectral variability patterns in the shock-in-jet synchrotron mirror model. The multi-wavelength light curves at 5 representative frequencies (high-frequency radio, optical, X-rays, high-energy [HE, 200 MeV], and very-high-energy [VHE, 200 GeV] γ-rays) are shown in Figure 2. All light curves in the Compton SED component (X-rays to VHE γ-rays) show a flare due to the synchrotron-mirror Compton emission. Note that the VHE γ-ray light curve had to be scaled up by a factor of 10^{10} to be visible on this plot. Thus, the apparently large VHE flare is actually at undetectably low flux levels for the parameters chosen here. In contrast, the 230 GHz radio and optical light curves show a dip due to increased radiative cooling during the synchrotron mirror action. The radio dip is significantly delayed compared to the optical due to the longer cooling time scales of electrons emitting in the radio band.

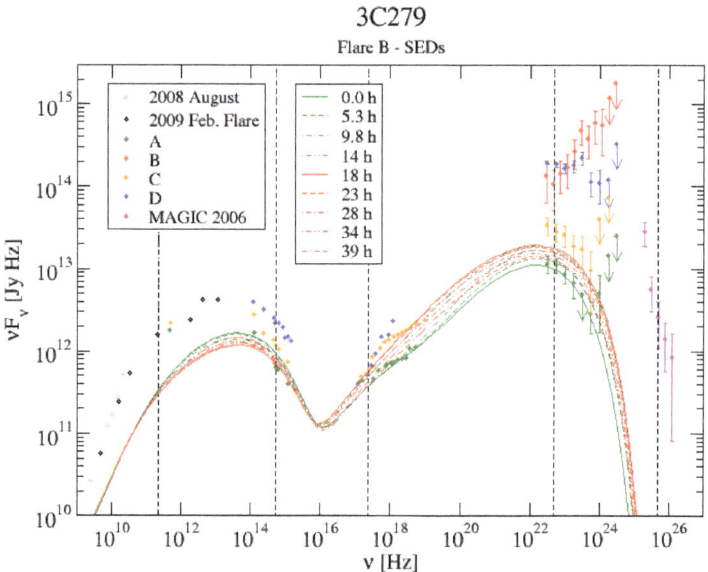

Figure 1. Spectral energy distributions (SEDs) of 3C279 in 2013–2014, from [36], along with snap-shot model SEDs from the shock-in-jet synchrotron-mirror model. The dashed vertical lines indicate the frequencies at which light curves and hardness-intensity relations were extracted. The legend follows the nomenclature of different periods from Hayashida et al. (2015) [36].

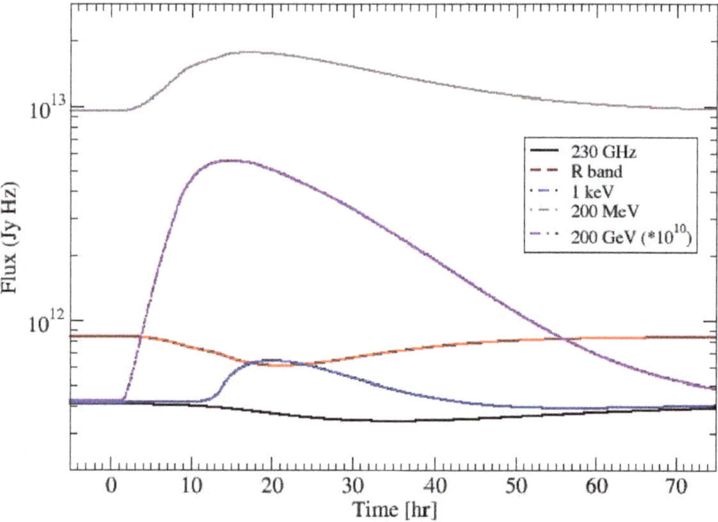

Figure 2. Model light curves in various frequency/energy bands resulting from the synchrotron mirror simulation illustrated in Figure 1 at the 5 representative frequencies/energies marked by the vertical dashed lines. Note that the very-high-energy (VHE, 200 GeV) γ-ray flux is scaled up by a factor of 10^{10} in order to be visible on the plot.

Cross-correlation functions between the various light curves from Figure 2 are shown in Figure 3. As expected from inspection of the light curves, significant positive correlations between X-rays and the 2 γ-ray bands with only small time lags (γ-rays leading X-rays by a few hours) and between the radio and optical band, with optical leading the radio by $\sim$15 h, are seen. The synchrotron (radio and optical) light curves are anti-correlated with the Compton (X-rays and γ-rays) ones, again with a significant lag of the radio emission by $\sim$15 h.

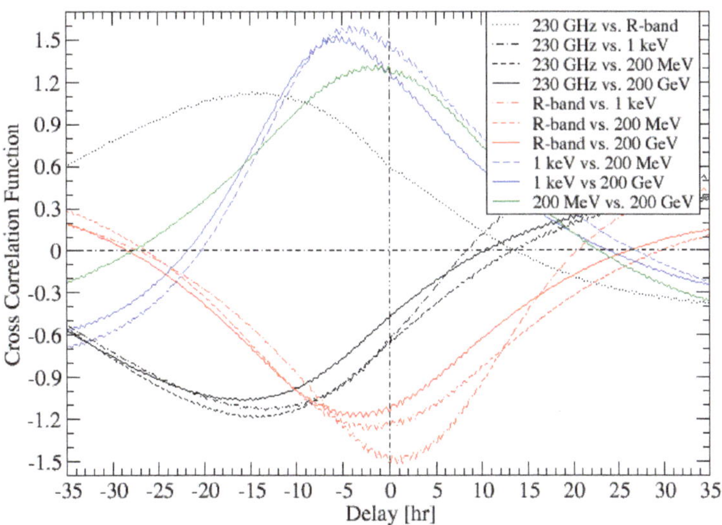

Figure 3. Cross-correlation functions between the model light curves in various energy / frequency bands.

Figure 4 shows the hardness-intensity diagrams for the 5 selected frequencies / energies, i.e., the evolution of the local spectral index (a, defined by $F_\nu \propto \nu^{-a}$) vs. differential flux. Generally, all bands, except the optical, exhibit the frequently observed harder-when-brighter trend. Only the radio and X-ray bands show very moderate spectral hysteresis. The dip in the optical R-band) light curve is accompanied by a very slight redder-when-brighter trend, likely as the consequence of the increased radiative cooling during the synchrotron-mirror action.

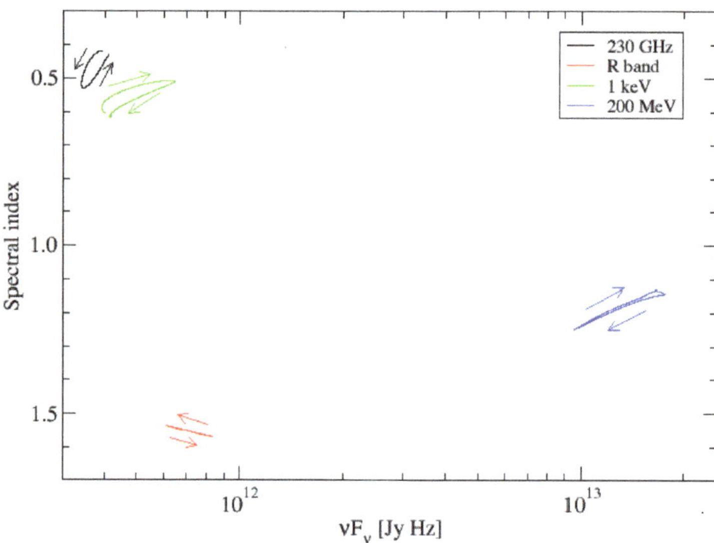

Figure 4. Model hardness-intensity diagrams at the selected frequencies/energies. The spectral index a is defined by $F_\nu \propto \nu^{-a}$, so that a smaller value indicates a harder spectrum. The VHE band has been omitted here due to its unobservably low flux level and very steep local spectral index. Arrow indicate the evolution in time.

4. Summary, Discussion, and Conclusions

In this paper, the leptonic shock-in-jet blazar model of [31] is extended with the addition of a self-consistent synchrotron mirror component. This was motivated by the difficulty in modeling orphan γ-ray flares with such an effectively single-zone model. A particularly high-amplitude (factor of $\sim$10) orphan γ-ray flare of the blazar 3C279 from December 2013 was chosen as a case study. However, the attempt to model this flare with the shock-in-jet synchrotron mirror model developed here, failed because the maximum γ-ray flux was limited by the (fixed) amount of power injected into shock-accelerated electrons, allowing for orphan flares with amplitudes of at most $\sim$2–3. Higher-amplitude flares would require an enhanced energy injection into relativistic electrons, in addition to more efficient pitch-angle scattering, leading to a harder electron spectrum. However, this would cause the same difficulties of having to decrease the magnetic field, followed by a fine-tuned recovery to its quiescent state, as were encountered in [31].

More successful model representations of this particular flare of 3C279 were presented by several authors. Hayashida et al. [36] use the model of Nalewajko et al. [50] to reproduce this orphan γ-ray flare by introducing an extreme hardening of the electron spectrum, along with a location of the emission region much closer to the BH and accretion disk. A hard electron spectrum $n_e(\gamma) \propto \gamma^{-1}$ up to a cut-off energy of a few 1000 is invoked, which may be difficult, but not impossible, to achieve with standard particle acceleration mechanisms. Asano and Hayashida [51] employ a time-dependent one-zone model with second-order Fermi acceleration, where an enhanced acceleration efficiency leads to a hardening of the electron spectrum, and a significant reduction of the magnetic field is required to suppress a simultaneous optical flare. While their model represents the γ-ray spectrum during the flare well, it does predict a non-negligible optical synchrotron flare accompanying the γ-ray flare. A similar strategy, based on an analytical solution to the steady-state electron distribution, was adopted by Lewis et al. [52], also requiring a significant reduction of the magnetic field to suppress a simultaneous optical synchrotron flare. Yan et al. [53] modelled the orphan-flare SED using a time-dependent single-zone model with rapid electron cooling. However, it is unclear whether a transition from the quiescent to this flaring state may be produced in a natural way. Lepto-hadronic models naturally de-couple the (proton-initiated) high-energy emission from the (electron-initiated) synchrotron radiation and therefore offer an alternative way of reproducing orphan γ-ray flares. Paliya et al. [54] used the time-dependent lepto-hadronic model of Diltz et al. [55] to model the December 2013 orphan γ-ray flare of 3C279. They also considered the possibility of a two-zone model, with a small emission region emitting an SED with very large Compton dominance, emerging during the time of the orphan γ-ray flare.

A representative simulation of the shock-in-jet synchrotron mirror scenario was then used for a generic study of the expected spectral variability patterns. X-ray and γ-ray light curves as well as radio and optical light curves are significantly correlated with each other, while radio and optical light curves are significantly anti-correlated, with radio and optical (synchrotron) dips accompanying the high-energy flare resulting from more efficient radiative cooling of the electrons. The response in the radio light curves is found to be delayed by $\sim$15 h with respect to other bands. In the scenario investigated here, where no changes to the diffusive shock acceleration process along the entire evolution of the flare are assumed, significant spectral hysteresis is not expected, but a mild harder-when-brighter trend in most wavebands is found.

While it is found that the specific December 2013 orphan γ-ray flare of 3C279 can not be successfully reproduced with this scenario, it may be applicable to other, more moderate orphan γ-ray flares. Especially the expected anti-correlation between Compton and synchrotron wavebands may serve as a smoking-gun signature of this scenario. The failure of the shock-in-jet synchrotron mirror model for the December 2013 flare of 3C279 was primarily caused by the fact that the shock-accelerated, relativistic electrons were already in the fast-cooling regime and radiating very efficiently, as was required by the fit to the quiescent state of 3C279. If the quiescent emission of a blazar can be produced by a less radiatively efficient configuration, then the increase in radiative efficiency in the synchrotron mirror scenario may lead to substantially higher-amplitude flares. A systematic study of different scenarios and applications to other sources will be presented in a forthcoming publication.

Funding: The work of M.B. is supported by the South African Researc Chairs Initiative of the National Research Foundation (any opinion, finding, and conclusion or recommendation expressed in this material is that of the authors, and the NRF does not accept any liability in this regard) and the Department of Science and Innovation of South Africa through SARChI grant no. 64789.

Conflicts of Interest: The author declares no conflict of interest.

References

1. Schlickeiser, R. Models of high-energy emission from active galactic nuclei. *Astron. Astrophys. Suppl. Ser.* **1996**, *120*, 481–489. Available online: https://adsabs.harvard.edu/pdf/1996A%26AS..120C.481S (accessed on 1 November 2021).
2. Crusius, A.; Schlickeiser, R. Synchrotron radiation in random magnetic fields. *Astron. Astrophys.* **1986**, *164*, L16–L18. Available online: https://articles.adsabs.harvard.edu/pdf/1986A%26A...164L..16C (accessed on 1 November 2021).
3. Crusius, A.; Schlickeiser, R. Synchrotron radiation in a thermal plasma with large-scale random magnetic fields. *Astron. Astrophys.* **1998**, *196*, 327–337. Available online: https://adsabs.harvard.edu/pdf/1988A%26A...196..327C (accessed on 1 November 2021).
4. Schlickeiser, R. Nonthermal radiation from jets of active galactic nuclei: Electrostatic bremsstrahlung as alternative to synchrotron radiation. *Astron. Astrophys.* **2003**, *419*, 397–414. [CrossRef]
5. Schlickeiser, R.; Vainio, R.; Böttcher, M.; Lerche, I.; Pohl, M.; Schuster, C. Conversion of relativistic pair energy into radiation in the jets of active galactic nuclei. *Astron. Astrophys.* **2002**, *393*, 69–87. [CrossRef]
6. Schlickeiser, R. Astrophysical gamma-ray production by inverse Compton interactions of relativistic electrons. *Astrophys. J.* **1979**, *233*, 294–301. [CrossRef]
7. Schlickeiser, R. Astrophysical gamma-ray production by inverse Compton interactions of relativistic electrons.II. Constraints on Compton emission models for NGC 4151, NP 0532, and PSR 0833-45 from gamma-ray data. *Astrophys. J.* **1980**, *236*, 945–950. [CrossRef]
8. Schlickeiser, R. Astrophysical gamma-ray production by inverse Compton interactions of relativistic electrons. III. Cutoff effect for inverse Compton spectra applied to the case of the hard X-ray and gamma-ray emission of NGC 4151. *Astrophys. J.* **1980**, *240*, 636–641. [CrossRef]
9. Maraschi, L.; Ghisellini, G.; Celotti, A. A jet model for the gamma-ray-emitting blazar 3C279. *Astrophys. J.* **1992**, *397*, L5–L9. [CrossRef]

10. Bloom, S.D.; Marscher, A.P. An analysis of the synchrotron self-Compton model for the multi-wave band spectra of blazars. *Astrophys. J.* **1996**, *461*, 657–663. [CrossRef]
11. Dermer, C.D.; Schlickeiser, R.; Mastichiadis, A. High-energy gamma radiation from extragalactic radio sources. *Astron. Astrophys.* **1992**, *256*, L21–L30.
12. Dermer, C.D.; Schlickeiser, R. Model for the high-energy emission from blazars. *Astrophys. J.* **1993**, *416*, 458–484. [CrossRef]
13. Sikora, M.; Begelmann, M.C.; Rees, M.J. Comptonization of diffuse ambient radiation by a relativistic jet: The source of gamma rays from blazars? *Astrophys. J.* **1994**, *421*, 153–162. [CrossRef]
14. Bla´zejowski, M.; Sikora, M.; Moderski, R.; Madejski, G.M. Comptonization of infrared radiation from hot dust by relativistic jets in quasars. *Astrophys. J.* **2000**, *545*, 107–116. [CrossRef]
15. Georganopoulos, M.; Kazanas, D. Decelerating flows in TeV blazars: A resolution to the BL Lacertae–FR I unification problem. *Astrophys. J.* **2003**, *594*, L27–L30. [CrossRef]
16. Tavecchio, F.; Ghisellini, G. Spine-sheath layer radiative interplay in subparsec-scale jets and the TeV emission from M87. *Mon. Not. R. Astron. Soc.* **2008**, *385*, L98–L102. [CrossRef]
17. Dermer, C.D.; Schlickeiser, R. Transformation properties of external radiation fields, energy-loss rates, scattered spectra, and a model for blazar variability. *Astrophys. J.* **2002**, *575*, 667–686. [CrossRef]
18. Dermer, C.D.; Schlickeiser, R. On the location of the acceleration and emission sites in gamma-ray blazars. In *International Astronomical Union Colloquium*; Cambridge University Press: Cambridge, UK, 1994; Volume 142, pp. 945–948.
19. Schlickeiser, R. Cosmic-ray transport and acceleration. I. Derivation of the kinetic equation and application to cosmic rays in static cold media. *Astrophys. J.* **1989**, *336*, 243–263. [CrossRef]
20. Schlickeiser, R. Cosmic-ray transport and acceleration. II. Cosmic rays in moving cold media with application to diffusive shock wave acceleration. *Astrophys. J.* **1989**, *336*, 264–293. [CrossRef]
21. Schlickeiser, R.; Dermer, C.D. Proton and electron acceleration through magnetic turbulence in relativistic outflows. *Astron. Astrophys.* **2000**, *360*, 789–794.
22. Peacock, J.A. Fermi acceleration by relativistic shock waves. *Mon. Not. R. Astron. Soc.* **1981**, *196*, 135–152. [CrossRef]
23. Kirk, J.G.; Heavens, A.F. Particle acceleration at oblique shock fronts. *Mon. Not. R. Astron. Soc.* **1989**, *239*, 995–1011. [CrossRef]
24. Kirk, J.G.; Guthmann, A.W.; Gallant, Y.A.; Achterberg, A. Particle acceleration at ultrarelativistic shocks: An eigenfunction method. *Astrophys. J.* **2000**, *542*, 235–242. [CrossRef]
25. Ellison, D.C.; Jones, F.C.; Reynolds, S.P. First-order Fermi particle acceleration by relativistic shocks. *Astrophys. J.* **1990**, *360*, 702–714. [CrossRef]
26. Bednarz, J.; Ostrowski, M. Energy spectra of cosmic rays accelerated at ultrarelativistic shock waves. *Phys. Rev. Lett.* **1998**, *80*, 3911–3914. [CrossRef]
27. Ellison, D.C.; Double, G.P. Diffusive shock acceleration in unmodified relativistic, oblique shocks. *Astropart. Phys.* **2004**, *22*, 323–338. [CrossRef]
28. Niemiec, J.; Ostrowski, M. Cosmic-ray acceleration at relativistic shock waves with a "realistic" magnetic field structure. *Astrophys. J.* **2004**, *610*, 851–867. [CrossRef]
29. Summerlin, E.J.; Baring, M.G. Diffusive acceleration of particles at oblique, relativistic magnetohydrodynamic shocks. *Astrophys. J.* **2012**, *745*, 63–85. [CrossRef]
30. Baring, M.G.; Böttcher, M.; Summerlin, E.J. Probing acceleration and turbulence at relativistic shocks in blazar jets. *Mon. Not. R. Astron. Soc.* **2017**, *464*, 4875–4894. [CrossRef]
31. Böttcher, M.; Baring, M.G. Multi-wavelength variability signatures of relativistic shocks in blazar jets. *Astrophys. J.* **2019**, *887*, 133. [CrossRef]
32. Böttcher, M.; Mause, H.; Schlickeiser, R. γ-ray emission and spectral evolution of pair plasmas in AGN jets. I. General theory and a prediction for the GeV–TeV emission from ultrarelativistic jets. *Astron. Astrophys.* **1997**, *324*, 395–409.
33. Böttcher, M.; Chiang, J. X-ray Spectral Variability Signatures of Flares in BL Lacertae Objects. *Astrophys. J.* **2002**, *581*, 127–142. [CrossRef]
34. Böttcher, M.; Reimer, A.; Sweeney, K.; Prakash, A. Leptonic and hadronic modeling of Fermi-detected blazars. *Astrophys. J.* **2013**, *768*, 54. [CrossRef]
35. Ch, ra, S.; Böttcher, M.; Goswami, P.; Singh, K.P.; Zacharias, M.; Kaur, N.; Bhattacharyya, S.; Ganesh, S.; Chandra, D. X-ray observations of 1ES 1959 + 650 in its high activity state in 2016–2017 with AstroSat and Swift. *arXiv* **2021**, arXiv:2105.08119.
36. Hayashida, M.; Nalewajko, K.; Madejski, G.M.; Sikora, M.; Itoh, R.; Ajello, M.; Blandford, R.D.; Buson, S.; Chiang, J.; Fukazawa, Y.; et al. Rapid variability of blazar 3C279 during flaring states in 2013–2014 with joint Fermi-LAT, NuSTAR, Swift, and ground-based multiwavelength observations. *Astrophys. J.* **2015**, *807*, 79. [CrossRef]
37. Baring, M.G.; Ogilvie, W.; Ellison, C.; Forsyth, R.J. Acceleration of solar wind ions by nearby interplanetary shocks: Comparison of Monte Carlo simulations with Ulysses observations. *Astrophys. J.* **1997**, *476*, 889–902. [CrossRef]
38. Ghisellini, G.; Madau, P. On the origin of the γ-ray emission in blazars. *Mon. Not. R. Astron. Soc.* **1996**, *280*, 67–76. [CrossRef]
39. Böttcher, M.; Dermer, C.D. On Comptonscattering scenarios for blazar flares. *Astrophys. J.* **1998**, *501*, L51–L54. [CrossRef]
40. Bedmarek, W. On the application of the mirror model for the gamma-ray flare in 3C279. *Astron. Astrophys.* **1998**, *336*, 123–129.
41. Vittorini, V.; Tavani, M.; Cavaliere, A.; Striani, E.; Vercellone, S. The blob crashes into the mirror: Modeling the exceptional γ-ray flaring activity of 3C454.3 in 2010 November. *Astrophys. J.* **2014**, *793*, 98. [CrossRef]

42. Tavani, M.; Vittorini, V.; Cavaliere, A. An emerging class of gamma-ray flares from blazars: Beyond one-zone models. *Astrophys. J.* **2015**, *814*, 51–63. [CrossRef]

43. MacDonald, N.R.; Marscher, A.P.; Jorstad, S.G.; Joshi, M. Through the ring of fire: γ-ray variabilit in blazars by a moving plasmoid passing a local source of seed photons. *Astrophys. J.* **2015**, *804*, 111. [CrossRef]

44. MacDonald, N.R.; Jorstad, S.G.; Marscher, A.P. "Orphan" γ-ray flares and stationary sheaths of blazar jets. *Astrophys. J.* **2017**, *850*, 87. [CrossRef]

45. Barkov, M.V.; Aharonian, F.A.; Bogovalov, S.V.; Kelner, S.R.; Khangulyan, D. Rapid TeV variability in blazars as a result of jet-star interaction. *Astrophys. J.* **2012**, *749*, 119. [CrossRef]

46. Araudo, A.T.; Bosch-Ramon, V.; Romero, G.E. Gamma-ray emission from massive stars interacting with active galactic nuclei jets. *Mon. Not. R. Astron. Soc.* **2013**, *436*, 3626–3639. [CrossRef]

47. Zacharias, M.; Böttcher, M.; Jankowsky, F.; Lenain, J.-P.; Wagner, S.J.; Wierzcholska, A. Cloud ablation by a relativistic jet and the extended flare in CTA 102 in 2016 and 2017. *Astrophys. J.* **2017**, *851*, 72. [CrossRef]

48. Zacharias, M.; Böttcher, M.; Jankowsky, F.; Lenain, J.-P.; Wagner, S.J.; Wierzcholska, A. The extended flare in CTA 102 in 2016 and 2017 within a hadronic model through cloud ablation by the relativistic jet. *Astrophys. J.* **2019**, *871*, 19. [CrossRef]

49. Dermer, C.D.; Menon, G. *High Energy Radiation from Black Holes: Gamma Rays, Cosmic Rays, and Neutrinos*; Princeton University Press: Princeton, NJ, USA, 2009.

50. Nalewajko, K.; Begelman, M.C.; Sikora, M. Constraining the location of gamma-ray flares in luminous blazars. *Astrophys. J.* **2014**, *789*, 161 [CrossRef]

51. Asano, K.; Hayashida, M. The most intensive gamma-ray flare of quasar 3C 279 with the second-order *Fermi* acceleration. *Astrophys. J. Lett.* **2015**, *808*, L18. [CrossRef]

52. Lewis, T.R.; Finke, J.D.; Becker, P.A. Electron Acceleration in Blazars: Application to the 3C 279 Flare on 2013 December 20. *Astrophys. J.* **2019**, *884*, 116. [CrossRef]

53. Yan, D.; Zhang, L.; Zhang, S.-N. Formation of very hard electron and gamma-ray spectra of flat-spectrum radio quasars in the fast-cooling regime. *Mon. Not. R. Astron. Soc.* **2016**, *459*, 3175–3181. [CrossRef]

54. Paliya, V.S.; Diltz, C.D.; Böttcher, M.; Stalin, C.S.; Buckley, D.A.H. A hard gamma-ray flare from 3C 279 in 2013 December. *Astrophys. J.* **2016**, *817*, 61. [CrossRef]

55. Diltz, C.S.; Böttcher, M.; Fossati, G. Time dependent hadronic modeling of flat spectrum radio quasars. *Astrophys. J.* **2015**, *802*, 133. [CrossRef]

Article

Proton-Alpha Drift Instability of Electromagnetic Ion-Cyclotron Modes: Quasilinear Development

Shaaban M. Shaaban [1,*], Marian Lazar [2,3], Peter H. Yoon [4], Stefaan Poedts [2,5] and Rodrigo A. López [6]

[1] Theoretical Physics Research Group, Physics Department, Faculty of Science, Mansoura University, Mansoura 35516, Egypt

[2] Centre for Mathematical Plasma Astrophysics, KU Leuven, Celestijnenlaan 200B, B-3001 Leuven, Belgium; marian.lazar@kuleuven.be (M.L.); stefaan.poedts@kuleuven.be (S.P.)

[3] Institut für Theoretische Physik IV, Ruhr-Universität Bochum, D-44780 Bochum, Germany

[4] Institute for Physical Science and Technology, University of Maryland, College Park, MD 20742, USA; yoonp@umd.edu

[5] Institute of Physics, University of Maria Curie-Skłodowska, Pl. M. Curie-Skłodowska 5, 20-031 Lublin, Poland

[6] Departamento de Física, Facultad de Ciencias, Universidad de Santiago de Chile, Santiago 8320000, Chile; rodrigo.lopez.h@usach.cl

* Correspondence: s.m.shaaban88@gmail.com

Abstract: The ability of space plasmas to self-regulate through mechanisms involving self-generated fluctuations is a topic of high interest. This paper presents the results of a new advanced quasilinear (QL) approach for the instability of electromagnetic ion-cyclotron modes driven by the relative alpha-proton drift observed in solar wind. For an extended parametric analysis, the present QL approach includes also the effects of intrinsic anisotropic temperatures of these populations. The enhanced fluctuations contribute to an exchange of energy between proton and alpha particles, leading to important variations of the anisotropies, the proton-alpha drift and the temperature contrast. The results presented here can help understand the observational data, in particular, those revealing the local variations associated with the properties of protons and alpha particles as well as the spatial profiles in the expanding solar wind.

Keywords: solar wind; plasmas; instabilities; waves

Citation: Shaaban, S.M.; Lazar, M.; Yoon, P.H.; Poedts, S.; López, R.A. Proton-Alpha Drift Instability of Electromagnetic Ion-Cyclotron Modes: Quasilinear Development. *Physics* **2021**, *3*, 1175–1189. https://doi.org/10.3390/physics3040075

Received: 16 September 2021
Accepted: 23 November 2021
Published: 1 December 2021

Publisher's Note: MDPI stays neutral with regard to jurisdictional claims in published maps and institutional affiliations.

1. Introduction

In collision-poor plasmas in space, e.g., solar wind and planetary magnetospheric environments, the dynamics of plasma particles, and implicitly their macroscopic properties, are expected to be constrained by the wave turbulence and the enhanced fluctuations, which are important components of these hot and dilute plasmas [1–3]. Very intriguing is the ability of these natural plasmas to self-regulate any deviation from kinetic isotropy, e.g., drifts, beams or anisotropic temperatures in velocity space, most likely, due to the inexorable implication of the self-generated instabilities [4–6].

In solar plasma outflows, protons (subscript p) are dominant, with a very high relative density $n_p > 90\%$, while alpha particles (subscript α) are highly contrasting, with only $n_\alpha < 8\%$, and a drift relative to protons of the order of local Alfvén speed [7–9]. These relative drifts may ignite the so-called ion-beam instabilities of various electrostatic (ion-acoustic) or electromagnetic (EM) modes (Alfvénic, magnetosonic). The resulting enhanced fluctuations can evolve fast enough so as to affect the regulation of ion drifts and their anisotropies [10,11], but may also contribute to the preferential heating of minor ions [5,12–14]. Their number densities, and the relative proton-alpha drift are indeed observed to decline with heliospheric distance [15,16], most probably due to scattering of the beaming particles involving self-generated instabilities.

The evolution of growing fluctuations, as well as the effects of their interaction with anisotropic plasma particles, cannot be captured by a linear theory of dispersion and stability, but needs more elaborate modeling and investigation using quasilinear (QL) approaches or/and numerical simulations [10,17,18]. In this paper, an advanced quasilinear (QL) approach of the EM ion-cyclotron (EMIC) instabilities, driven by the proton-alpha relative drifts, is presented as recently predicted by linear theory [11]. Such an extensive QL analysis is apt to characterize the amplification and saturation of the growing fluctuations in time, but also the relaxation effects on macroscopic properties of ion populations, including their relative drift speed [10,19]. A QL approach based on velocity moments, provides a reliable and straightforward description of the EM instabilities driven by the kinetic anisotropies of plasma particles [20,21], i.e., combinations of ion beams and intrinsic temperature anisotropies, typically observed in solar wind. The evolution of the main velocity moments, such as drifting or beaming speeds, and temperature components (parallel and perpendicular to the magnetic field) is confirmed by the numerical simulations, which also show that transient deformations of the distributions fade over time, while their initial shape (e.g., bi-Maxwellian with lower drifts for drifting components) is mainly restored during the relaxation [10,21–24]. More elaborate QL diffusion theories attempting to reproduce transient deformations of the anisotropic distribution [25] are however complicated and restricted thus far to a limited approximation of treating the wave spectral intensity as fixed and not evolving in time, which make their implementation to fully describe the saturation of the fluctuations and relaxation of the anisotropic distribution not yet feasible.

2. Quasilinear Kinetic Theory of Proton-Alpha Drift and Anisotropy Instabilities

2.1. Dispersion Relation and Wave Properties

Below, the EMIC-like instabilities, driven either by temperature anisotropies of ion populations (Section 3.1), or an alpha-proton (small) drift (Section 3.3), or the interplay of this drift and intrinsic temperature anisotropy (Section 3.2), are examined. Let us start by overviewing the linear wave properties associated with a plasma in which the ions are made of majority protons and alpha particles as the minor species. The basic theoretical framework was already discussed in a recent paper by Rehman et al. [11], but we hereby give a brief overview for the sake of completeness. The low-frequency waves of interest satisfy the cold-plasma dispersion relation specified by

$$\frac{c^2 k^2}{\omega_{pp}^2} = \frac{\epsilon_+ \epsilon_-}{\epsilon},$$

$$\epsilon_\pm = \frac{\omega}{\omega \pm \Omega_p} + \frac{n_\alpha}{n_p} \frac{\omega \mp \Omega_p \pm \Omega_\alpha}{\omega \pm \Omega_\alpha},$$

$$\epsilon = \frac{\Omega_p^2}{\omega^2 - \Omega_p^2} + \frac{n_\alpha}{n_p} \frac{\Omega_p^2}{\omega^2 - \Omega_\alpha^2}, \tag{1}$$

where $\omega_{pp} = (4\pi n_p e^2/m_p)^{1/2}$, $\Omega_p = eB_0/m_p c$, and $\Omega_\alpha = \Omega_p/2$ represent the proton plasma frequency, proton cyclotron frequency, and the cyclotron frequency associated with the alpha-particles, respectively, e, n_p, m_p, B_0, and c denoting the unit electric charge, proton number density, proton mass, ambient magnetic field intensity, and the speed of light in vacuum, respectively. In Equation (1), ω and k stand for the angular frequency and the wave number, respectively. Assuming that the ambient magnetic field lies along z axis, $\mathbf{B}_0 = \hat{\mathbf{z}} B_0$, the wave vector may be assumed to lie in xz plane without loss of generality, $\mathbf{k} = \hat{\mathbf{x}} k_\perp + \hat{\mathbf{z}} k_\parallel = \hat{\mathbf{x}} k \sin\theta + \hat{\mathbf{z}} k \cos\theta$, where $k_\perp$ and $k_\parallel$ are perpendicular and parallel wave vector components with respect to the ambient magnetic field vector, $k = (k_\perp^2 + k_\parallel^2)^{1/2}$ and $\theta = \cos^{-1}(k_\parallel/k)$ being the magnitude of the wave vector and the wave propagation angle, respectively. Note that the dispersion relation (1) supports the proton-cyclotron resonance ($\omega \sim \Omega_p$) and the alpha-cyclotron resonance ($\omega \sim \Omega_\alpha$). Instabilities may take place in the vicinity of these cyclotron frequencies when the appropriate free energies are available.

Among the useful properties of low-frequency waves is the polarization vector. The unit electric field vector $\hat{\mathbf{e}} = \delta\mathbf{E}/|\delta\mathbf{E}|$, where $\delta\mathbf{E}$ denotes the perturbed wave electric field, may be defined with respect to three orthogonal vectors, $\mathbf{e}_1 = [(\hat{\mathbf{b}} \times \hat{\mathbf{k}}) \times \hat{\mathbf{k}}]/|\hat{\mathbf{k}} \times \hat{\mathbf{b}}|$, $\hat{\mathbf{e}}_2 = (\hat{\mathbf{b}} \times \hat{\mathbf{k}})/|\hat{\mathbf{k}} \times \hat{\mathbf{b}}|$, and $\hat{\mathbf{e}}_3 = \hat{\mathbf{k}}$, where $\hat{\mathbf{k}} = \mathbf{k}/|\mathbf{k}|$ and $\hat{\mathbf{b}} = \mathbf{B}_0/|\mathbf{B}_0|$. For the geometry of interest, namely, $\mathbf{B}_0 = \hat{\mathbf{z}} B_0$ and $\mathbf{k} = \hat{\mathbf{x}} k \sin\theta + \hat{\mathbf{z}} k \cos\theta$, one can express $\mathbf{e}_1 = \hat{\mathbf{t}}(\sin\theta/|\sin\theta|)$, $\hat{\mathbf{e}}_2 = \hat{\mathbf{a}}(\sin\theta/|\sin\theta|)$, and $\hat{\mathbf{e}}_3 = \hat{\kappa}$, where $\hat{\kappa} = \hat{\mathbf{x}} \sin\theta + \hat{\mathbf{z}} \cos\theta$, $\hat{\mathbf{a}} = \hat{\mathbf{y}}$, and $\hat{\mathbf{t}} = \hat{\mathbf{x}} \cos\theta - \hat{\mathbf{z}} \sin\theta$. In short, the unit electric field vector can be expressed by

$$\hat{\mathbf{e}}(\mathbf{k}) = \frac{\delta \mathbf{E_k}}{|\delta \mathbf{E_k}|} = \frac{K\hat{\kappa} + T\hat{\mathbf{t}} + i\hat{\mathbf{a}}}{(1 + K^2 + T^2)^{1/2}}. \tag{2}$$

Thus, the polarization of the wave electric field is determined through the coefficients K and T. If $K = \infty$, then the wave is characterized as the longitudinal mode. If either $K = 0$ or $T = \infty$, then the mode is a transverse mode. Rehman et al. [11] show that these constants are determined from the dispersion relation as follows:

$$K = -M \sin\theta \frac{\omega}{\Omega_p},$$

$$T = -M \cos\theta \frac{\omega}{\Omega_p},$$

$$M = \left(\frac{\omega^2}{\Omega_p^2} \epsilon + \frac{c^2 k_\parallel^2}{\omega_{pp}^2} \right)^{-1} \left(\frac{\omega^2}{\omega^2 - \Omega_p^2} + \frac{n_\alpha}{n_p} \frac{\omega^2 + \Omega_\alpha^2}{\omega - \Omega_\alpha^2} \right), \tag{3}$$

where ϵ is defined in Equation (1).

Another useful linear wave property is the magnitude of the group velocity,

$$\frac{v_g}{c} = \frac{2ck}{\omega_{pp}^2} \frac{1}{(\partial/\partial\omega)(ck/\omega_{pp})^2}. \tag{4}$$

Rehman et al. [11] show that the quantity $(\partial/\partial\omega)(ck/\omega_{pp})^2$ is given by

$$R = \omega_p \frac{\partial}{\partial\omega} \left(\frac{c^2 k^2}{\omega_{pp}^2} \right) = \frac{\omega \Omega_p^3}{(\omega^2 - \Omega_p^2)(\omega^2 - \Omega_\alpha^2)} \left(\frac{2c^2 k^2}{\omega_{pp}^2} \cos^2\theta + \frac{\omega^2}{\Omega_p^2} \epsilon(1 + \cos^2\theta) \right)^{-1}$$

$$\times \left[\frac{2\omega^2}{\omega^2 - \Omega_p^2} \frac{(\omega^2 - 2\Omega_p^2)(\omega^2 - \Omega_\alpha^2)}{\Omega_p^4} \right.$$

$$+ \frac{2c^2 k^2 (1 + \cos^2\theta)}{\omega_{pp}^2} \left(\frac{(\omega^2 - \Omega_\alpha^2)}{\omega^2 - \Omega_p^2} + \frac{n_\alpha}{4n_p} \frac{\omega^2 - \Omega_p^2}{\omega^2 - \Omega_\alpha^2} \right) \tag{5}$$

$$\left. + \frac{n_\alpha}{4n_p} \frac{\omega^2}{\Omega_p^2} \frac{8\omega^4(2\omega^2 - 5\Omega_p^2) + 3\Omega_p^4(9\omega^2 - 2\Omega_p^2)}{\Omega_p^2(\omega^2 - \Omega_p^2)(\omega^2 - \Omega_\alpha^2)} + \frac{2n_\alpha^2}{n_p^2} \frac{(\omega^2 - \Omega_p^2)(\omega^2 - \Omega_\alpha^2)}{\Omega_p^4} \right].$$

2.2. Proton-Alpha Drift and Anisotropy Instability Growth Rate

The (quasi)-linear growth/damping rate is generally given by [26]

$$\gamma = \frac{\pi}{2\omega} \sum_{j=p,\alpha} \frac{n_j}{n_0} \frac{\omega_{pj}^2}{(1 + T^2)R} \int dv v_\perp^2 \sum_{n=-\infty}^{\infty}$$

$$\times \left\{ \frac{\omega}{\Omega_j} \left[K\sin\theta + T\left(\cos\theta - \frac{kv_\parallel}{\omega} \right) \right] \frac{J_n(b_j)}{b_j} - J_n'(b_j) \right\}^2$$

$$\times \delta(\omega - n\Omega_j - kv_\parallel \cos\theta) \left(\frac{n\Omega_j}{v_\perp} \frac{\partial}{\partial v_\perp} + k\cos\theta \frac{\partial}{\partial v_\parallel} \right) f_j,$$

$$b_j = \frac{kv_\perp \sin\theta}{\Omega_j}, \tag{6}$$

where $n_0 = n_p + n_\alpha$ denotes the net ambient plasma density, $J_n(x)$ is the Bessel function of the first kind of order n, and the quantities K, T, and R are given in Equations (3) and (5).

In ref. [11], it is assumed that both the alpha particles ("$j = \alpha$") and protons ("$j = p$") are described by drifting Maxwellian distribution functions, but in the present analysis, we extend the model to include temperature anisotropy. Hence, we keep the electrons cold, minimizing their influence, and assume that both protons and alpha particles are described by the drifting bi-Maxwellian distribution functions [27],

$$f_j = \frac{1}{\pi^{3/2} \alpha_{\perp j}^2 \alpha_{\|j}} \exp\left(-\frac{v_\perp^2}{\alpha_{\perp j}^2} - \frac{(v_\| - V_j)^2}{\alpha_{\|j}^2} \right), \tag{7}$$

where f_j is normalized to unity ($\int d\mathbf{v} f_j = 1$); V_j is the drifting velocity along the background magnetic field, and thermal velocities $\alpha_{\perp,\|j}$ (which may evolve in time) defined in terms of the corresponding kinetic temperatures $T_{\perp,\|j}$ are

$$\alpha_{\perp j} = \sqrt{\frac{2k_B T_{\perp j}}{m_j}}, \quad \alpha_{\|j} = \sqrt{\frac{2k_B T_{\|j}}{m_j}}. \tag{8}$$

The assumption of drifting bi-Maxwellian distribution functions is well supported by observations [28]. In the present discussion the temperature is defined in the unit of energy. As such, the Boltzmann constant can be set equal to unity $k_B = 1$.

Under the assumption of drifting isotropic Maxwellian model, in ref. [11], the linear growth/damping rate (6) is derived by carrying out the velocity integration. Under the more general model (7), the same calculation as that carried out in [11], is repeated. The result is a straightforward generalization,

$$\frac{\gamma}{\Omega_p} = -\sum_j \frac{n_\alpha}{n_p} \frac{m_p}{m_\alpha} \frac{\pi^{1/2}}{[1 + (M\cos\theta)^2(\omega/\Omega_p)^2]R} \sum_{n=-\infty}^{\infty} H_j^n \eta_j^n \exp\left[-\left(\zeta_j^n\right)^2\right], \tag{9}$$

where $m_\alpha = 4m_p$ is the alpha-particle mass; M and R are defined in Equations (3) and (5), respectively, and

$$\eta_j^n = \frac{1}{k_\| \alpha_{\|j}} \left[\frac{T_{\perp j}}{T_{\|j}} (\omega - k_\| V_j) - \left(\frac{T_{\perp j}}{T_{\|j}} - 1 \right) n\Omega_j \right],$$

$$\zeta_j^n = \frac{\omega - n\Omega_j - k_\| V_j}{k_\| \alpha_{\|j}},$$

$$H_j^n = \left(1 + \frac{M^2\omega^2}{\Omega_p^2} \right) \frac{n^2 \Lambda_n(\lambda_j)}{\lambda_j} - 2\left(\lambda_j + M\frac{n\omega}{\Omega_p} \right) \Lambda_n'(\lambda_j),$$

$$\Lambda_n(x) = I_n(x)e^{-x}, \quad \lambda_j = \frac{k_\perp^2 \alpha_{\perp j}^2}{2\Omega_j^2}. \tag{10}$$

Here, $I_n(x)$ is the modified Bessel function of the first kind of order n.

In the present analysis, it is assumed that the proton and alpha-particle distribution functions essentially retain their drifting bi-Maxwellian form throughout the time evolution of the instability whether it be driven by the proton-alpha relative drift or the temperature anisotropies. This is, of course, an approximation, but as already discussed in the Introduction, such an approach is well supported by validation against the numerical simulation. The time evolution of the particle distributions are controlled by the dynamical evolution of the temperature and drift velocities, $T_{\perp,\|j}$ and V_j, which is discussed below. We note that the wave energy density evolution is dictated by the QL wave kinetic equation,

$$\frac{\partial}{\partial t}\left(\frac{\langle \delta E^2\rangle_{\mathbf{k}}}{\langle \delta B^2\rangle_{\mathbf{k}}}\right) = 2\gamma\left(\frac{\langle \delta E^2\rangle_{\mathbf{k}}}{\langle \delta B^2\rangle_{\mathbf{k}}}\right). \tag{11}$$

2.3. Quasilinear Particle Kinetic Equation and Velocity Moment Equations

The QL evolution of the particle velocity distribution function f_j can be described by the general velocity space diffusion equation for the particles as given by [19,21,26]

$$\frac{\partial f_j}{\partial t} = \frac{1}{v_\perp}\frac{\partial}{\partial v_\perp} v_\perp\left(D_{\perp\perp}\frac{\partial f_j}{\partial v_\perp} + D_{\perp\parallel}\frac{\partial f_j}{\partial v_\parallel}\right) + \frac{\partial}{\partial v_\parallel}\left(D_{\parallel\perp}\frac{\partial f_j}{\partial v_\perp} + D_{\parallel\parallel}\frac{\partial f_j}{\partial v_\parallel}\right), \tag{12}$$

with

$$D_{ab} = \frac{\pi e_j^2}{m_j^2}\int d\mathbf{k}\,\frac{\langle \delta E^2\rangle_{\mathbf{k}}}{1+K^2+T^2}\sum_{n=-\infty}^{\infty}\left\{\frac{\omega}{\Omega_j}\left[K\sin\theta + T\left(\cos\theta - \frac{kv_\parallel}{\omega}\right)\right]\frac{J_n(b_j)}{b_j} - J_n'(b_j)\right\}^2$$

$$\times\ \delta(\omega - k_\parallel v_\parallel - n\Omega_j)\,\Delta_a\Delta_b, \qquad (a,b=\perp,\parallel),$$

$$\Delta_\perp = \frac{n\Omega_j}{\omega}, \qquad \Delta_\parallel = \frac{v_\perp}{v_\parallel}\frac{\omega - n\Omega_j}{\omega}. \tag{13}$$

The temporal evolution of velocity moments of the distribution function, such as drift velocities of the species j and their temperature components $T_{\perp j}$ and $T_{\parallel j}$, are given by

$$\frac{dV_j}{dt} = \frac{\partial}{\partial t}\int d\mathbf{v}\,v_\parallel f_j,$$

$$\frac{dT_{\perp j}}{dt} = \frac{m_j}{2}\frac{\partial}{\partial t}\int d\mathbf{v}\,v_\perp^2 f_j, \tag{14}$$

$$\frac{dT_{\parallel j}}{dt} = m_j\frac{\partial}{\partial t}\int d\mathbf{v}\,(v_\parallel - V_j)^2 f_j.$$

Upon substituting the kinetic Equation (12) to the right-hand sides of Equation (14), and taking partial derivatives, one gets:

$$\frac{dV_j}{dt} = -\int d\mathbf{v}\left(D_{\parallel\perp}\frac{\partial f_j}{\partial v_\perp} + D_{\parallel\parallel}\frac{\partial f_j}{\partial v_\parallel}\right),$$

$$\frac{dT_{\perp j}}{dt} = -m_j\int d\mathbf{v}\,v_\perp\left(D_{\perp\perp}\frac{\partial f_j}{\partial v_\perp} + D_{\perp\parallel}\frac{\partial f_j}{\partial v_\parallel}\right), \tag{15}$$

$$\frac{dT_{\parallel j}}{dt} = -2m_j\int d\mathbf{v}\,(v_\parallel - V_j)\left(D_{\parallel\perp}\frac{\partial f_j}{\partial v_\perp} + D_{\parallel\parallel}\frac{\partial f_j}{\partial v_\parallel}\right).$$

Making explicit use of the definitions for diffusion coefficients (13) as well as the drifting bi-Maxwellian distribution for f_j, defined by Equation (7), it is possible to show, after some straightforward albeit tedious mathematical manipulations, that the velocity moment kinetic equations that describe the time evolution of V_j, $T_{\perp j}$ and $T_{\parallel j}$ are given by

$$\frac{dV_j}{dt} = \frac{2\pi^{3/2}e_j^2}{m_j^2 c^2} \sum_{n=-\infty}^{\infty} \int_0^\infty dk\, k \int_{-1}^1 d(\cos\theta) \frac{|\cos\theta|\langle\delta B^2\rangle_{\mathbf{k}}}{1+M^2(\omega/\Omega_p)^2} H_j^n \eta_j^n e^{-(\zeta_j^n)^2},$$

$$\frac{dT_{\perp j}}{dt} = \frac{2\pi^{3/2}e_j^2}{m_j c^2} \sum_{n=-\infty}^{\infty} n\Omega_j \int_0^\infty dk \int_{-1}^1 d(\cos\theta)$$

$$\times \frac{|\cos\theta|}{\cos\theta} \frac{\langle\delta B^2\rangle_{\mathbf{k}}}{1+M^2(\omega/\Omega_p)^2} H_j^n \eta_j^n e^{-(\zeta_j^n)^2}, \tag{16}$$

$$\frac{dT_{\parallel j}}{dt} = \frac{4\pi^{3/2}e_j^2}{m_j c^2} \sum_{n=-\infty}^{\infty} \int_0^\infty dk \int_{-1}^1 d(\cos\theta)$$

$$\times \frac{|\cos\theta|}{\cos\theta} \frac{(\omega - n\Omega_j - k_\parallel V_j)\langle\delta B^2\rangle_{\mathbf{k}}}{1+M^2(\omega/\Omega_p)^2} H_j^n \eta_j^n e^{-(\zeta_j^n)^2},$$

where the dynamical Equation (16) are now expressed in terms of perturbed magnetic field energy density,

$$\left\langle \delta B^2 \right\rangle_{\mathbf{k}} = \frac{c^2 k^2}{\omega^2} \left\langle \delta E^2 \right\rangle_{\mathbf{k}}. \tag{17}$$

Equation (16) together with the wave kinetic Equation (11) describe the dynamics of the instability, whose growth rate is given by Equation (9) at each time step.

Here we should caution the readers that the assumption of drifting bi-Maxwellian velocity distribution functions for all time for both protons and alpha particles while simply calculating for the time evolution of velocity moments that define these distributions is a highly idealized approach. It is known that for an electrostatic bump-on-tail instability problem under a one-dimensional approximation, the quasilinear theory predicts a local deformation of the resonant range of velocity space that leads to the velocity space plateau formation. For an electromagnetic instability driven by the temperature anisotropy, on the other hand, it is also known that the quasilinear diffusion takes place, not along the parallel velocity space, but rather along a circularly path centered around the wave phase speed, which leads to the pitch-angle space diffusion and the resultant isotropization of the initial anisotropic temperatures [29,30]. For the present problem of EMIC instability driven by either the proton-alpha relative drift or the temperature anisotropies, the situation is more akin to the pitch-angle diffusion saturation picture as discussed by the above-referenced early literature rather than the velocity space plateau formation. Besides, recent papers by Harding et al. [31], Melrose et al. [32] discuss that the velocity plateau formation is relevant only for strictly one-dimensional problems and that for three-dimensional situations, the quasilinear process involved in the bump-on-tail instability also leads to the pitch-angle diffusion, which leads to the isotropic velocity distribution function. We thus believe that our assumption of the drifting bi-Maxwellian velocity distribution functions with time-varying velocity moments are justified.

3. Numerical Results

In order to aid the numerical analysis Equations (11) and (16) are considered in dimensionless form. Thus, the normalized quantities equations of relevance are:

$$\tau = \Omega_p t, \qquad x = \frac{\omega}{\Omega_p}, \qquad q = \frac{ck}{\omega_{pp}},$$

$$\beta_{\perp,\parallel j} = \frac{8\pi n_j T_{\perp,\parallel j}}{B_0^2}, \qquad \beta_{0j} = \frac{4\pi n_p m_p V_j^2}{B_0^2}, \qquad W(x,\theta)\, dq = \frac{4\pi n_p e^2}{m_p c^2} \frac{\langle\delta B^2\rangle_{\mathbf{k}}}{B_0^2}\, dk, \tag{18}$$

which stand for non-dimensional time, normalized wave frequency, normalized wave number, perpendicular and parallel beta's as well as the squared dimensionless drift speed for each species, and normalized wave magnetic field energy density. We solved the set of equations by numerical means. The basic time stepping method is the standard leap frog scheme. In the numerical analysis, the wave spectrum was separated into the forward

($\theta > 90°$) and backward ($90° < \theta < 180°$) components since the finite drift velocity breaks the symmetry associated with the forward versus backward wave propagation directions.

Below in this Section, a parametric analysis of the unstable modes, obtained as numerical exact solutions of the system of QL equations, are given. First, simple cases of instabilities, driven by temperature anisotropies of protons and alpha particles, are considered, and, then, the complexity of the study is gradually increased by introducing the cumulative effect of the alpha drift velocity.

3.1. Proton and Alpha Temperature Anisotropy-Driven Cyclotron Instabilities

In order to test the QL derivations obtained here, let us start with three cases describing cyclotron instabilities, driven either by the anisotropic protons with $A_p \equiv T_{\perp p}/T_{\parallel p} = 2$ (case 1). or by the anisotropic alpha particles $A_\alpha = 4$ (case 2), or, cumulatively, by the protons with $A_p = 2$ and alpha particles with $A_\alpha = 4$ (case 3). The input parameters are as follows:

- case 1: $A_p = 2$, $\beta_{\parallel p} = 2$, $A_\alpha = 1$, $\beta_{\parallel \alpha} = 0.2$;
- case 2: $A_p = 1$, $\beta_{\parallel p} = 2$, $A_\alpha = 4$, $\beta_{\parallel \alpha} = 0.2$;
- case 3: $A_p = 2$, $\beta_{\parallel p} = 2$, $A_\alpha = 4$, $\beta_{\parallel \alpha} = 0.2$.

Common plasma parameters in these cases are $n_\alpha/n_p = 0.05$, $T_{\parallel \alpha}/T_{\parallel p} = 2$, and $\beta_{0p,\alpha} = V_{p,\alpha}^2/v_A^2 = 0$.

Figure 1 presents the results of the QL temporal evolution for the enhanced magnetic fluctuations $\delta B^2/B_0^2 = \int d\mathbf{k}(\delta B_\mathbf{k}/B_0)^2$ (top row), and their effects on the macroscopic plasma parameters, i.e., proton plasma betas $\beta_{\perp,\parallel p}$ (middle), alpha plasma betas $\beta_{\perp,\parallel \alpha}$ (bottom), for the initial plasma parameters corresponding to case 1 (left), 2 (middle), and 3 (right).

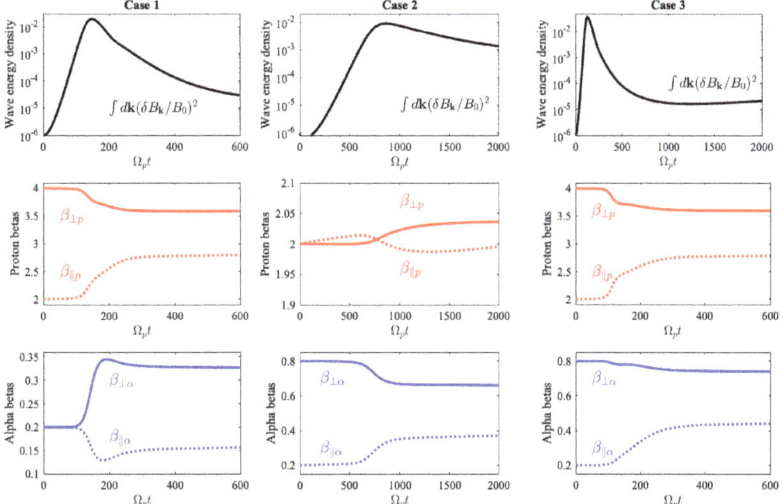

Figure 1. Quasilinear (QL) temporal variations for cases 1–3: the magnetic field energy density (**top row**), plasma beta parameters for protons (**middle row**), and alpha particles (**bottom row**). See text for details.

For case 1 (left column), the proton cyclotron instability is driven by the initially anisotropic protons $A_p(0) = 2$, and enhances the magnetic wave fluctuations $\delta B^2/B_0^2$. As a direct consequence, protons are subjected to perpendicular cooling and parallel heating, as indicated by, respectively, the perpendicular (red solid line) and parallel (red dashed line) components of the proton beta parameter $\beta_{\perp,\parallel p}$. On the other hand, initially isotropic $[A_\alpha(0) = 1]$ alpha particles are subjected to perpendicular heating and parallel cooling as shown by, respectively, the perpendicular (blue solid line) and parallel (blue dashed line) alpha beta parameters.

For case 2, the initially anisotropic alpha particles $[A_\alpha(0) = 4]$ excite the alpha cyclotron instability. The enhanced wave fluctuations $(\delta B^2 / B_0^2)$ determine a perpendicular cooling and parallel heating of alpha particles, as reflected by the perpendicular (blue solid line) and parallel (blue dashed line) alpha beta parameters $\beta_{\perp,\|\alpha}$, respectively. In this case plasma beta parameters of the initially isotropic protons experience perpendicular heating (red solid line) and parallel cooling (red dashed line).

In case 3, plasma beta parameters for protons and alphas are subjected to perpendicular cooling and parallel heating by the enhanced wave fluctuations. The interplay of protons and alpha particles temperature anisotropies inhibits the perpendicular cooling of alpha particles—see the bottom right panel. It is worth noting that all these results are in agreement with those reported in [10] for different plasma conditions.

Figure 2 summarizes the discussion on the cooling and heating of plasma ions by the means of temperature anisotropy for protons (red) and alpha particles (blue) $A_{p,\alpha} \equiv \beta_\perp / \beta_\|$ in the left panels, and alpha-proton parallel temperature ratio $T_{\|\alpha} / T_{\|p}$ in the right panels. For case 1, the temperature anisotropy of protons is relaxed as a function of time ($\tau = \Omega_p t$), while the initially isotropic alpha particles ended up with large temperature anisotropy in the perpendicular direction $A_\alpha(\tau_{\max}) > 1$. An opposite situation is observed in case 2, the temperature anisotropy of alpha particles is relaxed in time, while the initially isotropic protons gain small anisotropy in perpendicular direction $A_p(\tau_m) \gtrsim 1$. For case 3, both proton and alphas anisotropies are relaxed, under action of the enhanced magnetic wave fluctuations of the accumulated proton and alpha cyclotron instabilities. Moreover, a relaxation for the alpha-proton temperature ratio $T_{\|\alpha} / T_{\|p}$ is observed associated only with the excitation of the proton cyclotron instability (case 1), while in the other two cases (cases 2 and 3) one finds an enhancement of this temperature contrast mainly determined by the anisotropic alpha particles.

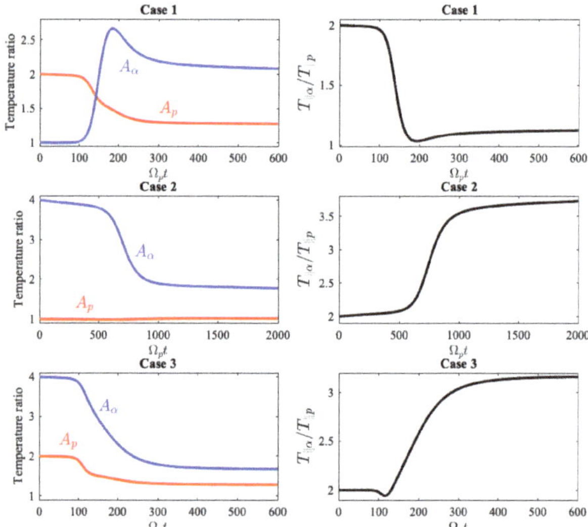

Figure 2. Temporal evolution for cases 1–3: temperature anisotropies (**left column**) and alpha-proton temperature ratio (**right column**). See text for details.

3.2. Interplay of Temperature Anisotropies and Alpha-Proton Drifts

Here, the complexity of the analysis is increased by considering a finite drift velocity of alpha particles (parallel to the background magnetic field) as an additional source of free energy. In order to show the effects induced by the drift, the same initial plasma parameters as in cases 1, 2 and 3 are considered, but with finite alpha drifts, which here are named cases 4, 5, and 6, respectively:

- case 4: $A_p = 2$, $A_\alpha = 1$, $\beta_{0\alpha} = \dfrac{V_\alpha^2}{v_A^2} = 1$;

- case 5: $A_p = 1$, $A_\alpha = 4$, $\beta_{0\alpha} = \dfrac{V_\alpha^2}{v_A^2} = 0.25$;

- case 6: $A_p = 2$, $A_\alpha = 4$, $\beta_{0\alpha} = \dfrac{V_\alpha^2}{v_A^2} = 1$.

Figure 3 shows the effects of the drift velocity of alpha particles on the temporal profiles of the macroscopic plasma parameters associated with the excitation of the proton cyclotron instability (case 4, left column), alpha cyclotron instability (case 5, middle column), and proton and alpha cyclotron cumulative instabilities (case 6, right column). Similar to case 1, the temporal evolution of the proton plasma beta parameters in case 4 are not affected by the drift velocity of alpha particles. Both alpha plasma beta parameters are increased in time, but the alphas are heated more in perpendicular direction, and become anisotropic at the final stage, that is, $A_\alpha(\tau_m) > 1$. For case 5, only parallel plasma beta parameter for protons is subjected to heating for a finite alpha drift. The interplay of the drift velocity and anisotropy of alpha particles enhance the cooling and heating mechanisms for the perpendicular and parallel alpha plasma betas, respectively; and alpha particles become less anisotropic, compared to case 2. For case 6, the interplay of three sources of free energies is considered, i.e., temperature anisotropies of protons and alphas, and the drift velocity of alpha particles. In this case, there are two opposite effects on the temporal profiles of the alpha plasma parameters (the first effect is already shown in Figure 1), such that temperature anisotropy inhibits the relaxation of the temperature anisotropy of alpha particles in the perpendicular direction, while the second effect is induced by the alpha drift velocity, markedly stimulating the cooling and heating mechanisms on alpha particles, which become isotropic at later time, i.e., $A_\alpha(\tau) \sim 1$. By comparing cases 5 and 6, one can state that the isotropization of the alpha particles is markedly enhanced with increasing the drift velocity, i.e., for $V_\alpha/v_A = 0.5$ in case 5, and $V_\alpha/v_A = 1$ in case 6.

Figure 4 summarizes the relaxation of the relative drift and the induction of temperature anisotropies of proton and alpha particles (left), and the alpha-proton temperature ratio (right). By contrasting with results in Figure 2, one can extract some effects induced by a finite alpha drift velocity. In case 4, the relaxation of proton anisotropy is not affected by the drift velocity of alpha particles, but the induced temperature anisotropy of alpha particles is much lower in case 4 than that in case 1. Furthermore, the relaxation of the alpha-proton temperature ratio is lower than that in case 1. For case 5, the alpha drift velocity slightly enhances the relaxation of alpha particles compared to case 2 (with $V_\alpha = 0$). In case 5, the initially isotropic protons remain nearly isotropic over time $A_p(\tau_m) \lesssim 1$. In case 6, one can see the relaxation of the proton anisotropy is comparable to that in case 3. However, the presence of alpha drift velocity markedly enhances the relaxation of the anisotropy of alpha particles, which become isotropic at final stage $A_\alpha(\tau_m) \sim 1$. In cases 5 and 6, the induced alpha-proton temperature ratios are enhanced by the initial drift of alpha particles. In all cases drift velocities are relaxed in time. It is worth to note that relaxation of the proton anisotropy and alpha drift velocity, and the induction of the alpha temperature anisotropy in case 4 (top left panel) are in good agreement with those obtained from the two-dimensional hybrid simulation reported by [33] for different plasma conditions, see Figure 1 therein.

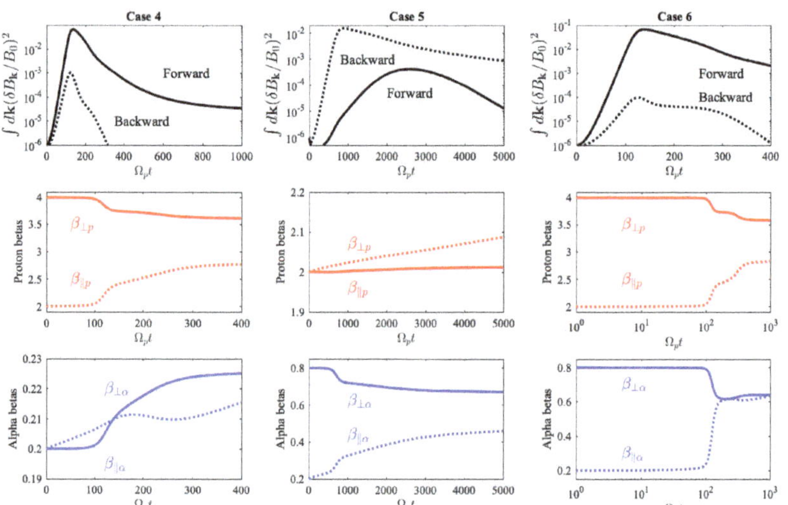

Figure 3. QL dynamical evolution for cases 4–6: the magnetic wave energy density (**top row**), plasma beta parameters for protons (**middle row**), and alpha particles (**bottom row**). See text for details.

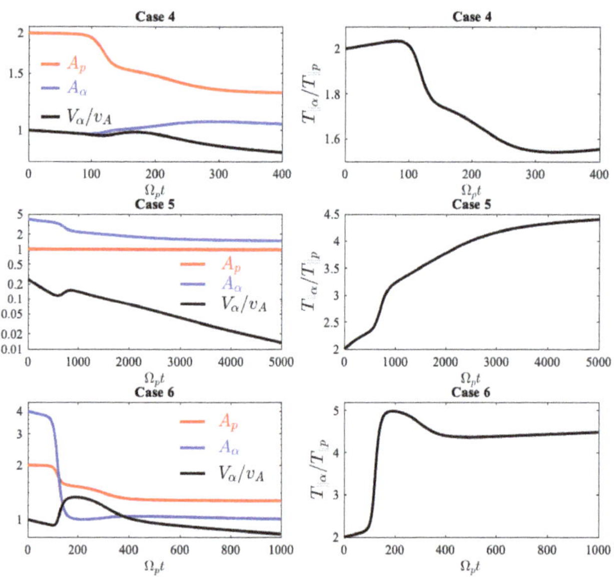

Figure 4. The relaxation of the drift velocity and the induced temperature anisotropies (**left column**), and alpha-proton temperature ratio (**right column**) for cases 5 and 6.

3.3. Instability Driven by the Alpha-Proton Drift (Isotropic Temperatures)

In this Section, it is assumed that the protons and alpha particles are initially isotropic, i.e., $A_{\alpha,p}(0) = 1$, and study the alpha cyclotron instability driven by the alpha drift velocity, and its consequences on the plasma particles. This case is summarized by the following initial parameters:

- case 7: $A_p = 1$, $\beta_{\|p} = 0.05$, $A_\alpha = 1$, $\beta_{\|\alpha} = 0.005$, $\beta_{0p} = \dfrac{V_p^2}{v_A^2} = 0$, $\beta_{0\alpha} = \dfrac{V_\alpha^2}{v_A^2} = 4$.

Figure 5 displays the temporal profiles for wave energy associated with the enhanced fluctuations (top-left). In this case the beaming cyclotron instability is driven by finite alpha-proton drift. Figure 5 also plots the plasma beta parameters of protons (bottom-left) and alpha particles (top-right), as well as their temperature anisotropies and drift velocity (bottom-right). The enhanced fluctuations are evident only for the forward propagating modes, meaning that the beam alpha cyclotron instability is excited in the direction parallel to the background magnetic field vector. Both components of plasma betas for proton and alpha particles are increased in time, i.e., as $\tau = \Omega_p t$ increases, but both species are heated more along the perpendicular direction. Bottom-right panel of Figure 5 summarizes the consequences of the enhanced fluctuations, presenting the temperature anisotropies of protons (red) and alpha particles (blue), and alpha drift velocity (black). Both species gain energy in perpendicular direction and become anisotropic at later stages with $A_{\alpha,p}(t_m) > 1$. However, the induced temperature anisotropy of alpha particles is much larger than that gained by the protons. The drift velocity is relaxed as τ increases. These results are in good agreement with those obtained recently from 2.5D and 3D hybrid simulations for different plasma conditions [22]; see Figure 1 in [22].

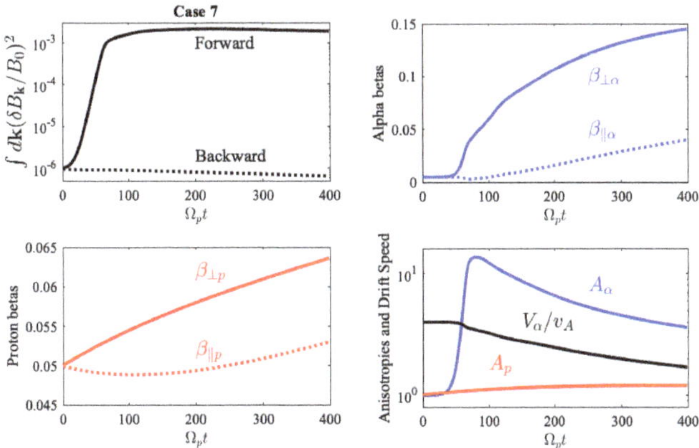

Figure 5. Case 7: QL dynamical evolution of the magnetic wave energy density (**top left panel**), plasma beta parameters for protons (**bottom left panel**), and alpha particles (**top right panel**), and for relaxing the drift velocity and inducing temperature anisotropies (**bottom right panel**). See text for details.

For the same plasma parameters as in case 7, but for different initial plasma beta parameters, the QL analysis is carried out further on:

- case 8: $\beta_p = 0.1$, $\beta_\alpha = 0.01$, $\beta_{0\alpha} \in [1, 2, d\beta_{0\alpha} = 0.25]$;
- case 9: $\beta_p = 0.5$, $\beta_\alpha = 0.05$, $\beta_{0\alpha} \in [1, 2, d\beta_{0\alpha} = 0.25]$;
- case 10: $\beta_p = 1.5$, $\beta_\alpha = 0.15$, $\beta_{0\alpha} \in [1, 2, d\beta_{0\alpha} = 0.25]$.

The result is given Figure 6, where the QL dynamical evolution of the alpha drift velocity (top), temperature anisotropies of protons (middle row) and alpha particles (bottom row) as functions of $\beta_{\|p}(\tau)$ (or $\beta_{\|\alpha}(\tau)$) for different initial plasma betas $\beta_{\|p}(0) = 0.1$ (case 8, left column), $\beta_{\|p}(0) = 0.5$ (case 9, middle column), and $\beta_{\|p}(0) = 1.5$ (case 10, right column) are displayed. For each case the evolutions for five different initial values of the alpha drift velocity $\beta_{0\alpha} \in [1, 2, d\beta_{0\alpha} = 0.25]$ are computed.

Top panels of Figure 6 show the relaxation of alpha drift velocity as a function of $\beta_{\|p}(\tau)$. The alpha-proton relative drift speed decreases to a very low level, i.e., $V_\alpha/V_A \to 0$, suggesting a complete relaxation of ion beams. Physically, ions can indeed be scattered by the resulting fluctuations, contributing to their redistribution in velocity space. In all these cases, the relaxation of the drift velocity is associated with a heating of protons in parallel direction. Middle panels show the dynamical evolution of the proton temperature anisotropy as a function of $\beta_{\|p}(\tau)$.

These dynamical evolution can be divided into two regimes, the first regime corresponds to low plasma betas, $\beta_{\|p} \sim 0.1$, when protons are slightly heated in the perpendicular direction and become anisotropic with $A_p > 1$, confirming the results in Figure 5. The second regime starts by increasing $\beta_{\|p} > 0.1$ (cases 9 and 10), or when the dynamical evolution of beta exceeds $\beta_{\|p}(\tau) > 0.1$. Increasing $\beta_{\|p}$ enhances the proton heating in parallel direction, and protons become (parallel) anisotropic at later stages with $A_p(\tau_m) < 1$. The same effects are observed for alpha particles in the bottom panels.

The dynamical evolution of the temperature anisotropy of alpha particles as a function of $\beta_{\|\alpha}(\tau)$ can also be divided into two regimes. For low betas $\beta_{\|\alpha} < 0.1$, the alpha particles are subjected to heating and cooling in the perpendicular and parallel directions, respectively, and alphas become anisotropic at later stages with large anisotropy in perpendicular direction $A_\alpha > 1$, confirming the results in Figure 5. The induced temperature anisotropies of alpha particles $A_\alpha > 1$ are in general increasing as the drift velocity increases. The second regime starts beyond $\beta_\alpha = 0.1$, when alpha particles gain (induced) temperature anisotropies only in parallel direction, i.e., $A_\alpha < 1$.

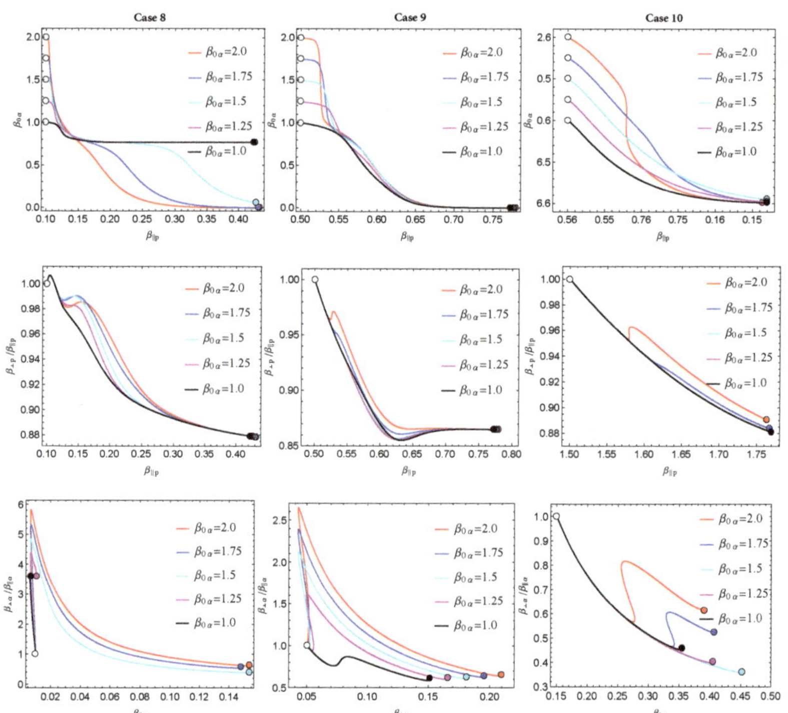

Figure 6. Dynamical evolution of the alpha drift velocity (**top row**), temperature anisotropies of protons (**middle row**) and alpha particles (**bottom row**) as functions of $\beta_{\|p}(\tau)$ (or $\beta_{\|\alpha}(\tau)$) for cases 8 (**left column**), 9 (**middle column**), and 10 (**right column**). See text for details.

4. Conclusions

In this paper, an advanced quasilinear (QL) analysis of the electromagnetic ion-cyclotron (EMIC) instabilities, driven by the kinetic anisotropies of protons and alpha particles, i.e., their relative drift, combined with or without the intrinsic temperature anisotropies, is presented. Such plasma conditions are specific to the solar atmosphere at short heliosphere distances, in the outer corona and solar wind. The long term evolution of the growing fluctuations, triggered by cyclotron instabilities, and also the consequences of their interaction were characterized with plasma particles, i.e., protons and alpha particles.

The paper presents for the first time the consequences of both the forward and backward fluctuations on the relaxation of the drift and temperature anisotropy of alpha and protons. Comparing the results obtained here with those of [33], where the analysis refers to total fluctuations, we were able to study the consequences of the backward fluctuations. The non-uniform relaxation of the macroscopic quantities in Figures 4 and 6 can be explained as a result of the competition between the backward and forward propagating modes and their enhanced fluctuations, as explained in [33] based on the hybrid simulations, e.g., in [22].

The results of the parametric study in Section 3 describe the effects from the interplay of the alpha-proton relative drift velocity and their temperature anisotropies on the saturation of the self-generated cyclotron instabilities and the relaxation of the non-thermal distributions. For a non-drifting scenario (Figures 1 and 2), the enhanced EMIC fluctuations are triggered by temperature anisotropy of protons or alphas, and on long term, show multiple effects on the particles. In addition to a relaxation of the anisotropy to quasi-stable states (below the thresholds of instability), the induction of a temperature anisotropy (in a perpendicular direction) to the other initially isotropic species are also observed. The induced temperature anisotropy of alpha particles is in general much larger than that of the protons. It is also found that the interplay of temperature anisotropies of protons and alpha particles has an inhibiting effect on the perpendicular cooling of alpha particles during the relaxation. Moreover, the alpha-proton temperature contrast ($T_{\|\alpha}/T_{\|p}$) is reduced only during the excitation of the proton cyclotron instability, but it is enhanced in the presence of the alpha instability fluctuations.

Comparing to the non-drifting scenario, an initial, relatively small alpha drift velocity stimulates the enhanced fluctuations and implicitly the relaxation of the alpha temperature anisotropy (Figures 3 and 4). The relaxation of temperature anisotropy is markedly enhanced with the increasing drift velocity of alpha particles, whose temperature becomes isotropic at final stage $A_\alpha \sim 1$. One the other hand, in the generation of the proton cyclotron fluctuations the relaxation of the proton anisotropy is not affected by the alpha drift velocity, but the induced temperature anisotropy of alpha particles is much lower than that in the non-drifting scenario. In all the cases drift velocities are relaxed as time evolves, and temporal profiles of the alpha-proton temperature contrast ($T_{\|\alpha}/T_{\|\alpha}$) are in general similar to those obtained for the non-drifting populations. However, a finite alpha drift may lead to an increase of this contrast in the excitation of the alpha cyclotron instability.

In Section 3.3, the QL temporal evolution of the beam alpha cyclotron instability, driven by the alpha-proton relative drift, is described with both protons and alpha particles considered initially isotropic (Figure 5). The enhanced fluctuations are associated only with the forward propagating modes, parallel to the background magnetic field. These fluctuations act back on the particles, reducing their drift velocity as time evolves. Concomitantly, the instability effectively leads to the perpendicular heating of the protons and alphas, so that at later stages, both species exhibit the perpendicular temperature anisotropies, $A_{p,\alpha} > 1$. The induced temperature anisotropy of alpha particles is in general much larger than that gained by protons. Figure 6 displays the QL dynamical evolution of the macroscopic plasma parameters, i.e., alpha drift velocity, temperature anisotropies of alphas and protons, and their temperature contrast, as a function of the parallel plasma betas $\beta_{\|\alpha,p}$. The dynamical evolution shows the relaxation of the drift velocity being associated with a heating of protons in parallel direction. One can distinguish two regimes conditioned by the parallel plasma betas $\beta_{\|\alpha,p}$ in this dynamical evolution of the temperature anisotropies of protons and alpha particles. First, for $\beta_{\|\alpha,p} < 0.1$, protons and alpha particles are heated in the perpendicular direction and become anisotropic with $A_{p,\alpha} > 1$. The second regime starts beyond $\beta_{\|\alpha,p} = 0.1$, when both protons and alpha particles gain induced anisotropy in parallel direction, i.e., $A_{p,\alpha} < 1$. These results are in good agreement with those obtained from hybrid and PIC simulations reported in the literature, see for instance [10,22,33].

To conclude, the study shows that the interplay of different sources of free energy present in solar wind has an important impact on the enhanced fluctuations, in particular those triggered by the EMIC instabilities, which in turn contribute to an exchange of energy between proton and alpha particles, leading to important variations of the temperature anisotropies, the proton-alpha drift and the temperature contrast. Future studies should also consider complementary conditions of more energetic beams (with higher drifts), which may excite different (e.g., firehose-like, or electrostatic) instabilities, possibly, with different consequences on the relaxation of the populations involved, especially, on their temperatures [34]. The results obtained here clearly show that self-generated EMIC instabilities can contribute to the regulation of drifts and anisotropies of ions present in the solar wind. Note that the QL theory, considered here, contains both parallel and perpendicular unstable solutions, although we do not give the detailed two-dimensional spectral characteristics associated with the unstable mode. Recently, Liu et al. [35] investigated the conditions when oblique instabilities are more competitive. In the future, it will be interesting to compare the QL theory here presented with the findings in [35] by analyzing the detailed 2D wave spectrum, but this is beyond the scope of the present study.

Author Contributions: The authors contributions are equal. All authors have read and agreed to the published version of the manuscript.

Funding: The authors acknowledge support from the Katholieke Universiteit Leuven, Ruhr-University Bochum, and Christian-Albrechts University Kiel. S. M. Shaaban acknowledges the Alexander-von-Humboldt Research Fellowship. R.A. López acknowledges the support of ANID Chile through FONDECyT grant No. 11201048. These results were also obtained in the framework of the projects C14/19/089 (C1 project Internal Funds KU Leuven), G.0D07.19N (FWO-Vlaanderen), SIDC Data Exploitation (ESA Prodex-12), and Belspo projects BR/165/A2/CCSOM and B2/191/P1/SWiM. P.H. Yoon acknowledges NASA Grant NNH18ZDA001N-HSR and NSF Grant 1842643 to the University of Maryland.

Data Availability Statement: The data that support the findings of this study are available from the corresponding author upon request.

Conflicts of Interest: The authors declare no conflict of interest.

References

1. Alexandrova, O.; Chen, C.H.K.; Sorriso-Valvo, L.; Horbury, T.S.; Bale, S.D. Solar wind turbulence and the role of ion instabilities. *Space Sci. Rev.* **2013**, *178*, 101–139. [CrossRef]
2. Woodham, L.D.; Wicks, R.T.; Verscharen, D.; Owen, C.J.; Maruca, B.A.; Alterman, B.L. Parallel-propagating fluctuations at proton-kinetic scales in the solar wind are dominated by kinetic instabilities. *Astrophys. J.* **2019**, *884*, L53. [CrossRef]
3. Bowen, T.A.; Mallet, A.; Huang, J.; Klein, K.G.; Malaspina, D.M.; Stevens, M.; Bale, S.D.; Bonnell, J.W.; Case, A.W.; Chandran, B.D.G.; et al. Ion-scale electromagnetic waves in the inner heliosphere. *Astrophys. J. Suppl. Ser.* **2020**, *246*, 66. [CrossRef]
4. Bale, S.D.; Kasper, J.C.; Howes, G.G.; Quataert, E.; Salem, C.; Sundkvist, D. Magnetic fluctuation power near proton temperature anisotropy instability thresholds in the solar wind. *Phys. Rev. Lett.* **2009**, *103*, 211101. [CrossRef] [PubMed]
5. Matteini, L.; Hellinger, P.; Goldstein, B.E.; Landi, S.; Velli, M.; Neugebauer, M. Signatures of kinetic instabilities in the solar wind. *J. Geophys. Res. Space Phys.* **2013**, *118*, 2771–2782. [CrossRef]
6. Sun, H.; Zhao, J.; Xie, H.; Wu, D. On kinetic instabilities driven by ion temperature anisotropy and differential flow in the solar wind. *Astrophys. J.* **2019**, *884*, 44. [CrossRef]
7. Asbridge, J.R.; Bame, S.J.; Feldman, W.C.; Montgomery, M.D. Helium and hydrogen velocity differences in the solar wind. *J. Geophys. Res.* **1976**, *81*, 2719–2727. [CrossRef]
8. Neugebauer, M. Observations of solar-wind helium. *Fundam. Cosm. Phys.* **1981**, *7*, 131–199. [CrossRef]
9. Marsch, E. Kinetic physics of the solar corona and solar wind. *Living Rev. Sol. Phys.* **2006**, *3*, 1. [CrossRef]
10. Yoon, P.H.; Seough, J.; Hwang, J.; Nariyuki, Y. Macroscopic quasi-linear theory and particle-in-cell simulation of helium ion anisotropy instabilities. *J. Geophys. Res. Space Phys.* **2015**, *120*, 6071–6084. [CrossRef]
11. Rehman, M.A.; Shaaban, S.M.; Yoon, P.H.; Lazar, M.; Poedts, S. Electromagnetic instabilities of low-beta alpha/proton beams in space plasmas. *Astrophys. Space Sci.* **2020**, *365*, 107. [CrossRef]
12. Maneva, Y.G.; Araneda, J.A.; Marsch, E. Regulation of ion drifts and anisotropies by parametrically unstable finite-amplitude Alfvén-cyclotron waves in the fast solar wind. *Astrophys. J.* **2014**, *783*, 139. [CrossRef]
13. Verscharen, D.; Chandran, B.D.G.; Bourouaine, S.; Hollweg, J.V. Deceleration of alpha particles in the solar wind by instabilities and the rotational force: Implications for heating, azimuthal flow, and the Parker spiral magnetic field. *Astrophys. J.* **2015**, *806*, 157. [CrossRef]
14. Markovskii, S.A.; Chandran, B.D.G.; Vasquez, B.J. Ion heating resulting from the deceleration of alpha particles by a proton-alpha drift instability in a nonuniform solar-wind plasma. *Astrophys. J.* **2019**, *870*, 121. [CrossRef]

15. Marsch, E.; Mühlhäuser, K.H.; Rosenbauer, H.; Schwenn, R.; Neubauer, F.M. Solar wind helium ions: Observations of the Helios solar probes between 0.3 and 1 AU. *J. Geophys. Res. Space Phys.* **1982**, *87*, 35–51. [CrossRef]
16. Neugebauer, M.; Goldstein, B.E.; Smith, E.J.; Feldman, W.C. Ulysses observations of differential alpha-proton streaming in the solar wind. *J. Geophys. Res. Space Phys.* **1996**, *101*, 17047–17055. [CrossRef]
17. Yoon, P.H.; Schlickeiser, R. On the equilibrium between proton distribution and compressible kinetic Alfvénic fluctuations. *Mon. Not. R. Astron. Soc.* **2018**, *482*, 4279–4289. [CrossRef]
18. Shaaban, S.M.; Lazar, M.; Schlickeiser, R. Electromagnetic ion cyclotron instability stimulated by the suprathermal ions in space plasmas: A quasi-linear approach. *Phys. Plasmas* **2021**, *28*, 022103. [CrossRef]
19. Yoon, P.H.; Seough, J. Quasilinear theory of anisotropy-beta relation for combined mirror and proton cyclotron instabilities. *J. Geophys. Res. Space Phys.* **2012**, *117*, A08102. [CrossRef]
20. Davidson, R.C.; Völk, H.J. Macroscopic quasilinear theory of the Garden-Hose instability. *Phys. Fluids* **1968**, *11*, 2259–2264. [CrossRef]
21. Yoon, P.H. Kinetic instabilities in the solar wind driven by temperature anisotropies. *Rev. Mod. Plasma Phys.* **2017**, *1*, 4. [CrossRef]
22. Ofman, L. Nonlinear Evolution of ion kinetic instabilities in the solar wind. *Sol. Phys.* **2019**, *294*, 51. [CrossRef]
23. López, R.A.; Shaaban, S.M.; Lazar, M.; Poedts, S.; Yoon, P.H.; Micera, A.; Lapenta, G. Particle-in-cell simulations of the whistler heat-flux instability in solar wind conditions. *Astrophys. J. Lett.* **2019**, *882*, L8. [CrossRef]
24. Micera, A.; Zhukov, A.N.; López, R.A.; Innocenti, M.E.; Lazar, M.; Boella, E.; Lapenta, G. Particle-in-cell simulation of whistler heat-flux instabilities in the solar wind: Heat-flux regulation and electron halo formation. *Astrophys. J. Lett.* **2020**, *903*, L23.
25. Jeong, S.Y.; Verscharen, D.; Wicks, R.T.; Fazakerley, A.N. A Quasi-linear Diffusion Model for Resonant Wave–Particle Instability in Homogeneous Plasma. *Astrophys. J.* **2020**, *902*, 128. [CrossRef]
26. Melrose, D.B. *Instabilities in Space and Laboratory Plasmas*; Cambridge University Press: Cambridge, UK, 1986. [CrossRef]
27. Shaaban, S.M.; Lazar, M.; Yoon, P.H.; Poedts, S.; López, R.A. Quasi-linear approach of the whistler heat-flux instability in the solar wind. *Mon. Not. R. Astron. Soc.* **2019**, *486*, 4498–4507. [CrossRef]
28. Alterman, B.L.; Kasper, J.C.; Stevens, M.L.; Koval, A. Comparison of alpha particle and proton beam differential flows in collisionally young solar wind. *Astrophys. J.* **2018**, *864*, 112. [CrossRef]
29. Vedenov, A.A.; Velikhov, E.P.; Sagdeev, R.Z. Nonlinear oscillations of rarified plasma. *Nucl. Fusion* **1961**, *1*, 82. (In Russian) [CrossRef]
30. Kennel, C.F.; Engelmann, F. Velocity space diffusion from weak plasma turbulence in a magnetic field. *Phys. Fluids* **1966**, *9*, 2377. [CrossRef]
31. Harding, J.C.; Cairns, I.H.; Melrose, D.B. Electron-Langmuir wave resonance in three dimensions. *Phys. Plasmas* **2020**, *27*, 020702. [CrossRef]
32. Melrose, D.B.; Harding, J.C.; Cairns, I.H. Type-III electron beams: 3D quasilinear effects. *Sol. Phys.* **2021**, *296*, 42. [CrossRef]
33. Gary, S.P.; Yin, L.; Winske, D.; Ofman, L.; Goldstein, B.E.; Neugebauer, M. Consequences of proton and alpha anisotropics in the solar wind: Hybrid simulations. *J. Geophys. Res. Space Phys.* **2003**, *108*, 1–11. [CrossRef]
34. Ďurovcová, T.; Šafránková, J.; Němeček, Z. Evolution of relative drifts in the expanding solar wind: Helios observations. *Sol. Phys.* **2019**, *294*, 97. [CrossRef]
35. Liu, W.; Zhao, J.; Xie, H.; Yao, Y.; Wu, D.; Lee, L.C. Electromagnetic proton beam instabilities in the inner heliosphere: Energy transfer rate, radial distribution, and effective excitation. *Astrophys. J.* **2021**, *920*, 158. [CrossRef]

Article

A Perspective on the Solar Modulation of Cosmic Anti-Matter

Marius S. Potgieter [1],*, O. P. M. Aslam [2], Driaan Bisschoff [2,3] and Donald Ngobeni [2,3]

[1] Institute for Experimental and Applied Physics, Christian-Albrechts University in Kiel, 24118 Kiel, Germany
[2] Centre for Space Research, North-West University, Potchefstroom 2520, South Africa;
    aslamklr2003@gmail.com (O.P.M.A.); driaanb@gmail.com (D.B.); donald.ngobeni@nwu.ac.za (D.N.)
[3] School of Physical and Chemical Sciences, North-West University, Mmabatho 2735, South Africa
* Correspondence: marius.s.potgieter@gmail.com or potgieter@physik.uni-kiel.de

Abstract: Global modulation studies with comprehensive numerical models contribute meaningfully to the refinement of very local interstellar spectra (VLISs) for cosmic rays. Modulation of positrons and anti-protons are investigated to establish how the ratio of their intensity, and with respect to electrons and protons, are changing with solar activity. This includes the polarity reversal of the solar magnetic field which creates a 22-year modulation cycle. Modeling illustrates how they are modulated over time and the particle drift they experience which is significant at lower kinetic energy. The VLIS for anti-protons has a peculiar spectral shape in contrast to protons so that the total modulation of anti-protons is awkwardly different to that for protons. We find that the proton-to-anti-proton ratio between 1–2 GeV may change by a factor of 1.5 over a solar cycle and that the intensity for anti-protons may decrease by a factor of ~2 at 100 MeV during this cycle. A composition is presented of VLIS for protons, deuteron, helium isotopes, electrons, and particularly for positrons and anti-protons. Gaining knowledge of their respective 11 and 22 year modulation is useful to interpret observations of low-energy anti-nuclei at the Earth as tests of dark matter annihilation.

Keywords: cosmic rays; galactic anti-matter; solar modulation; solar cycle

Citation: Potgieter, M.S.; Aslam, O.P.M.; Bisschoff, D.; Ngobeni, D. A Perspective on the Solar Modulation of Cosmic Anti-Matter. *Physics* **2021**, *3*, 1190–1225. https://doi.org/10.3390/physics3040076

Received: 1 October 2021
Accepted: 19 November 2021
Published: 7 December 2021

Publisher's Note: MDPI stays neutral with regard to jurisdictional claims in published maps and institutional affiliations.

1. Introduction

Galactic anti-matter, specifically positrons and anti-protons, has been observed on Earth simultaneously, continuously and with high precision over prolonged periods of time by PAMELA [1–3] and AMS-02 [4,5]. Although these missions, deployed in geospace, were designed to study specific astrophysical characteristics related to galactic cosmic rays (GCRs), they both have contributed extensively to the study of solar modulation in the inner heliosphere. For solar modulation and numerical modelling studies, observations of GCRs lower than ~30 GV are essential, with modulation effects growing in significance the lower the rigidity is. In this context, PAMELA had recorded GCRs down to ~80 MV from 2006 to 2016, ending just after the recent maximum solar activity. AMS-02 has been recording GCRs down to ~0.5–1.0 GV since 2011, including the period of solar maximum activity, the reversal of the heliospheric magnetic field (HMF) and the recent solar minimum period with published spectra averaged over Bartels rotations until 2017. However, full spectra of anti-matter particles detected at regular intervals (e.g., every solar rotation) by these missions have not been published. This lower limit in energy of AMS-02 is not ideal for studying the exceptional features of solar modulation occurring at lower energy but the consistent preciseness of these observations over time has opened up other avenues that can be studied in conjunction with sophisticated numerical models. These recurrently measured spectra of such a variety of GCRs over an extensive period of time have made it possible to take also the investigation of the modulation of GCRs with comprehensive numerical models to a higher level of preciseness at a rigidity range not previously achievable. This encourages studying modulation effects occurring at Earth that are quite small and which on its part provides interesting challenges to numerical modeling attempting to explain these easily overlooked features and the underlying physics on which they are based.

Observations of GCRs made by the two Voyager spacecraft in the outer heliosphere have been of utmost importance for modeling efforts on a global scale. Voyager 1 and 2 crossed the heliopause (HP) in August 2012 at 121.6 AU and in November 2018 at 119 AU from the Sun, respectively, but far apart, about 167 AU, and at different heliographic latitudes [6,7]. It is now reasonably well known how wide the inner heliosheath is, what the average distance is between the termination shock (TS) and the HP and where the HP is located as the preferred modulation boundary (at least in the nose-direction of the heliosphere). Also, that none or very little modulation of GCRs (e.g., [8–10]) seems to occur beyond the HP (this region is considered to be the outer heliosheath, depending on the shape of the heliosphere and given that a heliospheric bow shock may exist; see [11,12]. Since no significant GCR spatial gradients in intensity were observed over this vast outer heliospheric region, the assumption that the local interstellar spectra (LISs) for GCRs are arriving isotropically at the heliosphere still seems reasonable (except at very high rigidity where it is less than 0.1% (e.g., [13,14]). The point is that three-dimensional (3D) simulations can now be done much less speculatively than before, both in terms of the spatial dimensions of the heliosphere and what spectra to utilize at the modulation boundary as very LISs (VLISs), an aspect that will be discussed in depth in the next section. It is fair to say that the global and total modulation of GCRs between the HP and the Earth can be described and simulated more convincingly than before. In this context a number of open publications by the late W.R. Webber and his colleagues can be found on the eprint archive arXiv. Also important to the aspired enhancement of global numerical modelling are observations that were made by the Ulysses mission in the inner heliosphere between 1990 and 2009 from which an interpretation could be formed of how GCR modulation may happen at high heliolatitudes (e.g., [15,16]; and the reviews in [17,18]).

In the context of establishing VLISs, both Voyager spacecraft have made measurements in the kinetic energy (KE) range of 3 MeV/nuc to 346 MeV/nuc for protons, total helium and other light GCR nuclei as well as for electrons (actually negatrons) from 2.7 to 74 MeV [19]. The energy range for these particles was specified somewhat differently in [20,21] also reported observations for the secondary GCR isotopes H-2 (^{2}H$_1$ deuteron), and He-3 (^{3}He$_2$), apart from protons and He-4 (^{4}He$_2$), as primary particles. However, anti-matter such as positrons and anti-protons are not measured by them, so that it remains an issue what to use at these low rigidities for their VLISs in global modeling. Fortunately, solar modulation can be considered negligible at an appropriate high enough rigidity where precise measurements now exist so that at least their VLISs are easily determined at these high rigidity values. The only point of contention here is at what rigidity the solar modulation of GCRs truly commences [22].

A way to test for dark matter annihilation may be found in observations of low-energy anti-nuclei in GCRs, such as anti-protons, anti-deuteron and anti-helium, making the proper numerical study of the modulation of these anti-particles in the heliosphere imminent. Rudimentary modulation models, such as the Force-Field model [23] and analytical variations there-of (e.g., [24]), are inadequate for this purpose and can easily yield misleading results when it comes to precise modeling as pointed out in [25–28].

The layout of this paper is as follows: First the essence of the basic transport theory used for numerical models and our modeling approach to solar modulation studies are briefly described, including global modulation parameters, as well as the diffusion coefficients in terms of their rigidity dependence. Particle drifts are discussed and what the drift related predictions for solar modulation are and how it manifests from the difference between the modulation of electrons and positrons, protons and anti-protons and what differences there are between positrons and anti-protons. In Section 3, the manner is discussed in which the relevant VLISs are established by using both observations and numerical modulation models. In Sections 4 and 5 simulations of the modulation of positrons and anti-protons are shown in particular and the differences between them are discussed. A brief discussion is given on the modulation of other types of anti-matter, with a summary and a compilation of computed VLISs in the conclusion.

2. Numerical Model and Modeling Approach to Solar Modulation Studies

2.1. Theory and Assumptions

The modeling here is based on solving numerically the transport equation (TPE) derived in [29]:

$$\frac{\partial f}{\partial t} = \nabla.(\boldsymbol{K}.\nabla f) - (\mathbf{V} + \langle \boldsymbol{v}_D \rangle).\nabla f + \frac{1}{3}(\nabla.\mathbf{V})\frac{\partial f}{\partial \ln p},\tag{1}$$

where $f(\mathbf{r}, p, t)$ is the omnidirectional GCR distribution function, p is particle momentum, $\mathbf{r}$ is the heliocentric position vector and t is time. The differential intensity of GCRs is given by $I = p^2 f$ or equivalently $I = R^2 f$, where R is rigidity in GV. The terms on the right-hand side represent diffusion, convection, gradient and curvature drifts, and adiabatic energy changes. The solar wind velocity is $\mathbf{V} = V(r, \theta)\mathbf{e}_r$. Adiabatic energy losses occur when $\nabla.\mathbf{V} > 0$ which applies to most of the heliosphere. The diffusion tensor $\boldsymbol{K}$ consists of three distinct diffusion coefficients (DCs), first, $K_{\parallel}$ which is parallel to the average HMF, $K_{\perp r}$ which is radially perpendicular to this field and $K_{\perp \theta}$ which is perpendicular to this field in the heliospheric polar direction. The averaged guiding center drift velocity for a near isotropic distribution is given by

$$\langle \mathbf{v}_D \rangle = \nabla \times (\mathbf{e}_B K_D)\tag{2}$$

with $\mathbf{e}_B = \mathbf{B}/B$. Here, K_D is the drift coefficient (in the literature it is also indicated with K_A) where B is the magnitude of the background HMF at a given position in the heliosphere, the value of which changes spatially and with solar activity.

The global HMF is assumed to have a basic Parkerian geometry in the equatorial plane, but modified in the polar regions similar to the approach of [30]; for a full description of this particular aspect, see [31]. The global latitudinal dependence of the V is assumed to change from 430 km s^{-1} in the equatorial plane to 750 km s^{-1} in the polar regions only during solar minimum activity in accordance with Ulysses observations (see [32], and reference there-in; and reviews [17,18,33]). The HP position is assumed at 122 AU, with any asymmetry in the shape of the heliosphere considered to be negligible for GCR modulation studies (e.g., [34,35]). The position of TS is changed according to solar activity levels [36] from 88 AU (e.g., in 2006) to 80 AU (e.g., in 2009). The shock compression ratio of 2.5 is assumed for the TS consistent with observations [37,38] in order to account for the drop in V beyond the TS as assumed in the model. For details on this modeling approach, see [39] and references there-in. However, the re-acceleration of GCRs at the TS is not considered for this study; for details on such a modeling approach; see [40,41].

This 3D numerical model and all relevant assumptions were comprehensively described in [42–46]. The main features of this model and several applications can also be found in these publications as well as those recently published [47–51]. For general reviews on solar modulation, see [27,33,52–54] and for reviews on deriving transport equations for GCRs, see [26,55,56] for those applicable to heliospheric transport.

2.2. Modeling Approach

The method and procedures followed here consist of the following: The VLIS for electrons and protons were established as the first step and then used in the 3D modulation model to reproduce the observed electron and proton spectra at the Earth from PAMELA for 2006 to 2009, thus focussing at first on solar minimum conditions. This was extended to include both PAMELA and AMS-02 spectra for after solar minimum to full maximum conditions, including the reversal of the HMF polarity during solar maximum and afterwards up to the present solar minimum. In this manner the model, with its assumptions about the global heliosphere as well as all the modulation parameters, DCs and the drift coefficient were tested and vindicated. The set of proton modulation parameters was applied to other GCR nuclei, e.g., helium isotopes as well as anti-protons over time (solar activity) and the electron parameter set was applied to positron modulation. This assures that the modeling parameters are not changed in an ad hoc and inconsistent manner for the purpose of simply fitting observational data sets at the Earth, including those for GCR anti-matter.

The only differences in the modulation of protons and anti-protons are their global drift directions in the heliosphere, and of course, their VLISs; similarly the only differences in the modulation of electrons and positrons are their VLISs and their global drift directions. The latter is of course responsible for the charge-sign dependent modulation effect over a 22-year cycle related to the reversal of the HMF polarity as will be discussed further in what follows.

Reproducing the precise observational spectra on Earth, especially at higher R, is being studied to determine refinements to the rigidity dependence of the DCs over time. This approach allows that the mentioned three DCs and the drift coefficient can be quantified in terms of their rigidity so that the 'free parameter space' in these models has become significantly reduced. How the 11-year and specifically the 22-year modulation cycles come about are explained, including charge-sign effects, the physics of which is not described by the Force-Field modulation approach (e.g., [27,57–59]).

The 3D numerical model for this modulation study is for a steady-state, so that $\partial f/\partial t = 0$ in Equation (1), so that modulation effects shorter than 27-days and all transient events such as Forbush decreases or co-rotating interaction regions cannot be adequately described with this model. Determination of the averaged values for modulation parameters that vary with solar activity is an important part of the modeling approach, specifically the tilt angle α of the heliospheric current sheet (HCS) and the magnitude B of the HMF, both considered as good proxies for solar activity when it comes to charged particles; see also [60,61]. These variations in α and B are used in the numerical model to set up realistic modulation conditions for each epoch.

The top panel of Figure 1 shows α at Earth from 2005 to 2018 [62] together with the calculated 15-month smoothed moving averages (SMAs) as used in the model. The observed B values at the Earth are shown in the bottom panel for the same period [63] together with the corresponding SMA. Also shown in this figure is the period when no well-defined HMF polarity was reported, as indicated by the shaded band, and during which the HMF polarity reversal occurred [64]. There seems no consensus on the exact time span of this band as discussed in [65]. However, important is that before this period the polarity of the HMF was A < 0 and afterwards A > 0; see the detailed discussion in Section 2.5.1. The lowest values of both α and B occurred in 2009 as part of that prolonged solar minimum, of which the specific modulation effects were discussed in detail in [42]. It is expected that similar low values may occur during the present solar minimum period. The highest values for α occurred around the beginning of 2013, with extraordinary high but temporary values in 2011 and late in 2014, but for B the highest values occurred in late 2014, remaining relatively high until the end of 2015.

2.3. Rigidity Dependence of the Diffusion Coefficients

The DCs for GCR modulation in the heliosphere are related to the particles' mean free paths (MFPs), λ, through $K = (v/3)\lambda$ with $v = \beta c$ the particle's speed, c the speed of light so that β is the corresponding ratio. The relationship between, e.g., $\lambda_\parallel$ and $K_\parallel$ is given then by $K_\parallel = (v/3)\lambda_\parallel = (\beta c/3)\lambda_\parallel$, similarly for $K_{\perp r}, K_{\perp \theta}$ and K_D. The rigidity dependence of the three relevant MFPs $\lambda_\parallel, \lambda_{\perp r}, \lambda_{\perp \theta}$ and the drift scale λ_D are shown in Figure 2 for positrons in the left panel and for anti-protons in the right panel for yearly averages from 2006 to 2017 as established in [45,60,66,67] in order to reproduce the corresponding observed electron and positron spectra, and the proton and anti-proton spectra at Earth.

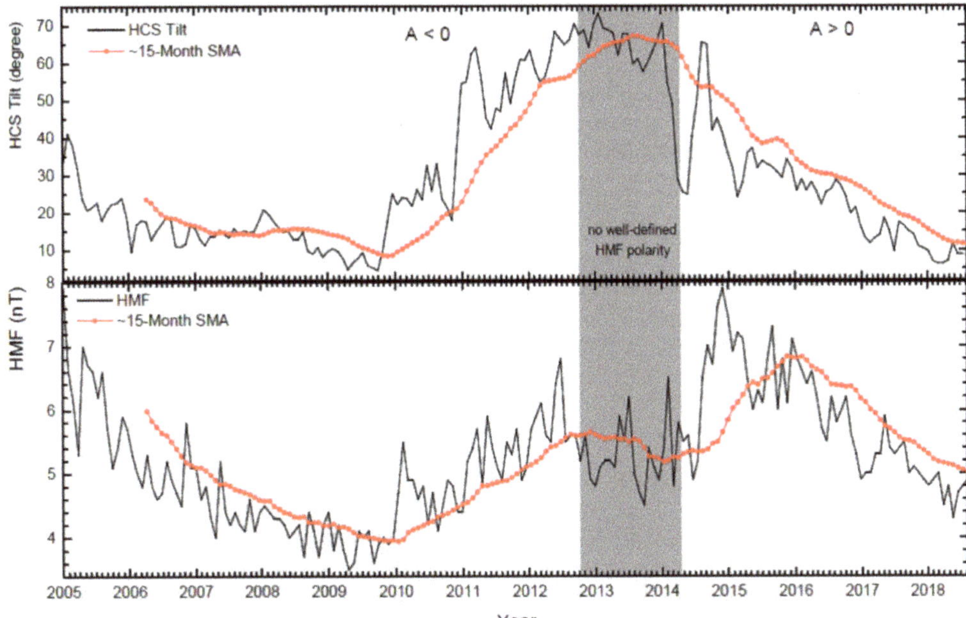

Figure 1. The 27-day averaged values for the heliospheric current sheet (HCS) tilt angle (**top** panel; black line) and magnitude of the heliospheric magnetic field (HMF) (**bottom** panel; black line) at the Earth from January 2005 to December 2018 along with the calculated 15-month smoothed moving averages (SMA; red dots) as applied in the model. The period with no well-defined HMF polarity is indicated by the shaded band during which the HMF polarity reversal took place changing from A < 0 to A > 0. These entities serve as good proxies for solar activity when it comes to charged cosmic particles.

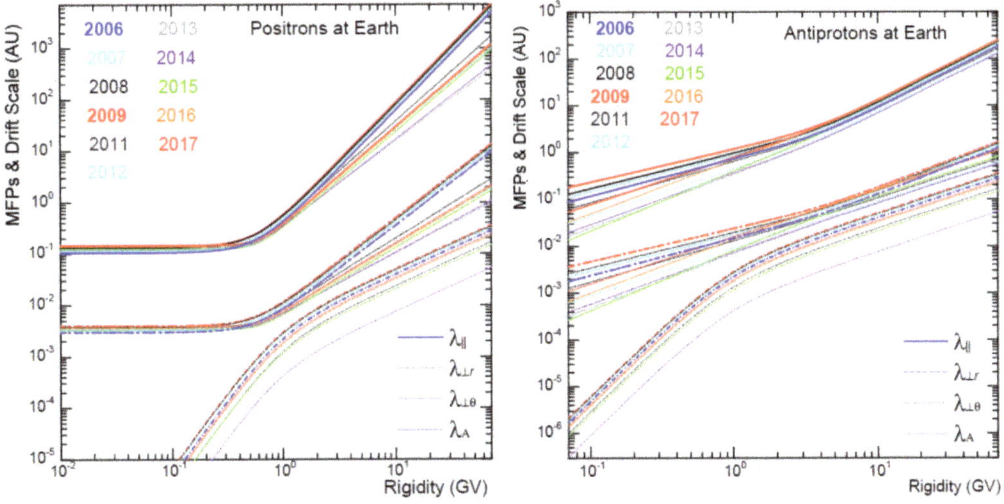

Figure 2. The particles' mean free paths (MFPs) $\lambda_\parallel$ $\lambda_{\perp r}$, $\lambda_{\perp \theta}$ and drift scale $\lambda_D (\equiv \lambda_A)$ as a function of rigidity at Earth, for positrons (**left** panel) and anti-protons (**right** panel), averaged yearly from 2006 to 2017; solid lines indicate the parallel MFPs, dashed-dotted lines indicate the perpendicular MFPs (middle set of curves) and the drift scale (lower set of curves), with different colours for each year as indicated; values of $\lambda_{\perp r}$ and $\lambda_{\perp \theta}$ are the same at Earth.

Figure 2 illustrates important aspects of how positrons and anti-protons (and electrons and protons) in terms of the R dependence of their MFPs are modulated, apart from the depicted time-dependence which is discussed later. Both these particle types have a different R dependence for the three MFPs at low compared to high R; for positrons this change in spectral slope occurs typically around ~0.4 GV and for anti-protons around ~4 GV, with steeper slopes consistently above these turn-point (tipping point) in R. The R-dependence of $\lambda_{\perp r}$ and $\lambda_{\perp \theta}$ are the same and modestly different from $\lambda_{\parallel}$ at high R but not at low R; overall $\lambda_{\perp r}$ and $\lambda_{\perp \theta}$ are only about 2% of the value of $\lambda_{\parallel}$ but their values differ away from the equatorial plane of the simulated heliosphere (see [45]). The MFPs for positrons (and electrons) have a characteristic slope, almost independent of R below ~0.4 GV, which means that the total value of the modulation of these particles between the HP and the Earth is the same below 0.4 GV because their modulation is diffusion-dominated in this rigidity range, not adiabatically-dominated as for protons and anti-protons. This aspect was elaborately illustrated with GCR spectra [45,60,66,67] and relates to what was discussed theoretically in [68–70] and from a turbulence theory point of view in [55,71] The drift scale also shows a difference at high R (where $\lambda_D \propto R$) compared to low R, with the change in spectral slope occurring typically between 1–2 GV for both positrons and anti-protons. This significantly steeper decrease in λ_D for $R < $ ~1 GV has been reported and studied comprehensively with numerical models (e.g., [72], and references there-in) and theoretically (e.g., [73,74], and references there-in).

In the modeling used here it is assumed that the DCs (and MFPs) in terms of their spatial dependence scale proportional to B_0/B with $B_0 = 1$ nT. This is the simplest and most pragmatic approximation to turbulence theory that can be assumed for DCs in numerical models and has turned out to be quite reasonable; for discussions and applications of more complicated assumptions regarding the relationship between these DCs and heliospheric turbulence; see [72,75–78]. Mathematical expressions for the R-dependence of the three MFPs, and the drift scale, as shown in Figure 2, are not repeated here (they are quite lengthy and would take up a substantial amount of space). They were given extensively in [48,60,72], and many were also utilized in [47,79–81].

2.4. Time Dependence of the Diffusion Coefficients

Another intriguing aspect of what is depicted in Figure 2 is the time variation found for the R dependence of the MFPs. For anti-protons their time variation is largest below 4 GV with a remarkable spread in values between 2006 and 2017, at the lowest R this spread is in the order of a factor of 10 but much less above 4 GV, about a factor of 3. This is a manifestation of the larger modulation that GCRs nuclei experience the lower the rigidity. For positrons (and electrons) this is different; the spread in time is significant for $R > 0.4$ GV, in the order of a factor of 10 but less significant below this value. The latter is based on comparing the model with published PAMELA observations for the solar minimum period before 2010; for electrons the observational cut off was established at ~80 MeV but originally higher for positrons at ~200 MeV because of statistics [2,82]; with AMS-02 observations from $R > 0.5$ GV but usually published for $R \geq 1$ GV ([5] and references there-in). This aspect, as well as the apparent steady behaviour in terms of rigidity, is therefore difficult to investigate more precisely, also complicated by Jovian electrons that dominate the spectrum at the Earth below ~50 MeV [83,84]. Although observations (detection) of electrons and positrons at lower R in geospace are intricate [85,86], such data will be quite useful for future analysis. For λ_D this time dependence is about the same at all rigidities, although somewhat less for $R < $ ~1 GV than above; see also [81].

Emphasizing the significance of the mentioned time dependence, Figure 3 shows $\lambda_\parallel$, $\lambda_{\perp r}$, $\lambda_{\perp\theta}$ and λ_D as determined for protons (assumed to be valid also for anti-protons) but now averaged over three distinct periods, for solar minimum conditions from 2006 to 2008, for increasing solar activity from 2011 to 2013 and then for maximum activity conditions from 2013 to 2015, in order to reproduce both PAMELA and AMS02 proton and antiproton spectra. Qualitatively, similar behaviour is shown as in Figure 2 but now it clearly illustrates how the time dependence differ below and above ~10 GV; below this value the MFPs decrease progressively and significantly from 2006–08 to 2013–15 but the opposite, and at a much smaller scale, occurs above this value. This seems counterintuitive, but similar behaviour (trends) using precise spectra from AMS-02, was reported also in [80,81,87] and implies that GCRs with $R > 10$ GV have somewhat smaller MFPs at solar minimum than at solar maximum; as such quite an interesting aspect of solar modulation at higher rigidity. In this context, it is well known that time trends in modulation is occurring differently at low and high rigidity during A > 0 and A < 0 cycles; see the discussion in [88,89], and references there-in, and recent modeling studies (e.g., [81]).

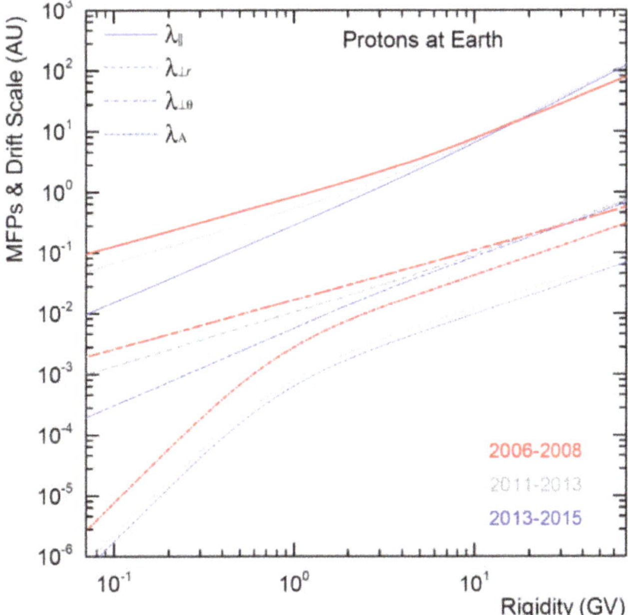

Figure 3. The MFPs, $\lambda_\parallel$, $\lambda_{\perp r}$, $\lambda_{\perp\theta}$, and the drift scale $\lambda_D (\equiv \lambda_A)$, from top to bottom, are shown for protons (assumed to be the same for anti-protons) as a function of rigidity but now averaged over three distinctive periods for solar minimum conditions from 2006–2008 (red lines), for increasing solar activity from 2011–2013 (grey lines) and for maximum activity conditions from 2013–2015 (blue lines); at Earth $\lambda_{\perp\theta} = \lambda_{\perp r}$.

An example of the time dependence of $\lambda_\parallel$ at the Earth is shown in Figure 4 for protons and anti-protons at 1 GV from July 2006 to May 2017. The period with no well-defined HMF polarity is indicated again by the shaded band, during which the HMF polarity changed from A < 0 to A > 0. During the A > 0 solar minimum $\lambda_\parallel$ had reached a maximum value in late 2009 and then decreased systematically to reach a minimum value in early 2015, apparently several months after the reversal of the HMF polarity was completed. The value in 2017 is already close to that in 2006 which was the beginning of the previous prolonged solar minimum. This investigation will be continued when new AMS-02 spectra get published for the present solar minimum (before and after 2020). A similar time-dependence was found for positrons and electrons in [62] with qualitative similar results.

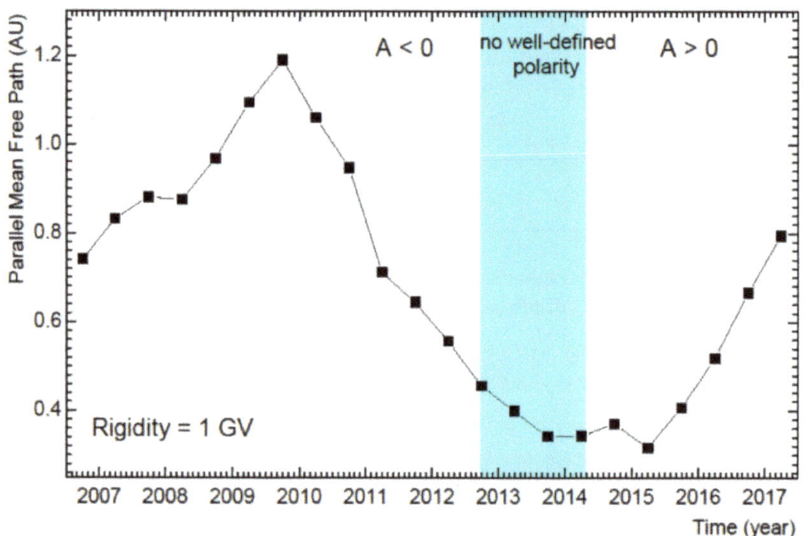

Figure 4. Parallel MFP $\lambda_{\parallel}$, at the Earth based on the simulations (this work) of 1 GV protons and anti-protons from July 2006 to May 2017. The period with no-well-defined HMF polarity is indicated by the shaded band during which the HMF polarity reversal occurred, changing from A < 0 to A > 0, the latter being the present modulation state.

2.5. Particle Drifts in the Heliosphere

As alluded to above, the solar modulation effects on GCRs of global gradient and curvature drifts, including the effect of the wavy HCS, are known and reported extensively since its introduction [90] and first modelled in 3D [91]. The main predictions are mentioned in the next section. Global particle drifts is one of the four main modulation processes as described in Equation (1). As numerical models have become more sophisticated, including the stochastic differential equation (SDE) approach to modulation modeling [9,31,92,93] and as new long-term and precise observations have become available from time to time, the insight on how large drift effects are and how this may change with rigidity, with space and with solar activity, as well as from one 11-year cycle to the next one, has significantly improved since the 1980's.

It has been realized that what is known theoretically as weak-scattering drift is an oversimplification. This is obtained by assuming that in the heliosphere $\omega\tau \gg 1$, where ω is the gyro-frequency of a charged particle in the HMF as a given location and with τ a time scale defined by its scattering so that the drift coefficient simplifies to:

$$K_D = \frac{\beta R}{3B}\frac{(\omega\tau)^2}{1+(\omega\tau)^2} = \frac{\beta R}{3B}. \tag{3}$$

This expression was used in earlier numerical models and obviously has an uncomplicated rigidity dependence, with its spatial dependence determined by what is assumed for B. During the Ulysses mission it became evident that this gives a far too large drift effect concerning the observed GCR latitudinal intensity gradients [17,33,39,54] at low rigidity, typically at $R < 1$ GV. This mission also indicated that Parker's descriptions of the global HMF structure is an oversimplification [94]. Nowadays, it is common practice in numerical modeling to modify K_D at lower rigidities as shown in Figures 2 and 3, and to modify the global geometry of the HMF to some degree; see [35,48,72] for the appropriate expressions and [31,94] for a study on HMF and HCS modifications. An extensive theoretical study on the reduction of particle drift, using various assumptions for heliospheric turbulence, was described in [74] (and references there-in) with previous related studies in, e.g., [73].

Apart from these explicit reductions of particle drift, the more subtle reduction of drift related modulation effects is done through changing diffusion which follows from the inspection of Equation (1) according to which the drift velocity is multiplied by ∇f (in effect the spatial gradients of the GCR intensity in the heliosphere at a given position and KE). This means that when these gradients are changed through changing any of the three DCs, drift effects consequently change. This is considered an implicit reduction of drift effects, not directly affecting the drift coefficient In this context, what is assumed, for example, for $K_{\perp\theta}$ is quite important because it affects directly the intensity gradients in the polar regions of the heliosphere and therefore the corresponding drift related modulation effects see also [72,84].

Another aspect of interest is how drift effects change with solar activity. At first, these effects were changed in simulation studies only through changing the waviness (tilt angles) of the HCS and later followed by changing B as it changes with solar activity. These aspects were simulated comprehensively in [95–97] who also introduced the concept of global merged interaction regions (GMIRs) into drift modeling; for more recent such simulations, see also, e.g., [98,99]. Simulations of these GMIRs showed that they could easily dominate the level of GCR modulation depending on how extensive in space they are and how many of them form beyond 5–10 AU during phases of higher solar activity. Particle drifts appeared to be dominant during solar minimum epochs (see review [100]) but how much particle drifts remains during periods of high solar activity could not be answered conclusively. However, afterwards it became clear that K_D (or λ_D) is required to be scaled with time in order to reproduce GCR observations over extended periods of time, specifically from the Ulysses mission; see reviews [17,18,33]. Examples of numerical simulation utilizing such an approach were given in [34,101–103] who also studied other modulation effects such as the role of the TS and inner heliosheath. Later, [104,105] applied a time-dependent scaling factor in their numerical model to reproduce GCRs at the Earth and especially along the trajectories of the two Voyager spacecraft. The first numerical calculations on specifically how K_D scales (effectively being reduced) over periods of increased solar activity up to solar maximum conditions was given in [106] when comparing their simulations to Ulysses observations. They found that relatively little particle drift (<10%) remained during the solar maximum periods of 1990–91 and 2000–02. Recently, [45,62] determined how much particle drift is needed at 1 GV to explain the time dependence from 2006 to 2015 related to the observed precise electron and positron spectra from PAMELA and AMS-02 during each solar activity phase, especially during the polarity reversal phase when no well-defined HMF polarity was present. The equivalent result based on our simulations of proton and anti-proton modulation is shown in Figure 5. Evidently, very little drift is found and required for the period of maximum solar activity.

2.5.1. Drift Related Modulation Effects

A main feature of drift related modulation is that the particle drift velocity field repeats its configuration every 22 years because it changes direction when the solar magnetic field flips over during every period of maximum solar activity. The northern and southern polar field then reverses sign so that if the direction in the northern hemisphere had been outwards it became inwards after the reversal. This process is known as the 'polarity' change of the solar magnetic field. The time it takes for the reversal to be completed differs from one solar maximum to the next one and how the process happens in time is always different for the northern polar field than for the southern field; see, e.g., [64]. From the view point of charged-particle drifts and solar modulation, the important aspect is that this period is characterized as a time with no well-defined HMF polarity and that it usually lasts more than a year. To briefly recap: The drift cycles are called A < 0 (e.g., 2001–2012) when positively charged GCRs (protons, positrons, GCR nuclei.) drift inward to the inner heliosphere mainly through the equatorial regions of the heliosphere, and in the process encounter the wavy HCS, and outward via the polar regions of the heliosphere. Negatively charged particles (electrons, anti-protons, anti-matter nuclei) then drift inwards and downwards to the Earth mostly from the polar regions and then outward mainly through the equatorial regions. During an A < 0 phase, it is expected that the changing wavy HCS plays an important even dominant role in the modulation of positively charged particles, whereas for an A > 0 polarity phase the drift directions become reversed so that negatively charged particles encounter the HCS during their entry.

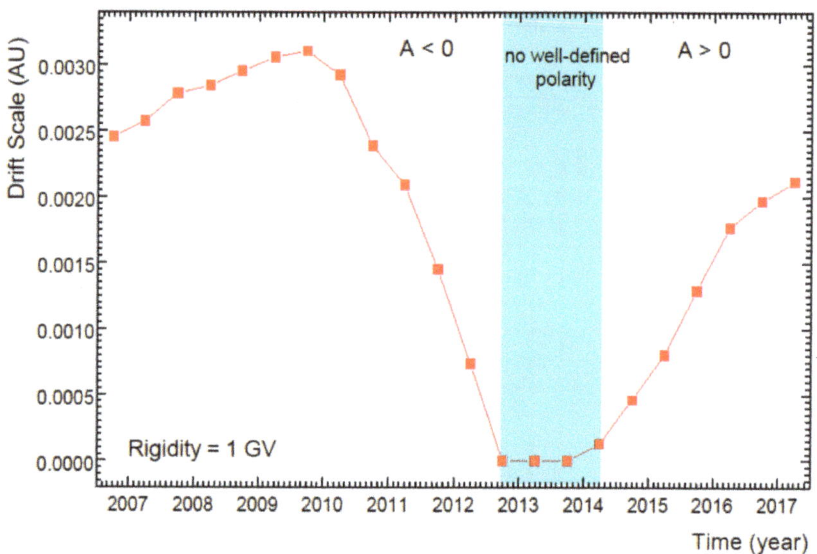

Figure 5. The drift scale, λ_D, at the Earth as a function of solar activity from July 2006 to May 2017, based on simulations of 1 GV proton and antiproton modulation; see also [60,66,67]; shaded band as in Figure 4.

The consequence of what is described above is that it produces a 22-year modulation cycle, not just in the GCR intensity-time profiles but also spatially in terms of the distribution of the radial and latitudinal intensity profiles in the heliosphere (and consequently the spatial gradients), and with the charge-sign dependent effect probably the most spectacular. Illustrative examples of the observation of this phenomenon with the Ulysses mission [15,107] and subsequent modeling was given in [108]. A full list of these drift related phenomena were given in the review [59].

A very subtle drift effect, not always mentioned, is the prediction that proton spectra for A < 0 solar minimum cycles compared to A > 0 minimum cycles should exhibit a spectral cross-over with means that below this cross-over rigidity the differential intensity is higher in the A > 0 than in the A < 0 cycles but not above this rigidity [109–111], as long as solar activity is at the same level. This phenomenon is discussed in detail in [88,89], especially how it is predicted with precise numerical modeling and how difficult it is to find and confirm observationally but surely possible now that precise long-term spectral observations have become available.

During the prolonged solar minimum before 2010, PAMELA had observed and confirmed the charge-sign dependent effect between electrons and protons conclusively [112] and later reported how this effect for electrons and positrons had changed with the reversal of the HMF polarity [2,113], although smaller than predicted by the leading drift models of that time [59]. Such an effect was also found for the very first time in the recover times of observed proton and electron Forbush decreases [114] and comprehensively modelled by [115]. The charge-sign dependent effect during the recent HMF polarity reversal was also observed and reported by AMS-02 [5], confirming the PAMELA observations, and is discussed further in Sections 4 and 5 comparison with our drift modeling.

3. The Question of VLISs at the Heliopause

The VLISs for GCRs are required as input spectra for the numerical modelling of their total and global modulation in the heliosphere [46,116–121]. For each type of charged cosmic particles such a specific input spectrum is considered to be unmodulated by solar activity beyond the HP. Global modulation modelling relates these VLISs to the corresponding observations at the Earth, through the physics contained in the assumed transport model applicable to the heliosphere. In order to reproduce (simulate) observed spectra over a wide rigidity range at the Earth for any type of GCRs, their associated VLISs must first be determined and evaluated against galactic propagation models such as GALPROP (e.g., [121–123]) and then tested and vindicated with comprehensive 3D modulation models as described here.

After Voyager 1 had observed GCRs at and beyond the HP, the question came up whether these spectra at ~122 AU, technically just heliopause spectra, could be used as VLISs and if the latter are truly the same as pristine LIS (say, 1000 AU away from the Sun) and if they were on their part truly the same as the averaged Galactic spectra (parsecs away). These questions have led to predictions done with SDE type modulation models that GCRs may be modulated even beyond the HP implying that the heliopause spectra were not the same as LISs and that the HP should not be considered the true modulation boundary. The original prediction [124] had come to such a conclusion but refined models later on predicted far less modulation (e.g., [125]) with [9] eventually illustrating with a comprehensive SDE study under what conditions at and just beyond the HP such modulation may happen or not, and why most recent Voyager 1 observations are consistent with no intensity gradient as mentioned above. These studies and also modeling of the TeV cosmic ray anisotropy [14] have been able to supply interstellar propagation parameter values. It is worth mentioning that disturbances related to solar activity have indeed been observed beyond the HP implying that the outer heliosheath is somewhat influenced by solar activity but apparently not to the extent that GCRs are impacted to show a subsequent modulation type response; see, e.g., the discussion in [10].

At first, establishing VLISs explicitly for solar modulation studies had been done by simply finding a statistical fit to the mentioned observations at low and at very high KE, but with values in between nothing more than estimates of the value and shape of the respective VLISs. This approach was followed for electrons [43,126] and for protons [42,44,127] and used in a comprehensive 3D modulation model; see also [27,116]. Similar approaches were followed in [47,119] with a different approach in [128]. As expected these attempts (statistically fitting observations) produce VLISs within about 10% of each other.

Precisely measured spectra at the Earth up to kinetic energy (KE) well above where modulation is assumed to become negligible [22] contribute significantly to the improved understanding of VLIS, providing meaningful constraints to galactic propagation models, and for the study of astrophysical concepts and explanations. However, the spectral value and shape of the VLIS above ~500 MeV up to whatever KE where solar modulation is considered negligible have remained uncertain to a large degree (see also the complementary discussion in [57]). Addressing this shortcoming, galactic propagation models are the main option, although there surely are other ways (e.g., [129,130]). Several such galactic propagation models of varying complexity have been developed over time; see, e.g., [123,131,132], including the SDE approach of [133]. However, the GALPROP code, being made available on-line, is an obvious choice to use and have been applied extensively to reproduce Voyager, PAMELA and AMS-02 observations also being used as constraints to these models. For typical examples of this approach where modulation models are used in conjunction with Galactic propagation models, see [19,46,117–119] also being discussed further below. Less sophisticated galactic propagation models, such as the so-called Leaky Box Model, had also been used; see, e.g., [21,134] where this model is applied extensively, and which nevertheless have provided useful insight into what VLISs may be. Of course, computing LISs with GALPROP for secondary GCRs are intrinsically more complicated than for GCR protons and nuclei. It is known from applying GALPROP that the spectral shape of the positron and anti-proton LISs are very different so it makes for quite interesting modulation differences between these anti-matter GCRs as shown below as the main focus of the present study.

This self-consistent approach has been followed in order to refine the computed LISs from GALPROP to transform them into VLISs. This means that the physics in GALPROP is attuned to comply with observational constraints from Voyager 1 and 2, PAMELA and AMS-02, and importantly, also providing in the process the shape of the VLISs in the middle range of KE, typically between 200 MeV/n to 30 GeV/n. We consider this approach as an improvement to using statistical fits to the mentioned observations. The VLISs obtained in this manner are then used as input (unmodulated) spectra in our 3D modulation model described above.

The next level of observational constraints comes forward when applying the 3D modulation model. Apart from having to reproduce spectra at the Earth over a wide range of KE, the model also has to reproduce spatial intensity gradients, both latitudinal as observed by Ulysses in the inner heliosphere, and radial as observed by the Voyagers from the Earth to the TS, and then from the TS to the HP, which is totally different from modulation up to the TS. The inner heliosheath causes what may be considered an additional modulation 'barrier' which has a completely different level of effects on GCRs of low KE than high KE. An example of this comprehensive approach to proton modulation is given in [39] and lately in [120]. The same model is then applied to GCR electrons as given in [111] and also positrons [135]. Once the model has reached this level of sophistication, it can be applied with confidence to predict the VLIS at the HP for positrons and anti-protons, in fact for all GCRs for which no observations exist beyond the HP as done in [121] with their HelMod approach. Utilizing the precise observations at the Earth, the VLISs for all GCRs may be fine-tuned based on the output of the modulation model at high rigidity. This procedure has been followed for protons [48], for total helium and separately for the two helium isotopes [48–51], as well as for electrons and positrons [45,60,67] and for protons and anti-protons [66].

In Figure 6 three examples of the VLISs obtained in this manner are shown, here depicted for protons, electrons and total helium (He-4 plus He-3) at the HP (122 AU) as specified in our modulation model. At high enough KE where solar modulation is assumed insignificant, these VLISs are required to match the observations at Earth. Evidently, as the KE decreases, the deviation of these observations at Earth from the corresponding VLIS increases as solar modulation becomes progressively larger.

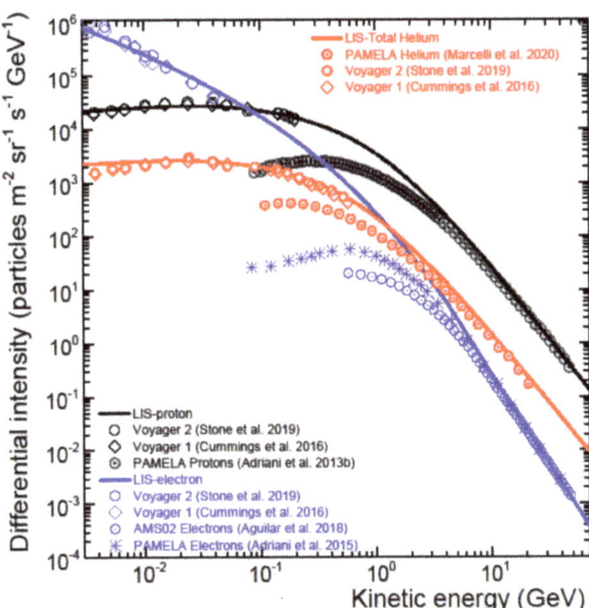

Figure 6. The very local interstellar spectra (VLISs) at the heliopause (HP) (122 AU) for galactic protons (black line), electrons (blue line) and total helium (red lines) with observations at the HP from Voyager 1 [19] and Voyager 2 [7], and corresponding spectra at the Earth from AMS-02 [5] and PAMELA [80,136,137].

4. Positron Modulation

Before the precise, simultaneous and long-term measurements of anti-matter at Earth, charge-sign dependent solar modulation was mostly studied using electrons (negatrons as the sum of electrons and positrons), protons and helium of the same rigidity (e.g., [15]). The first robust observational evidence of charge-sign dependent solar modulation was reported in [138] and modelled in [139] using a drift model which, today, is known as a first generation drift model. These type of drift models predicted that during A > 0 polarity cycles the intensity ratio of negative to positively charged particles as a function of time should exhibit a ∧ shape profile while during A < 0 cycles it should exhibit a ∨ shape around minimum modulation periods, but when only the waviness of the HCS was changed with time [140]; for updates and new insights, see Figures 2 and 3 in [141]. It was found that when other modulation parameters were also changed with time in later generation drift models these distinctive shapes fade and are been smoothed out (e.g., [106]).

Comprehensive modeling was done for electron and positron modulation in the global heliosphere [101,102] relying on the computations of galactic spectra [142–144] as alternative to the GALPROP calculations [121]. These modulation models were more advanced and also included the effects of the solar wind TS and the role of the inner heliosheath which was considered state-of-the-art at that time but not much attention was given to comparisons between these modulation predictions and observations at the Earth. This has changed since the first positron spectra from PAMELA [2,113,145] became available. Follow-up simulations done for positrons in [135] were based on new VLISs with the focussed on detailed aspects of the solar modulation of positrons during the extraordinary quiet solar modulation period from 2006–2009. For the first time, a meaningful modulation factor in the heliosphere was computed for positrons, from 50 GeV down to 1 MeV, as well as the electron to positron ratios as a function of time and rigidity for the mentioned period. In the next section, the focus is on contemporary modeling of these GCR positrons.

4.1. The VLIS and Modulated Spectra

A comprehensive numerical study of the modulation of galactic electrons and positrons was reported in [60]. First, they established the VLISs for electrons using the process described above [46] by adjusting the appropriate physics in the web version of the GALPROP code [146] (and references there-in) to reproduce both the observed Voyager 1 and 2 electron intensity levels from beyond the HP and the high KE observations from PAMELA where solar modulation is considered negligible. Since positron data is not available beyond the HP, the spectral shape of the positron VLIS at low KE is based on what GALPROP produced but empirically modified as explained in [45,67] to improve the reproduction of observed spectra at Earth. As mentioned, the only known differences in the heliospheric modulation of electrons and positrons are the particle drifts that they experience and their respective VLISs.

The computed VLIS for electrons and positrons are shown in Figure 7 together with observations for galactic electrons from PAMELA from January to December 2009 [82], from AMS-02 from May 2011 to May 2017 [5] and from Voyager 1 [6,19] and Voyager 2 [7], with positron observations from PAMELA [113] and AMS-02 [5]. As for Figure 6, solar modulation is assumed insignificant at high enough KE where these VLISs are required to match the Earth observations. Again, the deviation of these observations at the Earth from the corresponding VLISs increases as solar modulation becomes progressively larger with decreasing KE.

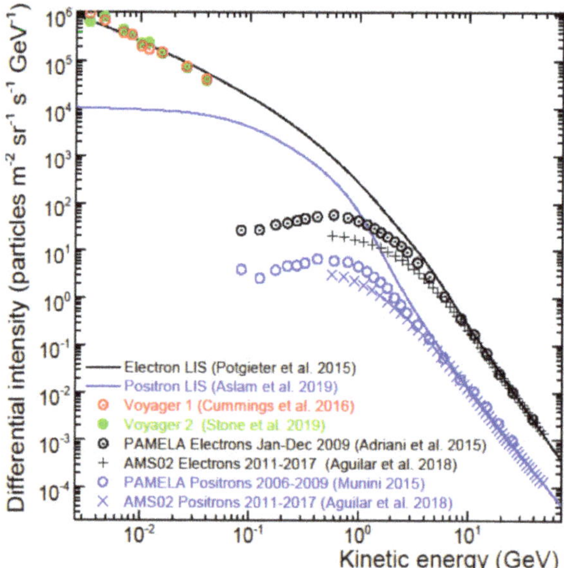

Figure 7. The computed VLIS for electrons (black line) from [43] and positrons (blue line) from [45] in comparison with observations from Voyager 1 [6,19] and Voyager 2 [7] at the HP, and observational averages from PAMELA [113] and AMS-02 [5] at the Earth as indicted in different colours. Figure is amended from [60].

In Figure 8, the modulation of positrons during extraordinary quite modulation conditions is shown as computed spectra at Earth compared to PAMELA data [113] for the period July to December 2009 (A > 0 cycle), and corresponding computed spectra throughout the heliosphere at 10 AU, 50 AU and 100 AU in the equatorial plane with respect to the VLIS at 122 AU. Examples of how large this modulation may be for an A > 0 cycle were given in [67], which can be tested once AMS-02 positron spectra are published for the present solar minimum period. Neutron monitors at ground level have already reached maximum counts in mid-2020 (see, e.g., [147]).

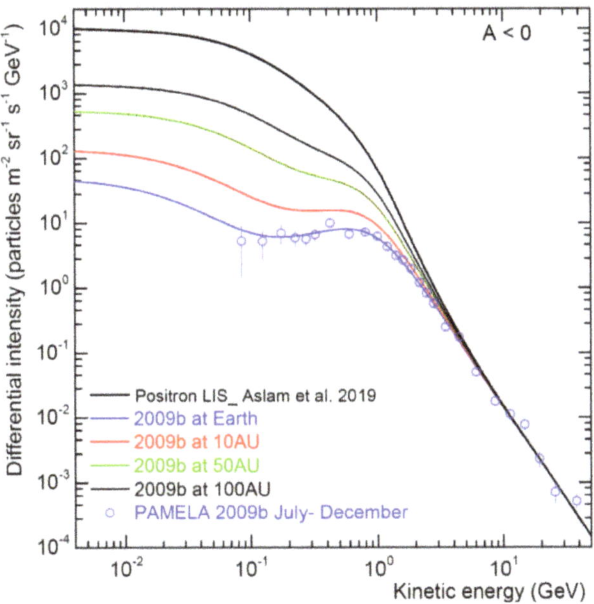

Figure 8. Computed positron spectra at Earth (blue line) compared to PAMELA data for the period July to December 2009 (A < 0 cycle), and corresponding spectra predicted for 10 AU, 50 AU and 100 AU with respect to the VLIS at 122 AU (upper black line) from [45,67].

4.2. Drift Effects

Figure 9 illustrates the predicted drift effects relevant to positrons displayed as computed spectra at Earth for modulation conditions applicable to the second halves of 2006 and 2009 (indicated as 2006b and 2009b) for the A < 0 cycle compared to the A > 0 cycle assuming the same modulation conditions. These spectra are shown with respect to the assumed VLIS at 122 AU. Evidently, focussing on the qualitative features, drift effects gradually subside with decreasing KE to vanish completely at low enough KE. For increasing and higher KE these effects gradually subside to become relatively small above a few GeV. This happens differently at higher KE than at the lower KE. Important is that the intensity for A > 0 cycles is predicted consistently higher than for A < 0 cycles, under the same modulation conditions, but not at all rigidities; note that the subtle phenomenon of the cross-over of these A > 0 and A < 0 spectra between 1–2 GeV is also present for these GCRs so that it is not obvious where these drift effects completely disappear with increasing KE. Modelling details related to this figure were reported in [45].

The quantitative features are displayed in Figure 10 by plotting the ratio of intensities for A < 0 to A < 0, first for the periods 2006b and 2009b for both electrons and positrons. In the left panel the ratios are shown at Earth related to Figure 9. The largest drift effects are occurring between 200 MV and 600 MV, progressively phasing out with decreasing rigidity to fade away below ~100 MV. Drift effects also subside with increasing rigidity (KE), but more strongly than at low rigidity, producing in the process the spectral cross-over phenomenon. As mentioned above, the inspection of the difference between A > 0 and A < 0 spectra above 1–2 GV in Figure 9 shows this cross-over, clearly evident in the ratios, so that A < 0 becomes higher than A > 0, in contrast to what happens at lower rigidity (KE). It appears that drift effects dissipate above ~10 GV but how small it gets before completely subsiding still needs to be investigated thoroughly (related to which rigidity solar modulation really commences at). In the right panel of Figure 10 the predicted ratios are shown at 10 AU, 50 AU and 100 AU in the equatorial plane, in addition to those at 1 AU. Evidently, drift effects decrease with increasing radial distance, from a maximum ratio of ~2.5 at 1 AU to ~1.2 at 100 AU; see also [35,148] who simulated drift effects for GCRs throughout the heliosphere. These results are consistent to what [84] reported for drift effects on electrons; see also predictions made for the two drift cycles for electrons and protons in [111].

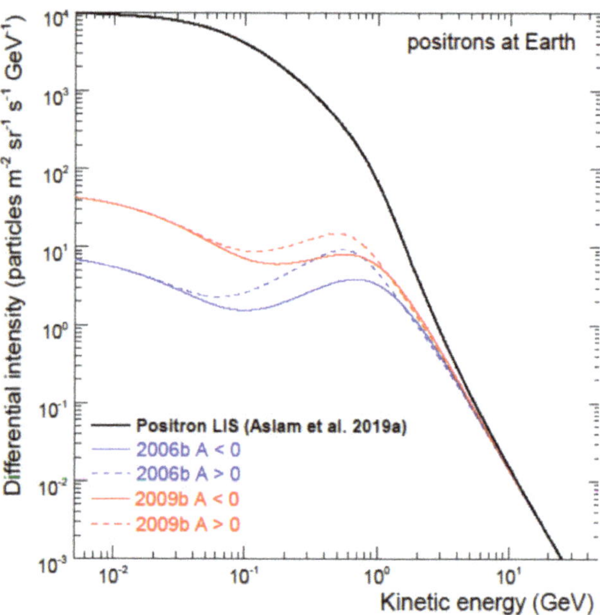

Figure 9. Predicted drift effects for positrons are displayed based on the differences in computed spectra between A > 0 (dashed lines) and A < 0 drift cycles (coloured solid lines). Modulation conditions applicable to the second halves of 2006 and 2009, indicated at 2006b and 2009b, are simulated for both cycles. The predicted spectra are shown at Earth with respect to the assumed VLIS] at 122 AU (black solid line). Figure is adapted from [45].

The charge-sign dependence displayed in Figure 10 is further highlighted in the next two figures by plotting the ratio of electron to positron intensities (e^-/e^+) as a function of KE, as well as for (e^+/e^-) in the bottom parts for illustrative purposes because it has been reported as such many times. These computed ratios are shown in Figure 11 at Earth with respect to the ratio of the two respective VLIS. The simulations are done for 2006b and 2009b modulation conditions so that a comparison can be made to the PAMELA observations [2,113] of electrons and positrons for these well-defined solar minimum periods. In Figure 12, these ratios are repeated for the A < 0 cycle as shown in Figure 11 but now also including the predicted ratios for an A > 0 drift cycle (the present solar epoch) assuming modulation conditions to be the same as in 2006b and 2009b. This serves to highlight the difference in these ratios between the two drift cycles.

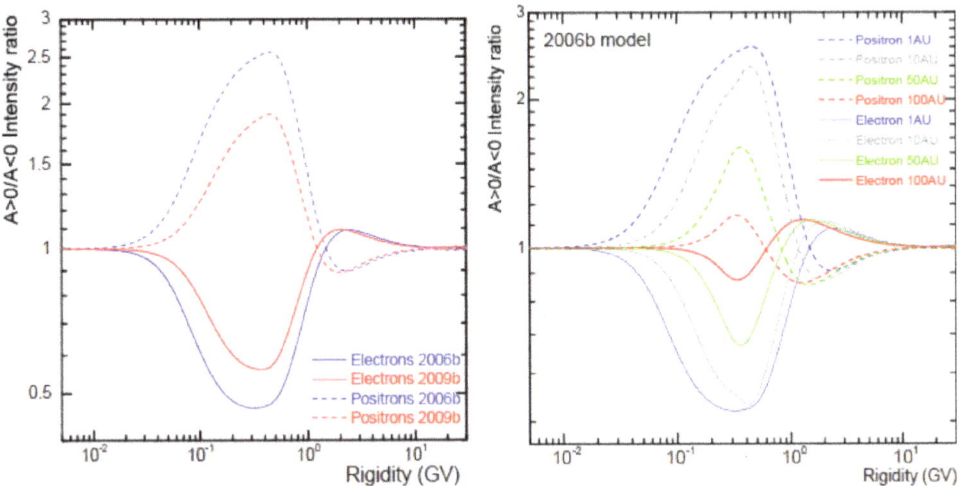

Figure 10. (**Left** panel) Predicted drift effects at Earth for positrons and electrons in terms of the intensity ratios between $A > 0$ and $A < 0$ drift cycles. This relates to what is shown in Figure 9 for positrons, again displayed for modulation conditions applicable to 2006b and 2009b. (**Right** panel) Drift effects for positrons and electrons at 1 AU, and predicted for 10 AU, 50 AU and 100 AU in the heliospheric equatorial plane. In this latter case modulation conditions for 2006b only are simulated.

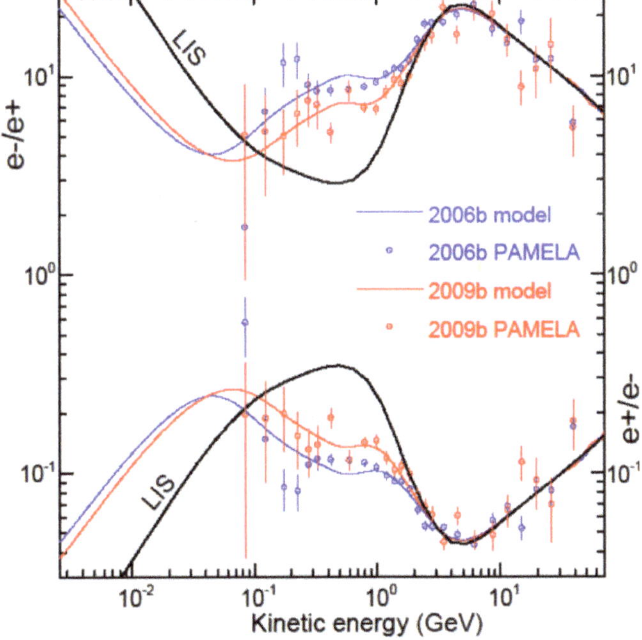

Figure 11. Ratio of electron to positron intensities (e^-/e^+) is shown as a function of kinetic energy (KE at Earth with respect to the ratio of the two respective VLIS (black lines) at 122 AU. In the lower part of the panel, the equivalent (e^+/e^-) is shown for illustrative purposes. Simulations are done for modulation conditions as in 2006b (blue lines) and 2009b (red lines) in comparison with PAMELA observations [2,82,113] during this solar minimum $A < 0$ drift cycle.

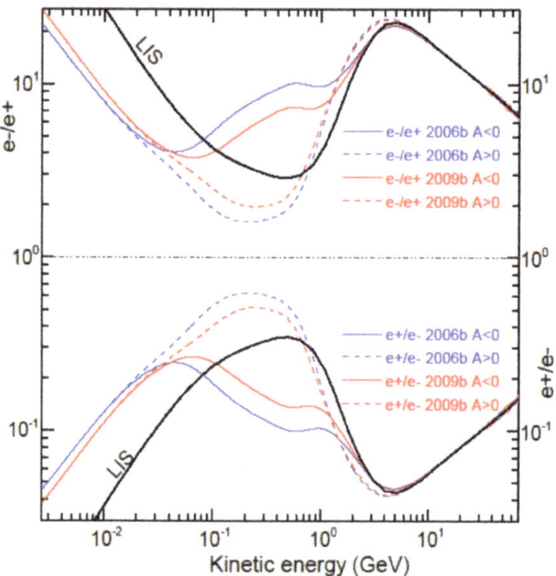

Figure 12. Same as Figure 11 (solid lines), but showing additionally ratios predicted for an A > 0 drift cycle, the present minimum epoch (dashed lines), assuming identical modulation conditions as in 2006b and 009b, respectively; following the colour coding as indicated in the legends.

An illustration of how charge-sign dependent modulation changes with time (solar activity), the observed and simulated ratios (e^+/e^-) are shown in Figure 13 for 1.01–1.46 GeV and for Bartels rotation numbers 2426 to 2506 (mid-2011 to mid-2017). The simulated ratios are based on reproducing the AMS-02 electron and positron spectra for these periods. Evidently, the simulations agree well with the observations displaying how the ratio changes from a low value before the reversal period to a significantly higher value afterwards. Figure 14 shows simulated ratios based on reproducing the observed PAMELA ratio [2] for 1.01–2.5 GeV over a longer period from 2006 to 2016. Modelling details related to this figure were reported in [60].

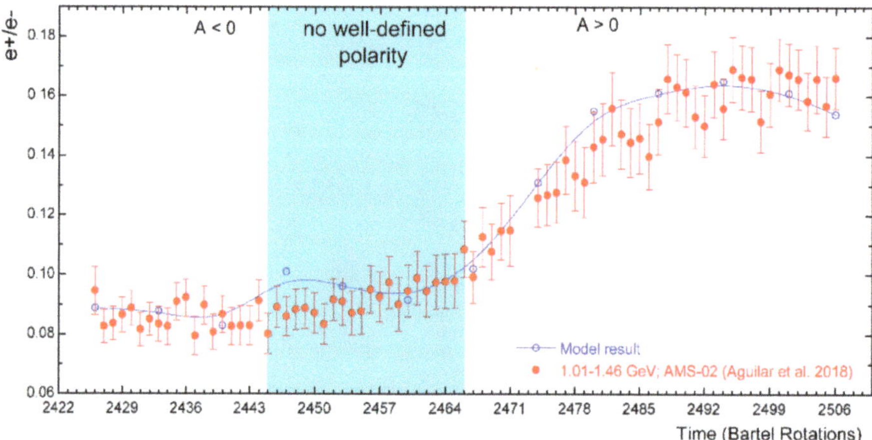

Figure 13. The simulated (e^+/e^-) ratio (blue circled point and smoothed line) as a function of time (solar activity) based on reproducing the AMS-02 electron and positron spectra [5] for this period from mid-2011 to mid-2017 shown here for 1.01–1.46 GeV in comparison with the observed ratio. Shaded band again indicates the period of no well-defined HMF polarity, with the A < 0 cycle before and the A > 0 cycle after this reversal period.

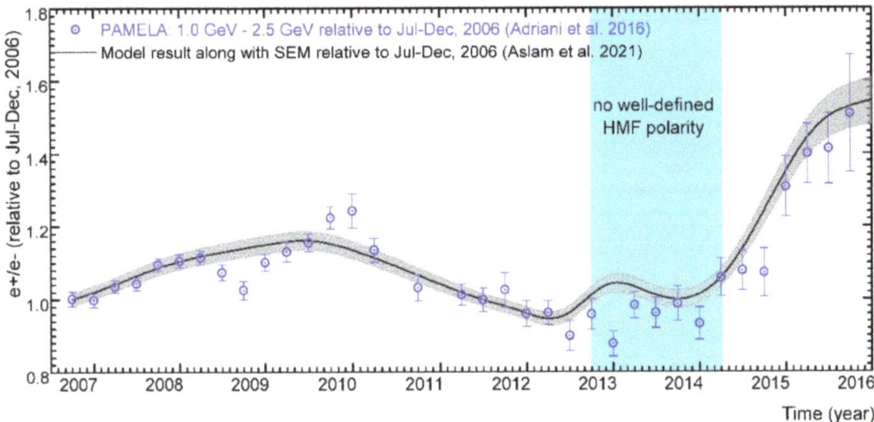

Figure 14. The simulated (e^+/e^-) ratio (black line with grey band) as a function of time (solar activity) based on reproducing the PAMELA observed ratio [2] here shown at 1.01–2.5 GeV (circled blue points with error bars) for this period. Vertical shaded band is as before. Modeling results are shown along with the standard error of mean (SEM; deviation/ $\sqrt{n}$). Figure amended from [60].

5. Anti-Proton Modulation

The first prediction of drift effects pertinent to anti-proton modulation was made in [149]. This early, first generation drift model, with the HP assumed at only 50 AU, predicted a change of a factor of ~8 in the anti-proton to proton ratio at 300 MeV between an A > 0 minimum and an A < 0 solar minima but about half of that for periods of increased solar activity. The LIS for protons and anti-protons were typical of what had been used then. The only time-dependent change in this modulation model was the tilt angle so what was predicted for solar maximum conditions was significantly over estimated. Comprehensive modeling was later done for anti-proton modulation in [101,102,150] relying on computations of galactic spectra with the GALPROP code [122]. Apart from drifts, their model also included the effects of the TS and the role of the outer heliosheath. These models predicted lesser drift effects than first generation models but they also consistently applied an overall reduction factor to the drift coefficient. The modulated anti-proton spectra at Earth reported in [102] differed by a factor of 1.5 below 100 MeV between the two drift cycles and also illustrated how the anti-proton to proton ratio changed from solar minimum to maximum.

Later, a 3D modulation model was used in [93] based on solving SDEs to study the modulation of protons and anti-protons and found a maximum drift effect of a factor of ~10 below ~100 MeV which gradually subsides with increasing KE to become negligible above 10 GeV. These predicted effects are large so that no attention was then given to smaller modulation effects above a few GeV as is done nowadays with the availability of precise GCR spectra. Recently, aspects of the modeling of anti-protons were reported in [28,119,151]; the latter raising additional questions about what the VLIS for anti-protons may be. In what follows the modelling study given here is described based on the approach outlined in this paper on how to obtain what can be considered a realistic anti-proton VLIS for solar modulation studies.

To recap concisely, the approach mentioned above is focus on gauging the numerical model and validating the modulation parameters on reproducing the observed protons and helium spectra at the HP at low KE and at Earth at much higher KE; see also [48–51]. The model is then applied to anti-protons with the assumption that there is no fundamental reason for anti-protons to be differently modulated than protons apart from the drift field directions that changes every 11 years. It is thus implied that anti-protons have the exact same MFPs as well as the drift scale than protons but oppositely directed drift velocity fields.

5.1. The Anti-Proton VLIS and Modulated Spectra

The numerical modeling described here is focussed on the PAMELA anti-proton spectra averaged from 2006 to 2009 to obtain improved statistics [145,152,153] which as such has less uncertainty than observations reported before this mission. The availability of very precise proton spectra for the same period and from the same detector makes the gauging process of the modeling much easier. The physics in the GALPROP code is adjusted to determine two possible anti-proton LISs which is then used in the modulation code at 122 AU to compute the corresponding modulated spectra at Earth based on the equivalent study of proton modulation. The procedure, followed here, including the adjustment of relevant physics in GALPROP, is described in [46,57]. It is worth mentioning that doing this for positrons and anti-protons is more complicated because they are secondary GCRs thus influenced by how they are created in galactic space through nuclear cross-sections. The two LISs computed with GALPROP are the plain diffusion (PD) and the reacceleration (REAC) approach, the latter being our preferred model for computing the LIS for protons. These two LISs are shown in Figure 15, the PD-model considered as the upper limit (black dotted line) and the REAC-model as the lower limit (black dashed line). The corresponding modulated spectra at Earth are computed based on the modulation parameters that worked optimally for protons for the solar minimum period 2009. These modulated spectra are compared to the mentioned PAMELA data. Inspection of these results shows that the REAC-model produces a LIS, specified in our modulation model as the HP (122 AU), that goes through the observational data at the Earth and thus fails to be considered as a realistic VLIS in our opinion. The PD-model gives a modulated spectrum significantly above the observations, which we anticipated because the same result was found for protons. This dilemma causes us to calculate an averaged VLIS (black solid line), using the two mentioned GALPROP spectra, and of which the corresponding modulated spectrum (red solid line) reproduces the observations reasonably well. We also repeated the modulation part using modulation conditions as found for protons for 2006 but found less agreement with these anti-proton observations; this modulated spectrum is found a factor 0.8 lower at its peak value. We therefore decided to settle on this averaged LIS (called LIS 3, Adjust, in this figure) as the optimal VLIS for further anti-proton modulation studies below 1 GeV. Adding earlier observations for anti-protons from balloon flights to this picture did not provide additional modulation insight and are not shown; see reviews of these earlier observations in, e.g., [145,154,155].

The modulation of anti-protons between the HP and the Earth, as illustrated in Figure 15, requires some additional remarks because it is rather peculiar in the sense that:

(1) The shape of the VLIS is quite different from GCR nuclei, having between 500 MeV and 1 GeV a spectral slope close to the slope evident in the modulated spectra below 500 MeV at Earth. This slope at Earth is caused by adiabatic energy losses inside the heliosphere and is a characteristic of modulated spectra for protons, anti-protons and all GCR nuclei (see also [48]). The shape of the VLIS of these particles below 200 MeV is therefore not reflected at Earth because of these energy losses inside the heliosphere; for illustrations of this modulation effect, see, e.g., [25,156].

(2) The total amount of modulation for anti-protons is far less than for protons of the same rigidity because of the very different shape of their VLISs, which for interest sake is compared to the VLIS for protons in Figure 16. As alluded to above, note where the peak in the anti-proton VLIS occurs compared to that for protons and how completely different the spectral slopes are at both low and high rigidities. The VLIS for protons is computed with GALPROP and normalized to PAMELA data at 100 GeV but the anti-proton spectrum is adjusted as described above.

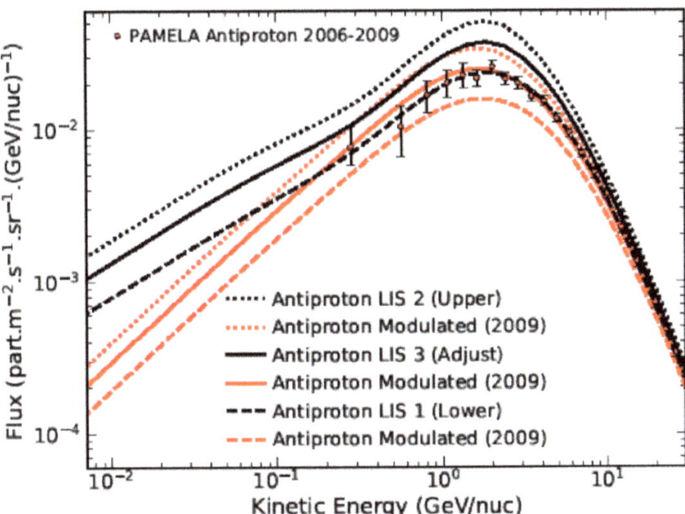

Figure 15. Computed LISs for anti-protons as a function of kinetic energy using two GALPROP models as described in the text, here indicated at 122 AU as LIS 1 Lower (dashed line) and LIS 2 Upper (dotted line). Corresponding modulated spectra at the Earth (red lines), based on modulation conditions for 2009, are found either too high or too low so that an averaged LIS is calculated as LIS 3 Adjust (black solid line) with the corresponding modulated spectrum (solid red line) reproducing optimally PAMELA observations averaged for 2006 to 2009 [152] (filled circles with error bars).

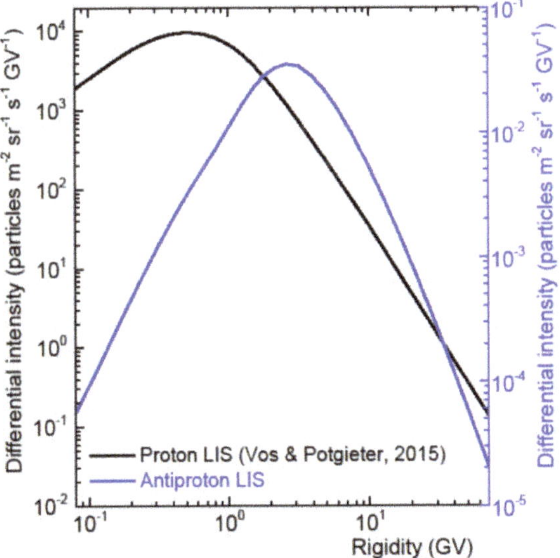

Figure 16. Computed VLISs for galactic protons [46,127] (left-side scale; black line) and for anti-protons (right-side scale; blue line) as a function of rigidity shown here for comparative reasons; apart from the peak values, note the vastly different spectral slopes at both low and high rigidity.

The computed, modulated anti-proton spectra at the Earth are shown in Figure 17 with respect to the assumed VLIS at 122 AU for anti-protons in both panels. Here, they are compared to AMS-02 observations averaged over the period 19 May 2011 to 26 May 2015 [4] in the left panel and to PAMELA observations on the right-side as in Figure 15, repeated here in a less cluttered panel. Although the modulated spectrum for 2006–2009 is in good agreement with the PAMELA data, the corresponding modulated spectrum for 2011–2015 is not agreeing at all rigidities with the AMS-02 observations. It is still quite reasonable in our view, given that the period after 2009 had experienced increased solar activity so that averaging over such a relatively long period makes the reproduction of observations more challenging. When applying our model for maximum solar activity conditions such as in early 2014, the peak value of the modulated anti-proton differential intensity in units as shown in Figure 17, drops to ~1.3 × 10^{-2} at 2 GeV whereas at a 100 MeV it drops to ~1.2 × 10^{-3} and at 10 MeV to ~1.2 × 10^{-4}.

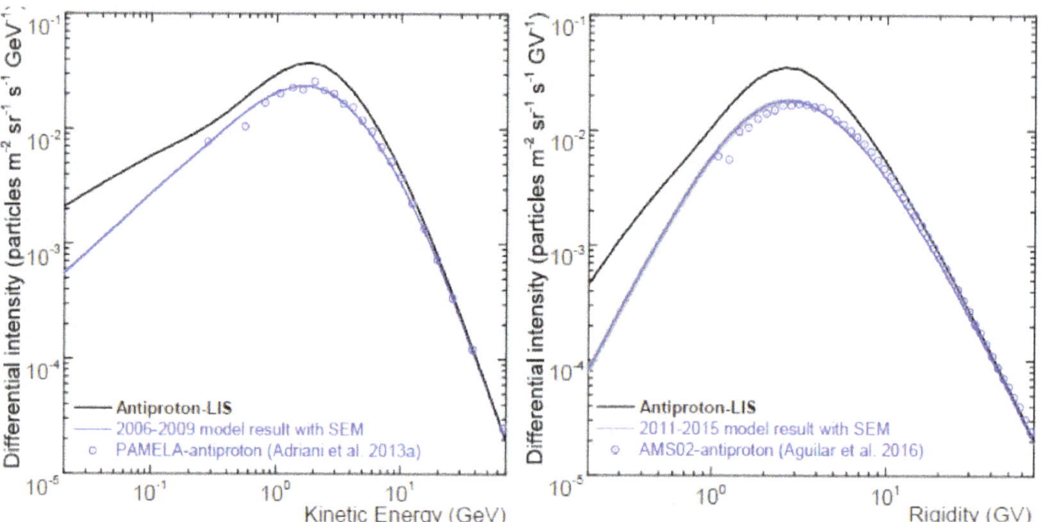

Figure 17. Modulated anti-proton spectra as computed with this work modulation model compared here with long-time averaged observations from PAMELA (July 2006–December 2009) [152] in the right-side panel and AMS-02 (May 2011–May 2015) [4] in the left-side panel as. The VLIS in each panel is shown as solid black lines and the computed, modulated spectra, averaged over corresponding periods, are shown by solid blue lines. SEM indicates standard error of mean (standard deviation/$\sqrt{n}$).

The total modulation factor for anti-protons is shown in Figure 18; this is the factor that the VLIS is reduced between the HP and Earth because of solar modulation as a function of KE and computed for the period 2015b (averaged over the last six months of 2015) when B at the Earth was observed to be at a maximum. This factor thus represents the total modulation over a distance of 121 AU during solar maximum modulation conditions. In this case solar modulation was assumed in the model to commence at 100 GeV. Compared to the simulated modulation factor for protons and electrons [27] and for positrons [45], this factor for anti-protons is much less at low KE and depicts a peculiar form (flatness) from ~300 MeV to around 2 GeV which is absent for protons.

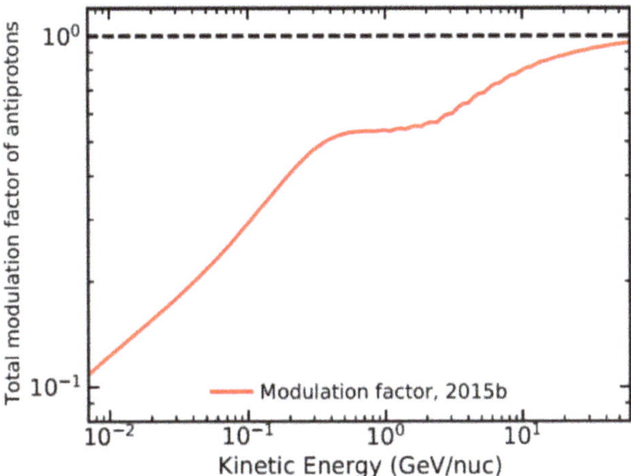

Figure 18. The total modulation factor for anti-protons as a function of KE at Earth with respect to the normalized VLIS specified at 122 AU; here computed for the period 2015b.

For the time being, we prefer to refrain from adjusting the VLIS for anti-protons simply to produce a better fit to the averaged AMS-02 data as shown in Figure 17; see [157]. The averaging of observational data over long periods is a complicating aspect when precise modeling is an objective. However, this issue remains important so we hope that observations which are analysed on a shorter observational time–scale will become available for the present solar minimum period.

5.2. Charge-Sign Dependence

After reproducing the AMS-02 proton and averaged anti-proton spectra, the subsequent anti-proton to proton ratio was calculated (simulated) as a function of time, from 2011–2017, for KE of 1.0–1.5 GeV. This is shown in Figure 19, with the ratio normalized to the value in 2006. It follows that this ratio increases gradually from 2011 to 2017, to reach a maximum in Nov-Dec 2013, and then decreases back to a minimum in Jan 2017. In order to reproduce the AMS-02 ratio, and in addition to varying the DCs (shown in Figure 4), we had to keep the drift scale (or drift coefficient) at a minimum level for the solar maximum period, and then had to increase it gradually to a maximum value as solar activity decreases to a minimum. What is shown here relates to what was shown in Figure 5 for the drift scale, at its maximum value during the 2006–2009 minimum epoch to decrease gradually up to its smallest value for the 2012–2014 reversal period, and then increasing gradually back to a maximum value in Dec 2016.

The numerical simulations for the anti-proton to proton variation with time reported in [158] produced large jumps during solar maximum period associated with the change in the HMF polarity because in their modeling approach drift was not changed with solar activity, similar to what was done with first generations drift models before the Ulysses mission [15–17,33]; see, e.g., [149]. The simulated anti-proton to proton ratio, not normalized and for a shorter time scale, is repeated in Figure 20 for the KE range 1.0–1.5 GeV as a function of time given by Bartels rotation numbers from 2426 to 2506 (May 2011 to May 2017) as reported by AMS-02.

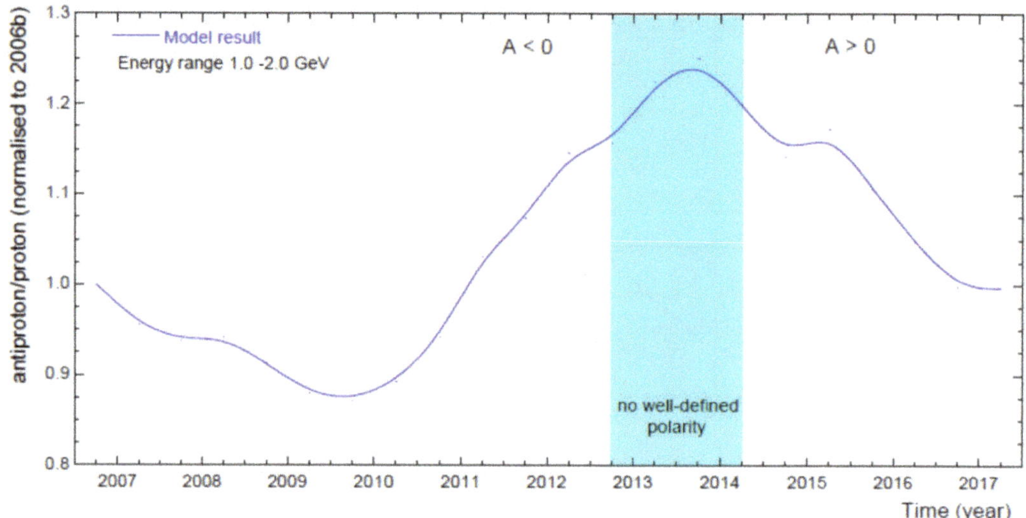

Figure 19. Computed anti-proton to proton ratio as a function of time for 1.0-2.0 GeV at the Earth, here normalized to the value in 2006b. Points indicate model results, interpolated and smoothed as indicated by the solid line. The HMF reversal period with no well-defined polarity is indicated as before. How solar activity has progressed over this period, is shown in Figure 1.

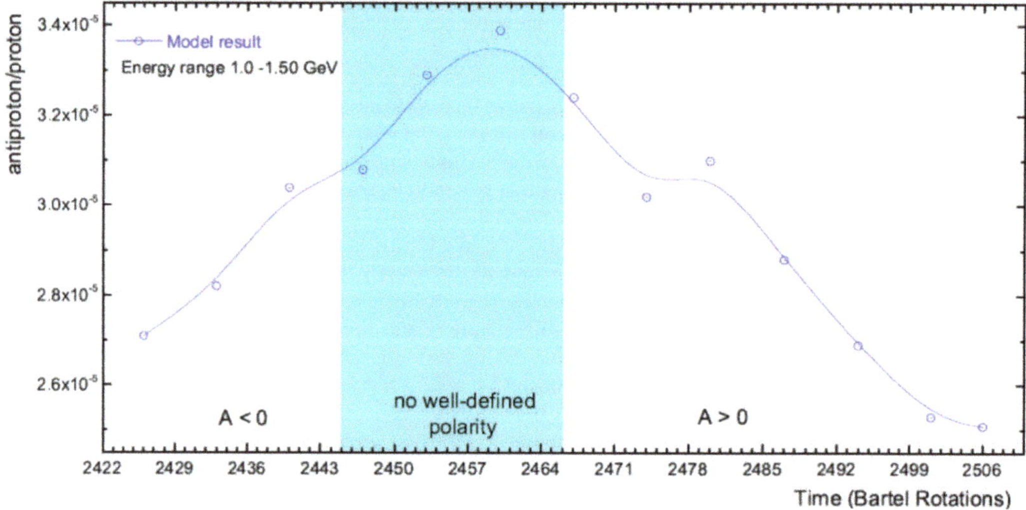

Figure 20. Similar to Figure 19, now enlarged and not normalized, showing the computed anti-proton to proton ratio for Bartels rotation numbers 2426–2506 (May 2011–May 2017), here for KE of 1.0–1.5 GeV. Circled points indicate model results, interpolated and smoothed as indicated by the solid line.

Let us emphasize that in order to obtain these ratios over time it is necessary to adjust the DCs and very specifically the drift coefficient systematically over time, with the latter becoming very small during the HMF polarity reversal period; see also Figures 4 and 5 above. This modeling study highlights the modulation differences between anti-protons and protons especially for the reversal period. In the next section, the focus will be on the difference between positron and anti-proton modulation.

6. Differences between Positron and Anti-Proton Modulation

A display of the main differences between positron and anti-proton modulation needs to start off with showing the enormous difference between the positron and anti-proton VLISs as used for this study. This is shown in Figure 21 in comparison with observations from PAMELA [113,152,153], illustrative of the observed positron and anti-proton spectra at the Earth based on the total modulation from 1 AU up to the HP at 122 AU during solar minimum conditions. Evidently, these two VLISs differ significantly the lower the KE, with the positron intensity exceeding the anti-proton intensity by a factor of up to 10^7 below 10^{-2} GeV. In contrast, the electron intensity exceeds the proton intensity by a factor of maximum 10 at this KE [111,135].

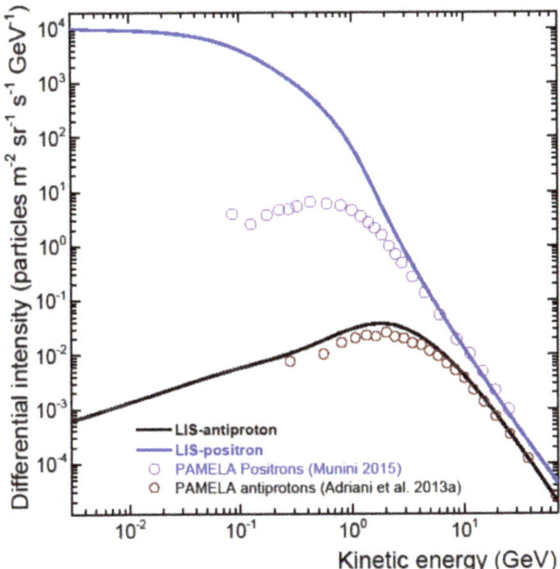

Figure 21. Computed VLISs for galactic positrons (solid blue line) and anti-protons (solid black line) along with observations from PAMELA [113,152] here averaged over 2006–2009, as examples of modulated positron and anti-proton spectra at the Earth based on the total modulation from 1 AU to the HP at 122 AU during solar minimum conditions.

6.1. Positron to Anti-Proton Ratio as a Function of Rigidity

Figure 22 displays the ratio of the VLISs of positrons and anti-protons (solid black line) together with the corresponding simulated positron to anti-proton modulated ratio at the Earth for modulation conditions as in 2006b (red lines) and 2009b (blue lines), for both drift cycles A < 0 and A > 0. The latter serves as predictions for the present solar minimum epoch. The observational ratio above 1 GV is from [4], matching the simulations reasonably well. The difference between the A < 0 and A > 0 cycles is indicative of the extent of drift effects for these GCRs, evidently becoming quite large below 1 GV. These differences relate to what was shown in Figure 2, according to which the positron modulation is dominated by diffusion at lower rigidity (with relatively large and steady MFPs) in contrast to anti-proton modulation which is dominated by adiabatic energy losses at low rigidities as the MFPs get systematically smaller with decreasing rigidity. The causes of these modulation effects are similar to the differences between electron and proton modulation [111] except for the charge-sign dependence which is not at play between positrons and anti-protons.

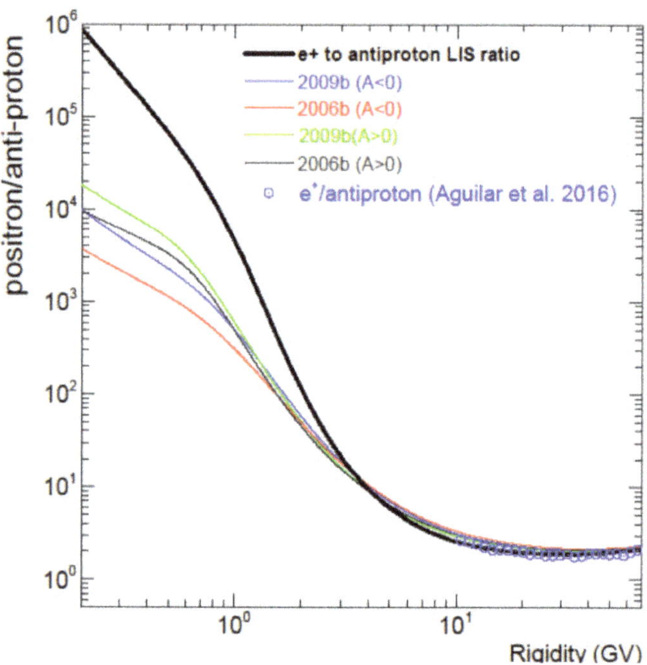

Figure 22. Ratio of the VLISs of positrons (e^+) and anti-protons (solid black line) as a function of rigidity together with the corresponding simulated positron to anti-proton ratio at the Earth for modulation conditions in 2006b (red lines) and 2009b (blue lines) and for both drift cycles A < 0 and A > 0. The latter, indicated as green and grey lines, serve as predictions for the present solar minimum epoch. Observational ratio (blue circles) above 10 GV is from [4].

6.2. Positron to Anti-Proton Ratio as a Function Time

The time dependence of the simulated positron to anti-proton ratio is shown in Figure 23 at two rigidity intervals: 1.0–2.0 GV (top panel) and 0.5–1.0 GV (bottom panel). The standard error of mean (SEM; standard deviation/ $\sqrt{n}$) is shown here for the modeling results (solid line with shaded grey band). Evidently, the largest values are obtained in 2009 with the smallest values during solar maximum activity, at the beginning of the reversal period. The variation with time is mainly because of the relatively larger modulation the positrons experience in contrast to the awkward behaviour of the anti-protons as described in relation to Figure 17. The large difference in the ratio values between these two adjacent rigidity intervals is conspicuous and illustrative of how the vast the difference become between the modulated spectra for these two types of anti-particles, also evident in Figures 21 and 22.

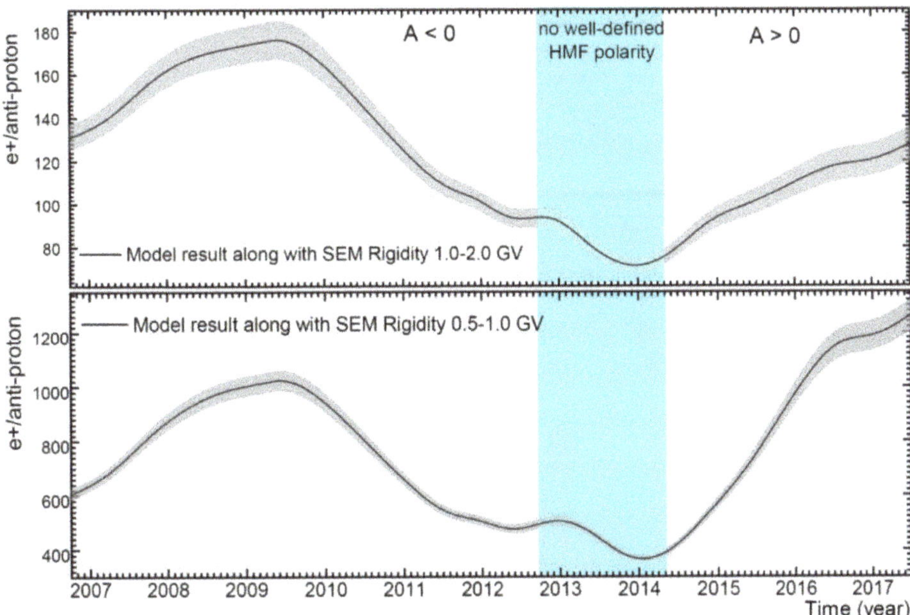

Figure 23. Examples of the simulated positron (*e*+) to anti-proton ratio as a function of time, from 2007 to 2017, at 1.0–2.0 GV (**top** panel) and 0.5–1.0 GV (**bottom** panel). Shaded grey bands indicated the standard error of mean (SEM; standard deviation/$\sqrt{n}$). Vertical shaded band is the same as used before. How solar activity has progressed over this period, is shown in Figure 1.

7. On the Modulation of Other Types of Anti-Matter

A way to test dark matter annihilation may be found in observations at the Earth of low-energy anti-nuclei in GCRs, such as anti-protons (see [154] describing BESS as one of the early ground-breaking anti-matter experiments), also anti-deuteron and anti-helium (e.g., [159]), which makes proper, realistic and precise modulation modeling eminent. This requires the study of all four major modulation processes including particle drifts that creates a 22-year modulation cycle, which is not described by the Force-Field approach to solar modulation. Applying a drift model to anti-deuteron (e.g., [160,161]) is in reach once the modulation of deuteron can be done properly with the modeling approach used here and when relevant precise spectral observations are published (e.g., [162,163]). Estimates exist for the LIS for deuteron and anti-deuteron (e.g., [164], and references there-in) and some effort has already been made for deuteron modulation modeling as shown in [57] based on deuteron observations at the HP from Voyager 1 [21] and from PAMELA [136,165] but only over a limited energy range. A next step in precise modeling should include the study of the solar modulation of anti-helium (e.g., [166,167]). However, this needs additional and dedicated investigation beyond the scope of the present study.

8. A Composition of VLISs

Here, a composition of computed VLISs is provided following the procedure described above for the purpose of pursuing precise global modelling of solar modulation for GCRs, similar to [57]. This is shown in Figure 24 which displays the VLISs for protons, deuteron H-2 ($^{2}H_{1}$) He-3 ($^{3}He_{2}$), He-4 ($^{4}He_{2}$), electrons, and particularly for positrons and anti-protons. It is interesting to note that the VLIS for electrons with KE < ~80 MeV has the largest flux of all GCRs so the question arises what the electron LIS may be down to 1 MeV and at even lower KE? Studying electron modulation to this very low energy at the Earth is difficult because adiabatic energy losses becomes large for electrons only with KE below their rest energy, apart from the dominating presence of Jovian electrons in the inner heliosphere below ~50 MeV [83,84,168,169]. The large flux of positrons below ~300 MeV, even larger than the flux for He-4 is worth noting whereas the H-2 flux is significantly larger than for He-3 below ~200 MeV/n. Not surprising is the low position of the anti-proton VLIS in this hierarchy of astrophysical charged particles.

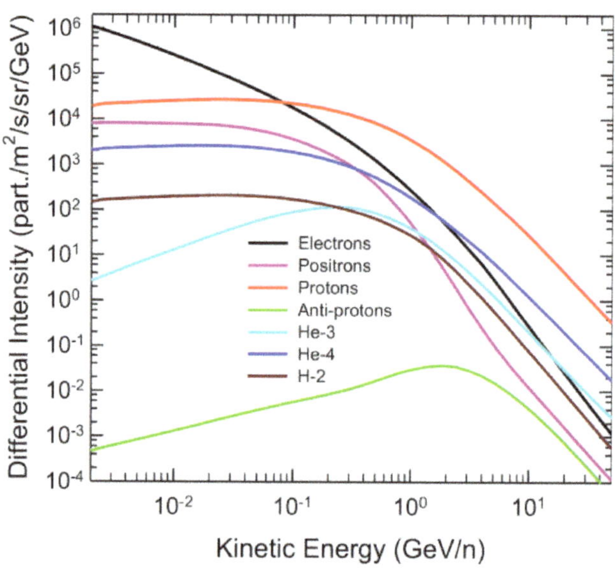

Figure 24. Compilation of computed VLISs shown here for protons, He-3, He-4, deuteron (H-2), electrons, positrons and anti-protons, following the colour coding as indicated in the legend.

The VLIS for deuteron as displayed in Figure 24 is not considered as the final word. However, we are confident that the VLISs for protons, anti-protons, electrons, positrons and total He (sum of both isotopes) are already close to what may be considered as optimal for solar modulation studies. Small adjustments to these VLISs at high rigidities, above a few GeV/n, may be required as the study of very small modulation effects at these high rigidities get better, again owing to these precise measurements being made at Earth and the accompanying precise modeling of the total modulation of GCRs in the heliosphere. Qualitatively, minor adjustments will not affect the obtained results on the computed charge-sign dependent ratios as a function of rigidity as shown in Figure 10 to Figure 16 for electrons and positrons and for protons and anti-protons in Figures 19 and 20.

Since we focus on the refinements of VLISs to be used in heliospheric modulation models, no discussion on how astrophysical aspects can change LISs and how that relates to the VLIS specified at the HP is done here. Examples of such details can be found, e.g., in [159,160,164,170–172]. If such studies would produce VLISs which were significantly different in terms of their spectral shapes (rigidity dependence) from what is presented here, they should be evaluated with comprehensive modulation models to determine their impact on the results obtained.

9. Summary and Conclusions

Determining the global and total modulation of GCRs in the heliosphere more meticulously than before has become possible because aspects such as the position of the TS and especially the HP are reasonably settled. Consensus on the actual shape of the heliosphere has not yet been reached (e.g., [12]) but so far there exists no conclusive evidence that the shape of the heliosphere [34,148] plays a real dominant role in the total modulation of GCRs observed at Earth.

Sophisticated solar modulation modelling of protons, electrons and total Helium has made it possible to refine the VLISs for all GCR nuclei, with the emphasis in this study on positrons and anti-protons. This approach, together with precise observations at Earth, contribute in restraining the uncertainties in the VLIS even for GCRs not observed by the Voyager spacecraft beyond the HP. Evidently, global modulation studies with comprehensive numerical models contribute meaningfully to the testing and refinement of the VLISs for GCRs.

The modeling presented here includes the effects of the solar magnetic field changing it's 'polarity' to create a 22-year modulation cycle and illustrates differences in how positrons and anti-protons are modulated over time and specifically how much particle drifts they experience which is significant at kinetic energies lower than a few GeV. Because the VLIS for anti-protons has a very peculiar spectral shape in contrast to the one for protons, the total modulation of anti-protons is awkwardly different compared to that for protons and other GCR nuclei.

The total modulation factor for anti-protons as a function of KE at Earth is computed with respect to the VLIS specified at the HP (122 AU); see Figure 18. The detailed anti-proton to proton ratio as a function of solar activity is computed for an 11-year period and serves as a prediction of what happens over periods of extreme solar activity and a change in the HMF polarity (Figures 19 and 20). This prediction deviates from other predictions on how this ratio changes during maximum solar activity; see, e.g., [149,158,173]. We found that this ratio decreases generally with decreasing solar activity and increases with increasing solar activity to reach a peak value during maximum solar activity, making a gradual change towards a peak value during this period. This ratio at 1–2 GeV may change as much as a factor of 1.5 over a solar cycle and the differential intensity for anti-protons may decrease by a factor of 2 at 100 MeV from solar minimum to solar maximum. Acquiring these charge-sign dependent ratios over time at the Earth it is necessary to adjust the heliospheric DCs and very specifically the drift coefficient systematically with solar activity, with the latter becoming quite small during the HMF polarity reversal period; see Figures 2–5.

The ratio of the VLISs of positrons to anti-protons are determined together with the corresponding simulated positron to anti-proton ratio at the Earth, based on modulation conditions that existed in 2006 and 2009, done for both drift cycles, A < 0 and A > 0. The latter serves as predictions for the present solar minimum epoch.

A composition of computed VLISs is given in Figure 24 for the purpose of pursuing precise global modelling of solar modulation for these GCRs. It follows that the VLIS for electrons with KE < ~80 MeV has the largest flux of all GCRs with the anti-proton flux by far the lowest.

Author Contributions: Conceptualization, M.S.P.; methodology, all authors; software, O.P.M.A., D.B., D.N.; validation, all authors; analysis, all authors; writing—original draft preparation, M.S.P.; writing—review and editing, all authors; visualization, all authors; supervision, M.S.P.; project administration, D.N., M.S.P.; funding acquisition, D.N. All authors have read and agreed to the published version of the manuscript.

Funding: M.D.N. thanks the South African (SA) National Research Foundation (NRF) for his partial financial support under Joint Science and Technology Research Collaboration between SA and Russia (Grant No: 118915) and BAAP (Grant No: 120642). He acknowledges that the opinions, findings, conclusions or recommendations expressed in publication generated by SA NRF supported research is that of the authors alone, and that the SA NRF accepts no liability whatsoever in this regard. D.B. and O.P.M.A. acknowledge the financial support they had received from the NWU's post-doctoral programme.

Conflicts of Interest: The authors declare no conflict of interest.

References

1.   Adriani, O.; Barbarino, G.C.; Bazilevskaya, G.A.; Bellotti, R.; Boezio, M.; Bogomolov, E.A.; Bonechi, L.; Bongi, M.; Bonvicini, V. Borisov, S.; et al. PAMELA results on the cosmic-ray antiproton flux from 60 MeV to 180 GeV in kinetic energy. *Phys. Rev. Lett* **2010**, *105*, 121101. [CrossRef] [PubMed]
2.   Adriani, O.; Barbarino, G.C.; Bazilevskaya, G.A.; Bellotti, R.; Boezio, M.; Bogomolov, A.E.; Bongi, M.; Bonvicini, V.; Bottai, S. Bruno, A.; et al. Time dependence of the electron and positron components of the cosmic radiation measured by the PAMELA experiment between July 2006 and December 2015. *Phys. Rev. Lett.* **2016**, *116*, 241105. [CrossRef] [PubMed]
3.   Adriani, O.; Barbarino, G.C.; Bazilevskaya, G.A.; Bellotti, R.; Boezio, M.; Bogomolov, A.E.; Bongi, M.; Bonvicini, V.; Bottai, S. Bruno, A.; et al. Ten years of PAMELA in space. *La Riv. Del Nuovo Cim.* **2017**, *40*, 473–522. [CrossRef]
4.   Aguilar, M.; Ali Cavasonza, L.; Alpat, B.; Ambrosi, G.; Arruda, L.; Attig, N.; Aupetit, S.; Azzarello, P.; Bachlechner, A.; Barao F.; et al. Antiproton flux, antiproton-to-proton flux ratio, and properties of elementary particle fluxes in primary cosmic rays measured with the Alpha Magnetic Spectrometer on the International Space Station. *Phys. Rev. Lett.* **2016**, *117*, 091103. [CrossRef] [PubMed]
5.   Aguilar, M.; Ali Cavasonza, L.; Ambrosi, G.; Arruda, L.; Attig, N.; Aupetit, S.; Azzarello, P.; Bachlechner, A.; Barao, F.; Barrau A.; et al. Observation of complex time structures in the Cosmic-ray electron and positron fluxes with the Alpha Magnetic Spectrometer on the International Space Station. *Phys. Rev. Lett.* **2018**, *121*, 051102. [CrossRef]
6.   Stone, E.C.; Cummings, A.C.; McDonald, F.B.; Heikkila, B.C.; Lal, N.; Webber, W.R. Voyager 1 observes low-energy galactic cosmic rays in a region depleted of heliospheric ions. *Science* **2013**, *341*, 150–153. [CrossRef] [PubMed]
7.   Stone, E.C.; Cummings, A.C.; Heikkila, B.C.; Lal, N. Cosmic ray measurements from Voyager 2 as it crossed into interstellar space *Nature Astron.* **2019**, *3*, 1013–1018. [CrossRef]
8.   Kóta, J.; Jokipii, J.R. Are cosmic rays modulated beyond the heliopause? *Astrophys. J.* **2014**, *782*, 1–6. [CrossRef]
9.   Luo, X.; Potgieter, M.S.; Zhang, M.; Pogorelov, N.; Feng, X.; Strauss, R.D. A numerical simulation of cosmic ray modulation near the heliopause: II. Some physical insights. *Astrophys. J.* **2016**, *826*, 182. [CrossRef]
10. Zhang, M.; Pogorelov, N.V. Modulation of galactic cosmic rays by plasma disturbances propagating through the Local Interstellar Medium in the outer heliosheath. *Astrophys. J.* **2020**, *895*, 1. [CrossRef]
11. Scherer, K.; Fichtner, H. The return of the bow shock. *Astrophys. J.* **2014**, *782*, 25. [CrossRef]
12. Pogorelov, N.; Fichtner, H.; Czechowski, A.; Lazarian, A.; Lembege, B.; le Roux, J.A.; Potgieter, M.S.; Scherer, K.; Stone, E.C.; Strauss, R.D.; et al. Heliosheath processes and the structure of the heliopause modeling energetic particles, cosmic rays, and magnetic fields. *Space Sci. Rev.* **2017**, *212*, 193–248. [CrossRef]
13. Schlickeiser, R.; Oppotsch, J.; Zhang, M.; Pogorelov, N.V. On the anisotropy of galactic cosmic rays. *Astrophys. J.* **2019**, *879*, 29 [CrossRef]
14. Zhang, M.; Pogorelov, N.V.; Zhang, Y.; Hu, H.B.; Schlickeiser, R. The original anisotropy of TeV Cosmic Rays in the Local Interstellar Medium. *Astrophys. J.* **2020**, *889*, 97. [CrossRef]
15. Heber, B.; Wibberenz, G.; Potgieter, M.S.; Burger, R.A.; Ferreira, S.; Müller-Mellon, R.; Kunow, H.; Ferrando, P.; Raviart, A.; Paizis, C.; et al. Ulysses cosmic ray and solar particle investigation/Kiel Electron Telescope observations: Charge sign dependence and spatial gradients during the 1990–2000 A > 0 solar magnetic cycle. *J. Geophys. Res.* **2002**, *107*, 1274. [CrossRef]
16. Heber, B.; Kopp, A.; Gieseler, J.; Müller-Mellin, R.; Fichtner, H.; Scherer, K.; Potgieter, M.S.; Ferreira, S.E.S. Modulation of galactic cosmic ray protons and electrons during an unusual solar minimum. *Astrophys. J.* **2009**, *699*, 1956–1963. [CrossRef]
17. Heber, B.; Potgieter, M.S. Cosmic rays at high heliolatitudes. *Space Sci. Rev.* **2006**, *127*, 117–194. [CrossRef]
18. Heber, B.; Potgieter, M.S. Galactic and anomalous cosmic rays through the solar cycle: New insights from Ulysses. In *The Heliosphere through the Solar Activity Cycle*; Balogh, A., Lanzerotti, L.J., Suess, S.J., Eds.; Springer: Berlin/Heidelberg, Germany, 2008; pp. 195–249. [CrossRef]
19. Cummings, A.C.; Stone, E.C.; Heikkila, B.C.; Lal, N.; Webber, W.R. Galactic cosmic rays in the local interstellar medium: Voyager 1 observations and model results. *Astrophys. J.* **2016**, *831*, 18. [CrossRef] [PubMed]
20. Webber, W.R.; Villa, T.L. A comparison of the galactic cosmic ray electron and proton intensities from 1 MeV/nuc to 1 TeV/nuc using Voyager and higher energy magnetic spectrometer measurements; Are there differences in the source spectra of these particles? *arXiv* **2018**, arXiv:1806.02808.
21. Webber, W.R.; Lal, N.; Heikkila, B. The spectra of 2H and 3He secondary cosmic ray isotopes from ~20-85 MeV/nuc as measured using the B-end HET Telescope on Voyager beyond the heliopause and a fit to these Interstellar Spectra using a Leaky Box Propagation Model. *arXiv* **2018**, arXiv:1802.08273.
22. Strauss, R.D.; Potgieter, M.S. Where does the heliospheric modulation of galactic cosmic rays start? *Adv. Space Res.* **2014**, *53*, 1015–1023. [CrossRef]
23. Gleeson, L.J.; Urch, I.A. A study of the force-field equation for the propagation of galactic cosmic rays. *Astrophys. Space Sci.* **1973**, *25*, 387–404. [CrossRef]
24. Cholis, I.; Hooper, D.; Linden, T. A predictive analytic model for the solar modulation of cosmic rays. *Phys. Rev. D.* **2016**, *93*, 043016. [CrossRef]
25. Caballero-Lopez, R.A.; Moraal, H. Limitations of the force field equation to describe cosmic ray modulation. *J. Geophys. Res.* **2004**, *109*, A01101. [CrossRef]
26. Moraal, H. Cosmic-ray modulation equations. *Space Sci. Rev.* **2013**, *176*, 299–319. [CrossRef]

27. Potgieter, M.S. The global modulation of cosmic rays during a quiet heliosphere: A modeling perspective. *Adv. Space Res.* **2017**, *60*, 848–864. [CrossRef]

28. Engelbrecht, N.E.; di Felice, V. Uncertainties implicit to the use of the force-field solutions to the Parker transport equation in analyses of observed cosmic ray antiproton intensities. *Phys. Rev. D* **2020**, *102*, 103007. [CrossRef]

29. Parker, E.N. The passage of energetic charged particles through interplanetary space. *Planet. Space Sci.* **1965**, *13*, 9–49. [CrossRef]

30. Smith, C.W.; Bieber, J.W. Solar cycle variation of the interplanetary magnetic field spiral. *Astrophys. J.* **1991**, *370*, 435. [CrossRef]

31. Raath, J.L.; Potgieter, M.S.; Strauss, R.D.; Kopp, A. The effects of magnetic field modifications on the solar modulation of cosmic rays with a SDE-based model. *Adv. Space Res.* **2016**, *57*, 1965–1977. [CrossRef]

32. McComas, D.J.; Elliot, H.A.; Gosling, J.T.; Reisenfield, D.B.; Skoung, R.M.; Goldstein, B.E.; Neugebauer, M.; Balogh, A. Ulysses' second fast latitude scan: Complexity near solar maximum and the reformation of polar coronal holes. *Geophys. Res. Lett.* **2002**, *29*, 1. [CrossRef]

33. Heber, B. Cosmic rays through the solar Hale Cycle. Insights from Ulysses. *Space Sci. Rev.* **2013**, *176*, 265–278. [CrossRef]

34. Langner, U.W.; Potgieter, M.S. The modulation of galactic protons in an asymmetrical heliosphere. *Astrophys. J.* **2005**, *630*, 1114–1124. [CrossRef]

35. Ngobeni, M.D.; Potgieter, M.S. Modulation of galactic cosmic rays in a north-south asymmetrical heliosphere. *Adv. Space Res.* **2011**, *48*, 300–307. [CrossRef]

36. Richardson, J.D.; Wang, C. Plasma in the heliosheath: 3.5 years of observations. *Astrophys. J.* **2011**, *734*, L21. [CrossRef]

37. Stone, E.C.; Cummings, A.C.; McDonald, F.B.; Heikkila, B.C.; Lal, N.; Webber, W.R. Voyager 1 explores the termination shock region and the heliosheath beyond. *Science* **2005**, *309*, 2017. [CrossRef] [PubMed]

38. Richardson, J.D.; Kasper, J.C.; Wang, C.; Belcher, W.; Lazarus, A.J. Cool heliosheath plasma and deceleration of the upstream solar wind at the termination shock. *Nature* **2008**, *454*, 63. [CrossRef] [PubMed]

39. Vos, E.E.; Potgieter, M.S. Global gradients for cosmic ray protons during the solar minimum of cycle 23/24. *Solar Phys.* **2016**, *291*, 2181–2195. [CrossRef]

40. Langner, U.W.; Potgieter, M.S.; Fichtner, H.; Borrmann, T. Effects of solar wind speed changes in the heliosheath on the modulation of cosmic ray protons. *Astrophys. J.* **2006**, *640*, 1119–1134. [CrossRef]

41. Ngobeni, M.D.; Potgieter, M.S. The heliospheric modulation of cosmic rays: Effects of a latitude dependent solar wind termination shock. *Adv. Space Res.* **2010**, *46*, 391–401. [CrossRef]

42. Potgieter, M.S.; Vos, E.E.; Boezio, M.; De Simone, N.; Di Felice, V.; Formato, V. Modulation of galactic protons in the heliosphere during the unusual solar minimum from 2006 to 2009: A modelling approach. *Solar Phys.* **2014**, *289*, 391–406. [CrossRef]

43. Potgieter, M.S.; Vos, E.E.; Munini, R.; Boezio, M.; Di Felice, V. Modulation of galactic electrons in the heliosphere during the unusual solar minimum of 2006 to 2009: A modelling approach. *Astrophys. J.* **2015**, *810*, 141. [CrossRef]

44. Adriani, O.; Barbarino, G.C.; Bazilevskaya, G.A.; Bellotti, R.; Boezio, M.; Bogomolov, E.A.; Bongi, M.; Bonvicini, V.; Borisov, S.; Bottai, S.; et al. Time dependence of the proton flux measured by PAMELA during the 2006 July-2009 December solar minimum. *Astrophys. J.* **2013**, *765*, 91. [CrossRef]

45. Aslam, O.P.M.; Bisschoff, D.; Potgieter, M.S.; Boezio, M.; Munini, R. Modeling of heliospheric modulation of cosmic-ray positrons in a very quiet heliosphere. *Astrophys. J.* **2019**, *873*, 70. [CrossRef]

46. Bisschoff, D.; Potgieter, M.S.; Aslam, O.P.M. New very local interstellar spectra for electrons, positrons, protons, and light cosmic ray nuclei. *Astrophys. J.* **2019**, *878*, 59. [CrossRef]

47. Corti, C.; Potgieter, M.S.; Bindi, V.; Consolandi, C.; Light, C.; Palermo, M.; Popkow, A. Numerical modeling of galactic cosmic ray proton and helium observed by AMS-02 during the solar maximum of Solar Cycle 24. *Astrophys. J.* **2019**, *871*, 253. [CrossRef]

48. Ngobeni, M.D.; Aslam, O.P.M.; Bisschoff, D.; Potgieter, M.S.; Ndiitwani, D.C.; Boezio, M.; Marcelli, N.; Munini, R.; Mikhailov, V.V.; Koldobskiy, S.A. The 3D numerical modeling of the solar modulation of galactic protons and helium nuclei related to observations by PAMELA between 2006 and 2009. *Astrophys. Space Sci.* **2020**, *365*, 182. [CrossRef]

49. Ngobeni, M.D.; Aslam, O.P.M.; Bisschoff, D.; Ndiitwani, D.C.; Potgieter, M.S.; Boezio, M.; Marcelli, N.; Munini, R.; Mikhailov, V.V.; Koldobskiy, S.A. Combined heliospheric modulation of galactic protons and Helium nuclei from solar minimum to maximum activity related to observations by PAMELA and AMS-02. *PoS* **2021**, *ICRC2021*, 1337. [CrossRef]

50. Ngobeni, M.D.; Potgieter, M.S.; Aslam, O.P.M.; Bisschoff, D.; Ramokgaba, I.I.; Ndiitwani, D.C. Numerical modeling of the solar modulation of helium isotopes in theiInner heliosphere. *PoS* **2021**, *ICRC2021*, 1338. [CrossRef]

51. Ngobeni, M.D.; Potgieter, M.S.; Aslam, O.P.M.; Bisschoff, D.; Ramokgaba, I.I.; Ndiitwani, D.C. Simulations of the solar modulation of helium isotopes constrained by observations. *Adv. Space Res.* **2021**. accepted.

52. Kóta, J. Theory and modeling of galactic cosmic rays: Trends and prospects. *Space Sci. Rev.* **2013**, *176*, 391–403. [CrossRef]

53. Potgieter, M.S. The modulation of galactic cosmic rays in the heliosphere: Theory and models. *Space Sci. Rev.* **1998**, *83*, 147–158. [CrossRef]

54. Potgieter, M.S. Solar modulation of cosmic rays. *Living Rev. Solar Phys.* **2013**, *10*, 3–66. [CrossRef]

55. Schlickeiser, R. *Cosmic Ray Astrophysics*; Springer: Berlin/Heidelberg, Germany, 2002. [CrossRef]

56. Quenby, J.J. The theory of cosmic ray modulation. *Space Sci. Rev.* **1984**, *37*, 201–267. [CrossRef]

57. Bisschoff, D.; Aslam, O.P.M.; Ngobeni, M.D.; Mikhailov, V.V.; Boezio, M.; Munini, R.; Potgieter, M.S. On the very local interstellar spectra for helium, positrons, anti-protons, deuteron and anti–deuteron. Proceedings of the 3rd International Symposium on Cosmic Rays & Astrophysics (ISCRA-2021), Moscow, Russia, 8–10 June 2021. *Phys. Atom. Nucl.* **2021**, *84*, 52–58. [CrossRef]

58. Zhu, C.-R.; Yuan, Q.; Wie, D.-M. Local interstellar spectra and solar modulation of cosmic ray electrons and positrons. *Astropart. Phys.* **2020**, *124*, 102495. [CrossRef]

59. Potgieter, M.S. The charge-sign dependent effect in the solar modulation of cosmic rays. *Adv. Space Res.* **2014**, *53*, 1415–1425 [CrossRef]

60. Aslam, O.P.M.; Bisschoff, D.; Ngobeni, M.D.; Potgieter, M.S.; Munin, I.R.; Boezio, M.; Mikhailov, V.V. Time and charge-sign dependence of the heliospheric modulation of cosmic rays. *Astrophys. J.* **2021**, *909*, 215. [CrossRef]

61. Kopp, A.; Raath, J.-L.; Fichtner, H.; Potgieter, M.S.; Ferreira, S.E.S.; Heber, B. Cosmic-ray transport in heliospheric magnetic structures. III. Implications of solar magnetograms for the drifts of cosmic rays. *Astrophys. J.* **2021**, *922*, 124. [CrossRef]

62. The Wilcox Solar Observatory. Available online: http://wso.stanford.edu (accessed on 1 July 2021).

63. Paths to Magnetic Field, Plasma, Energetic Particle Data Relevant to Heliospheric Studies and Resident at Goddard's Space Physics Data Facility. Available online: http://omniweb.gsfc.nasa.gov (accessed on 1 July 2021).

64. Sun, X.; Hoeksema, J.T.; Liu, Y.; Zhao, J. On polar magnetic field reversal and surface flux transport during solar cycle 24. *Astrophys. J.* **2015**, *798*, 114. [CrossRef]

65. Janardhan, P.; Fujiki, K.; Ingale, M.; Bisoi, S.K.; Rout, D. Solar cycle 24: An unusual polar field reversal. *Astron. Astrophys.* **2018**, *618*, A148. [CrossRef]

66. Aslam, O.P.M.; Bisschoff, D.; Potgieter, M.S. The solar modulation of protons and anti-protons. *PoS* **2019**, *ICRC2019*, 1054. [CrossRef]

67. Aslam, O.P.M.; Bisschoff, D.; Potgieter, M.S. The heliospheric modulation of electrons and positrons. *PoS* **2019**, *ICRC2019*, 1053. [CrossRef]

68. Bieber, J.W.; Matthaeus, W.H.; Smith, C.W.; Wanner, W.; Kallenrode, M.-B.; Wibberenz, G. Proton and electron mean free paths: The Palmer consensus revisited. *Astrophys. J.* **1994**, *420*, 294–306. [CrossRef]

69. Potgieter, M.S. The heliospheric modulation of galactic electrons: Consequences of new calculations for the mean free path of electrons between 1 MeV and $\sim$ 10 GeV. *J. Geophys. Res.* **1996**, *101*, 24411–24422. [CrossRef]

70. Teufel, A.; Schlickeiser, R. Analytical calculation of the parallel mean free path of heliospheric cosmic rays II. Dynamical magnetic slab turbulence and random sweeping slab turbulence with finite wave power at small wavenumbers. *Astron. Astrophys.* **2003**, *397*, 15. [CrossRef]

71. Shalchi, A. *Nonlinear Cosmic Ray Diffusion Theories*; Springer: Berlin/Heidelberg, Germany, 2009. [CrossRef]

72. Ngobeni, M.D.; Potgieter, M.S. Modelling the effects of scattering parameters on particle-drift in the solar modulation of galactic cosmic rays. *Adv. Space Res.* **2015**, *56*, 1525–1537. [CrossRef]

73. Tautz, R.C.; Shalchi, A. Drift coefficient of charged particles in turbulent magnetic fields. *Astrophys. J.* **2012**, *744*, 125. [CrossRef]

74. Engelbrecht, N.E.; Strauss, R.D.; le Roux, J.A.; Burger, R.A. Toward a greater understanding of the reduction of drift coefficients in the presence of turbulence. *Astrophys. J.* **2017**, *841*, 107. [CrossRef]

75. Manuel, R.; Ferreira, S.E.S.; Potgieter, M.S.; Strauss, R.D.; Engelbrecht, N.E. Time-dependent cosmic ray modulation. *Adv. Space Res.* **2011**, *47*, 1529–1537. [CrossRef]

76. Qin, G.; Shen, Z.-N. Modulation of galactic cosmic rays in the inner heliosphere, comparing with PAMELA measurements. *Astrophys. J.* **2017**, *846*, 56. [CrossRef]

77. Shen, Z.-N.; Qin, G. Modulation of galactic cosmic rays in the inner heliosphere over solar cycles. *Astrophys. J.* **2018**, *854*, 137. [CrossRef]

78. Moloto, K.D.; Engelbrecht, N.E. A fully time-dependent ab initio cosmic-ray modulation model applied to historical cosmic-ray modulation. *Astrophys. J.* **2020**, *894*, 121. [CrossRef]

79. Zhao, L.-L.; Qin, G.; Zhang, M.; Heber, B. Modulation of galactic cosmic rays during the unusual solar minimum between cycles 23 and 24. *J. Geophys. Res. Space Physics* **2014**, *119*, 1493–1506. [CrossRef]

80. Fiandrini, E.; Tomassetti, N.; Bertucci, B.; Donnini, F.; Graziani, M.; Khiali, B.; Reina Conde, A. Numerical modeling of cosmic rays in the heliosphere: Analysis of proton data from AMS-02 and PAMELA. *Phys. Rev. D.* **2021**, *104*, 023012. [CrossRef]

81. Song, X.; Luo, X.; Potgieter, M.S.; Liu, X.-M.; Geng, Z. A numerical study of the solar modulation of galactic protons and Helium from 2006 to 2017. *Astrophys. J. Supp. Ser.* **2021**, *257*, 48. [CrossRef]

82. Adriani, O.; Barbarino, G.C.; Bazilevskaya, G.A.; Bellotti, R.; Boezio, M.; Bogomolov, E.A.; Bongi, M.; Bonvicini, V.; Bottai, S.; Bruni, A.; et al. Time dependence of the e⁻ flux measured by PAMELA during the July 2006-December 2009 solar minimum. *Astrophys. J.* **2015**, *810*, 142. [CrossRef]

83. Vogt, A.; Heber, B.; Kopp, A.; Potgieter, M.S.; Strauss, R.D. Jovian electrons in the inner heliosphere: Proposing a new source spectrum based on 30 years of measurements. *Astron. Astrophys.* **2018**, *613*, A28. [CrossRef]

84. Nndanganeni, R.R.; Potgieter, M.S. The energy range of drift effects in the solar modulation of cosmic ray electrons. *Adv. Space Res.* **2016**, *58*, 453–463. [CrossRef]

85. Mechbal, S.; Mangeard, P.-S.; Clem, J.; Evenson, P.A.; Johnson, R.P.; Lucas, B.; Roth, J. Measurement of low-energy cosmic-ray electron and positron spectra at 1 AU with the AESOP-Lite spectrometer. *Astrophys. J.* **2020**, *903*, 21. [CrossRef]

86. Mikhailov, V.V.; Aleksandrin, S.Y.; Koldobskiy, S.A.; Boezio, M.; Munini, R.; Aslam, O.P.M.; Bisschoff, D.; Ngobeni, M.D.; Potgieter, M.S. Study of the modulation of galactic positrons and electrons in 2006–2016 with the PAMELA experiment. *PoS* **2021**, *ICRC2021*, 1307. [CrossRef]

37. Ndiitwani, D.C.; Ngobeni, M.D.; Aslam, O.P.M.; Bisschoff, D.; Potgieter, M.S.; Boezio, M.; Munini, R.; Mikhailov, V.V. A Simulation Study of galactic proton modulation from solar minimum to maximum conditions. *PoS* **2021**, *ICRC2021*, 1327. [CrossRef]

38. Krainev, M.; Kalinina, M.; Aslam, O.P.M.; Ngobeni, M.D.; Potgieter, M.S. On the dependence of maximum GCR intensity on heliospheric factors for the last five sunspot minima. *Adv. Space Res.* **2021**, *68*, 2953–2962. [CrossRef]

39. Krainev, M.; Gvozdevsky, B.B.; Kalinina, M.; Aslam, O.P.M.; Ngobeni, M.D.; Potgieter, M.S. On the solar poloidal magnetic field as one of the main factors for maximum GCR intensity for the last five sunspot minima. *PoS* **2021**, *ICRC2021*, 1322. [CrossRef]

90. Jokipii, J.R.; Levy, E.H.; Hubbard, W.B. Effects of particle drift on cosmic-ray transport. I. General properties, application to solar modulation. *Astrophys. J.* **1977**, *213*, 861–868. [CrossRef]

91. Kóta, J.; Jokipii, J.R. Effects of drift on the transport of cosmic rays. VI. A three-dimensional model including diffusion. *Astrophys. J.* **1983**, *265*, 573–581. [CrossRef]

92. Zhang, M. A Markov stochastic process theory of cosmic-ray modulation. *Astrophys. J.* **1999**, *513*, 409–420. [CrossRef]

93. Strauss, R.D.; Potgieter, M.S.; Boezio, M.; De Simone, N.; Di Felice, V.; Kopp, A.; Büsching, I. The heliospheric transport of protons and anti-protons: A stochastic modelling approach of PAMELA observations. In *Astroparticle, Particle, Space Physics and Detectors for Physics Applications. Proceedings of the 13th ICATPP Conference. Como, Italy, 3–7 October 2021*; Giani, S., Leroy, C., Price, L., Rancoita, P.-G., Ruchti, R., Eds.; World Scientific: Singapore, 2012; pp. 288–296.

94. Fisk, L.A. Motion of the footprints of heliospheric magnetic field lines at the Sun: Implications for recurrent energetic particle events at high heliographic latitudes. *J. Geophys. Res.* **1996**, *101*, 15547–15554. [CrossRef]

95. Le Roux, J.A.; Potgieter, M.S. A time-dependent drift model for the long-term modulation of cosmic rays with special reference to asymmetries with respect to the solar minimum of 1987. *Astrophys. J.* **1990**, *361*, 275–282. [CrossRef]

96. Le Roux, J.A.; Potgieter, M.S. The simulated features of heliospheric cosmic ray modulation with a time-dependent drift model. II. On the energy dependence of the onset of new modulation in 1987. *Astrophys. J.* **1992**, *390*, 661–667. [CrossRef]

97. le Roux, J.A.; Potgieter, M.S. The simulation of complete 11 and 22 year modulation cycles for cosmic rays in the heliosphere using a drift model with global interaction regions. *Astrophys. J.* **1995**, *442*, 847–851. [CrossRef]

98. Luo, X.; Zhang, M.; Rassoul, K.H.; Pogorelov, N. Cosmic-ray modulation by the global merged interaction region in the heliosheath. *Astrophys. J.* **2011**, *730*, 13. [CrossRef]

99. Luo, X.; Potgieter, M.S.; Zhang, M.; Feng, X. A numerically study of cosmic proton modulation using AMS02 observations. *Astrophys. J.* **2019**, *878*, 6. [CrossRef]

100. Potgieter, M.S. Time-dependent cosmic-ray modulation: Role of drifts and interaction regions. *Adv. Space Res.* **1993**, *13*, 6–239. [CrossRef]

101. Langner, U.W.; Potgieter, M.S. Solar wind termination shock and heliosheath effects on charge-sign dependent modulation for protons and anti-protons. *J. Geophys. Res.* **2004**, *109*, A01103. [CrossRef]

102. Langner, U.W.; Potgieter, M.S. Effects of the solar wind termination shock on charge-sign dependent cosmic ray modulation. *Adv. Space Res.* **2004**, *34*, 144–149. [CrossRef]

103. Langner, U.W.; Potgieter, M.S. The heliospheric modulation of cosmic ray protons during increased solar activity: Effects of the position of the solar wind termination shock and of the heliopause. *Ann. Geophys.* **2005**, *23*, 1–6. [CrossRef]

104. Manuel, R.; Ferreira, S.E.S.; Potgieter, M.S. Time-dependent modulation of cosmic rays in the heliosphere. *Solar Phys.* **2014**, *289*, 2207–2231. [CrossRef]

105. Manuel, R.; Ferreira, S.E.S.; Potgieter, M.S. The effect of a dynamic inner heliosheath thickness on cosmic ray modulation. *Astrophys. J.* **2015**, *799*, 223. [CrossRef]

106. Ndiitwani, D.C.; Ferreira, S.E.S.; Potgieter, M.S.; Heber, B. Modelling cosmic ray intensities along the Ulysses trajectory. *Ann. Geophys.* **2005**, *23*, 1–10. [CrossRef]

107. Heber, B.; Ferrando, P.; Raviart, A.; Paizis, C.; Posner, A.; Wibberenz, G.; Müller-Mellin, R.; Kunow, H.; Potgieter, M.S.; Ferreiera, S.; et al. 3–20 MeV electrons in the inner three-dimensional heliosphere at solar maximum: Ulysses COSPIN/KET observations. *Astrophys. J.* **2002**, *579*, 888–894. [CrossRef]

108. Ferreira, S.E.S.; Potgieter, M.S.; Heber, B.; Fichtner, H. Charge-sign dependent modulation over a 22-year cycle. *Ann. Geophys.* **2003**, *21*, 1359–1366. [CrossRef]

109. Reinecke, J.P.L.; Potgieter, M.S. An explanation for the difference in cosmic ray modulation at low and neutron monitor energies during consecutive solar minimum periods. *J. Geophys. Res.* **1994**, *99*, 14761–14768. [CrossRef]

110. Strauss, R.D.; Potgieter, M.S. Is the highest cosmic rays yet to come? *Sol. Phys.* **2014**, *289*, 3197–3205. [CrossRef]

111. Potgieter, M.S.; Vos, E.E. Difference in the heliospheric modulation of cosmic-ray protons and electrons during solar minimum of 2006 to 2009. *Astron. Astrophys.* **2017**, *601*, A23. [CrossRef]

112. Di Felice, V.; Munini, R.; Vos, E.E.; Potgieter, M.S. New evidence for charge-sign dependent modulation during the solar minimum of 2006 to 2009. *Astrophys. J.* **2017**, *834*, 89. [CrossRef]

113. Munini, R. Solar Modulation of Cosmic Ray Electrons and Positrons Measured by the PAMELA Experiment During the 23rd Solar Minimum. Ph.D. Thesis, University delgi Studi di Trieste, Trieste, Italy, 2015.

114. Munini, R.; Boezio, M.; Bruno, A.; Christian, E.C.; de Nolfo, G.A.; Di Felice, V.; Martucci, M.; Merge, M.; Richardson, I.G.; Ryan, J.M.; et al. Evidence of energy and charge sign dependence of the recovery time for the 2006 December Forbush event measured by the PAMELA Experiment. *Astrophys. J.* **2018**, *853*, 76. [CrossRef]

115. Luo, X.; Potgieter, M.S.; Zhang, M.; Feng, X. A study of electron Forbush decreases with a 3D SDE numerical model. *Astrophys. J.* **2018**, *860*, 160. [CrossRef]
116. Potgieter, M.S. Very local interstellar spectra for galactic electrons, protons and helium. *Brazilian J. Phys.* **2014**, *44*, 581–588 [CrossRef]
117. Bisschoff, D.; Potgieter, M.S. Implications of Voyager 1 observations outside the heliosphere for the local interstellar electron spectrum. *Astrophys. J.* **2014**, *794*, 166. [CrossRef]
118. Bisschoff, D.; Potgieter, M.S. New local interstellar spectra for protons, helium and carbon derived from PAMELA and Voyager 1 observations. *Astrophys Space Sci.* **2016**, *361*, 48. [CrossRef]
119. Boschini, M.J.; Della Torre, S.; Gervasi, M.; Grandi, D.; Jøhannesson, G.; Kachelriess, M.; La Vacca, G.; Masi, N.; Moskalenko, I.V.; Orlando, E.; et al. Solution of heliospheric propagation: Unveiling the local interstellar spectra of cosmic-ray species. *Astrophys. J.* **2017**, *840*, 115. [CrossRef] [PubMed]
120. Boschini, M.J.; Della Torre, S.; Gervasi, M.; Rancoita, P.G. The HELMOD model in the works for inner and outer heliosphere: From AMS to Voyager probes observations. *Adv. Space Res.* **2019**, *64*, 12–2476. [CrossRef]
121. Boschini, M.J.; Della Torre, S.; Gervasi, M.; Grandi, D.; Jøhannesson, G.; La Vacca, G.; Masi, N.; Moskalenko, I.V.; Pensotti, S.; Porter, T.A.; et al. Deciphering the local interstellar spectra of secondary nuclei with the Galprop/Helmod framework and a hint for primary lithium in cosmic rays. *Astrophys. J.* **2020**, *889*, 167. [CrossRef] [PubMed]
122. Moskalenko, I.V.; Strong, A.W.; Ormes, J.F.; Potgieter, M.S. Secondary antiprotons and propagation of cosmic rays in the Galaxy *Astrophys, J.* **2002**, *565*, 280–296. [CrossRef]
123. Strong, A.W.; Moskalenko, I.V.; Ptuskin, V.S. Cosmic-ray propagation and interactions in the galaxy. *Ann. Rev. Nucl. Part. Sci.* **2007**, *57*, 285–327. [CrossRef]
124. Scherer, K.; Fichtner, H.; Strauss, R.D.; Ferreira, S.E.S.; Potgieter, M.S.; Fahr, H.-J. On cosmic ray modulation beyond the heliopause: Where is the modulation boundary? *Astrophys. J.* **2011**, *735*, 128. [CrossRef]
125. Strauss, R.D.; Potgieter, M.S.; Ferreira, S.E.S.; Fichtner, H.; Scherer, K. Cosmic ray modulation beyond the heliopause: A hybrid modelling approach. *Astrophys. J. Lett.* **2013**, *765*, L18. [CrossRef]
126. Potgieter, M.S.; Nndanganeni, R.R. A local interstellar spectrum for galactic electrons. *Astropart. Phys.* **2013**, *48*, 25–29. [CrossRef]
127. Vos, E.E.; Potgieter, M.S. New modeling results of Galactic proton modulation during the minimum of solar cycle 23/24. *Astrophys. J.* **2015**, *815*, 119. [CrossRef]
128. Herbst, K.; Muscheler, R.; Heber, B. The new local interstellar spectra and their influence on the production rates of the cosmogenic radionuclides ^{10}Be and ^{14}C. *J. Geophys. Res. Space Physics* **2017**, *122*, 23–34. [CrossRef]
129. Schlickeiser, R.; Webber, W.R.; Kempf, A. Explanation of the local galactic cosmic ray energy spectra measured by Voyager 1. I. Protons. *Astrophys. J.* **2014**, *787*, 35. [CrossRef]
130. Drury, L.O.; Strong, A.W. Power requirements for cosmic ray propagation models involving diffusive reacceleration; estimates and implications for the damping of interstellar turbulence. *Astron. Astrophys.* **2017**, *597*, A117. [CrossRef]
131. Büsching, I.; Potgieter, M.S. The variability of the proton cosmic ray flux on the Sun's way around the galactic center. *Adv. Space Res.* **2008**, *42*, 504. [CrossRef]
132. Kissmann, R.; Werner, M.; Reimer, O.; Strong, A.W. Propagation in 3D spiral-arm cosmic-ray source distribution models and secondary particle production using PICARD. *Astropart. Phys.* **2015**, *70*, 39–53. [CrossRef]
133. Kopp, A.; Büsching, I.; Potgieter, M.S.; Strauss, R.D. A stochastic approach to Galactic proton propagation: Influence of the spiral arm structure. *New Astron.* **2014**, *30*, 32–37. [CrossRef]
134. Webber, W.R.; Higbie, P.R. Calculations of the propagated LIS electron spectrum which describe the cosmic ray electron spectrum below ~100 MeV measured beyond 122 AU at Voyager 1 and its relationship to the Pamela electron spectrum above 200 MeV. *arXiv* 2013.
135. Potgieter, M.S.; Vos, E.E.; Bisschoff, D.; Raath, J.L.; Boezio, M.; Munini, R.; Di Felice, V. Solar modulation of cosmic ray positrons in a very quiet heliosphere. In Proceedings of the International Cosimc Ray Conference (ICRC), Busan, Korea, 12–20 July 2017; Volume 44.
136. Adriani, O.; Barbarino, G.C.; Bazilevskaya, G.A.; Bellotti, R.; Boezio, M.; Bogomolov, E.A.; Bongi, M.; Bonvicini, V.; Borisov, S.; Bottai, S.; et al. Measurement of the isotopic composition of hydrogen and helium nuclei in cosmic rays with the PAMELA experiment. *Astrophys. J.* **2013**, *770*, 2. [CrossRef]
137. Marcelli, N.; Boezio, M.; Lenni, A.; Menn, W.; Aslam, O.P.M.; Bisschoff, D.; Ngobeni, M.D.; Potgieter, M.S.; Adriani, O.; Barbarino, G.C.; et al. Time dependence of the flux of Helium nuclei in cosmic rays measured by the PAMELA Experiment between 2006 July and 2009 December. *Astrophys. J.* **2020**, *893*, 145. [CrossRef]
138. Webber, W.R.; Kish, J.C.; Schrier, D.A. Asymmetries in the modulation of protons and helium nuclei over two solar cycles. In Proceedings of the 18th International Cosmic Ray Conference, Bangalore, India, 22 August–3 September 1983; Volume 3, pp. 35–38.
139. Potgieter, M.S.; Moraal, H. A drift model for the modulation of galactic cosmic rays. *Astrophys. J.* **1985**, *294*, 425–440. [CrossRef]
140. Potgieter, M.S.; Burger, R.A. The modulation of cosmic ray electrons, helium nuclei and positrons as predicted by a drift model with a simulated wavy neutral sheet. *Astron. Astrophys.* **1990**, *233*, 598–604. Available online: https://articles.adsabs.harvard.edu/pdf/1990A%26A...233..598P (accessed on 1 September 2021).

141. Raath, J.-L.; Potgieter, M.S. Charge-sign dependent modulation of cosmic ray electrons and positrons up to extreme solar maximum conditions. *PoS* **2017**, *ICRC2017*, 039. [CrossRef]

142. Langner, U.W.; de Jager, O.C.; Potgieter, M.S. On the local interstellar spectrum for cosmic ray electrons. *Adv. Space Res.* **2001**, *27*, 517–522. [CrossRef]

143. Langner, U.W.; Potgieter, M.S. Heliospheric modulation of cosmic rays computed with new local interstellar spectra. In Proceedings of the 27th International Cosmic Ray Conference, Hamburg, Germany, 7–15 August 2001; Volume 8, pp. 3686–3689.

144. Potgieter, M.S.; Langner, U.W. Heliospheric modulation of cosmic ray positrons and electrons: Effects of the heliosheath and solar wind termination shock. *Astrophys. J.* **2004**, *602*, 993–1001. [CrossRef]

145. Adriani, O.; Barbarino, G.C.; Bazilevskaya, G.A.; Bellotti, R.; Boezio, M.; Bogomolov, E.A.; Bongi, M.; Bonvicini, V.; Bottai, S.; Bruni, A.; et al. The PAMELA mission: Heralding a new era in precision cosmic ray physics. *Phys. Rep.* **2014**, *544*, 232–370. [CrossRef]

146. Vladimirov, A.E.; Digel, S.W.; Jóhannesson, G.; Michelson, P.F.; Moskalenko, I.V.; Nolan, P.L.; Orlando, E.; Porter, T.A.; Strong, A.W. GALPROP WebRun: An internet-based service for calculating galactic cosmic ray propagation and associated photon emissions. *Comp. Phys. Comm.* **2011**, *182*, 1156–1161. [CrossRef]

147. Cosmic Ray Database. Available online: https://cosmicrays.oulu.fi (accessed on 1 September 2021).

148. Ngobeni, M.D.; Potgieter, M.S. Modelling of galactic Carbon in an asymmetrical heliosphere: Effects of asymmetrical modulation conditions. *Adv. Space Res.* **2012**, *49*, 1660–1669. [CrossRef]

149. Webber, W.R.; Potgieter, M.S. A new calculation of the cosmic ray anti-proton spectrum in the galaxy and heliospheric modulation effects on this spectrum using a drift plus wavy current sheet model. *Astrophys. J.* **1989**, *344*, 779–785. [CrossRef]

150. Langner, U.W.; Potgieter, M.S. Differences in proton and antiproton modulation in the heliosphere. In Proceedings of the 27th International Cosmic Ray Conference, Hamburg, Germany, 7–15 August 2001; Volume 8, pp. 3657–3660.

151. Engelbrecht, N.E.; Moloto, K.D. An ab initio approach to antiproton modulation in the inner heliosphere. *Astrophys. J.* **2021**, *980*, 167. [CrossRef]

152. Adriani, O.; Bazilevskaya, G.A.; Barbarino, G.C.; Bellotti, R.; Boezio, M.; Bogomolov, E.A.; Bonvicini, V.; Bongi, M.; Bonechi, L.; Borisov, S.; et al. Measurement of the flux of primary cosmic ray antiprotons with energies of 60 MeV to 350 GeV in the PAMELA experiment. *JETP Lett.* **2013**, *96*, 621–627. [CrossRef]

153. Boezio, M.; Munini, R.; Adriani, O.; Barbarino, G.C.; Bazilevskaya, G.A.; Bellotti, R.; Bogomolov, E.A.; Bongi, M.; Bonvicini, V.; Bottai, S.; et al. The PAMELA experiment: A cosmic ray experiment deep inside the heliosphere. *PoS* **2017**, *ICRC2017*, 039. [CrossRef]

154. Mitchell, J.W.; Abe, K.; Anraku, K.; BESS Program. Precise measurements of the cosmic ray antiproton spectrum with BESS including the effects of solar modulation. *Adv. Space Res.* **2005**, *35*, 135–141. [CrossRef]

155. Menn, W.; Adriani, O.; Barbarno, G.C.; Bazilevskaya, G.A.; Bellotti, R.; Boezio, M.; Bogomolov, E.A.; Bonechi, L.; Bongi, M.; Bonvicini, V.; et al. The PAMELA space experiment. *Adv. Space Res.* **2013**, *51*, 209–218. [CrossRef]

156. Moraal, H.; Potgieter, M.S. Solutions of the spherically-symmetric cosmic-ray transport equation in interplanetary space. *Astrophys. Space Sci.* **1982**, *84*, 519. [CrossRef]

157. Munini, R.; Boezio, M.; Bisschoff, D.; Marcelli, N.; Ngobeni, M.D.; Aslam, O.P.M.; Potgieter, M.S. Solar modulation of galactic-cosmic ray antiprotons. *PoS* **2021**, *ICRC2021*, 1328. [CrossRef]

158. Tomassetti, N.; Orcinha, M.; Barão, F.; Bertucci, B. Evidence for a time lag in solar modulation of galactic cosmic rays. *Astrophys. J.* **2017**, *849*, 132. [CrossRef]

159. Korsmeier, M.; Donato, F.; Fornengo, N. Prospects to verify a possible dark matter hint in cosmic antiprotons with antideuterons and antihelium. *Phys Rev. D.* **2018**, *97*, 103011. [CrossRef]

160. Fuke, H.; Maeno, T.; Abe, K.; Haino, S.; Makida, Y.; Matsuda, S.; Matsumoto, H.; Mitchell, J.W.; Moiseev, A.A.; Nishimura, J.; et al. Search for cosmic-ray antideuterons. *Phys. Rev. Lett.* **2005**, *95*, 081101. [CrossRef]

161. Sakai, K.; Abe, K.; Fuke, H.; Haino, S.; Hams, T.; Hasegawa, M.; Kim, K.C.; Lee, M.H.; Makida, Y.; Mitchell, J.W.; et al. New result of antideuteron search in BESS-Polar II. *PoS* **2021**, *ICRC2021*, 0123. [CrossRef]

162. Dimiccoli, F.; Battiston, R.; Kanishchev, K.; Nozzoli, F.; Zuccon, P. Measurement of cosmic deuteron flux with the AMS-02 detector. *J. Phys. Conf. Ser.* **2020**, *1548*, 012034. [CrossRef]

163. Lenni, A.; Boezio, M.; Munini, R.; Menn, W.; Marcelli, N.; Potgieter, M.S.; Bisschoff, D.; Ngobeni, M.D.; Aslam, O.P.M. Study of the solar modulation for the cosmic ray isotopes with the PAMELA experiment. *PoS* **2021**, *ICRC2021*, 1310. [CrossRef]

164. Gomez Coral, D.M.; Menchaca-Rocha, A. SM antideuteron background to indirect dark matter signals in galactic cosmic rays. *J. Phys. Conf. Ser.* **2020**, *1602*, 012005. [CrossRef]

165. Adriani, O.; Barbarino, G.C.; Bazilevskaya, G.A.; Bellotti, R.; Boezio, M.; Bogomolov, E.A.; Bongi, M.; Bonvicini, V.; Bottai, S.; Bruno, A.; et al. Measurements of cosmic ray hydrogen and helium isotopes with the PAMELA experiment. *Astrophys. J.* **2016**, *818*, 68. [CrossRef]

166. Abe, K.; Fuke, H.; Haino, S.; Hams, T.; Hasegawa, M.; Horikoshi, A.; Itazaki, A.; Kim, K.C.; Kumazawa, T.; Kusumoto, K.A.; et al. (BESS program). Search for antihelium with the BESS-polar spectrometer. *Phys. Rev. Lett.* **2012**, *108*, 131301. [CrossRef]

167. von Doetinchem, P.; Perez, K.; Aramaki, T.; Baker, S.; Barwick, S.; Bird, R.; Boezio, M.; Boggs, S.E.; Cui, M.; Datta, A.; et al. Cosmic-ray antinuclei as messengers of new physics: Status and outlook for the new decade. *J. Cosm. Astrop. Phys.* **2020**, *8*, 35. [CrossRef]

168. Heber, B.; Dröge, W.; Ferrando, P.; Haasbroek, L.J.; Kunow, H.; Müller-Mellin, R.; Paizis, C.; Potgieter, M.S.; Raviart, A.; Wibberenz, G. Spatial variation of > 40 MeV/n nuclei fluxes observed during the Ulysses rapid latitude scan. *Astron. Astrophys.* **1996**, *316*, 538–546. Available online: https://articles.adsabs.harvard.edu/pdf/1996A%26A...316..538H (accessed on 1 September 2021).
169. Ferreira, S.E.S.; Potgieter, M.S.; Heber, B.; Fichtner, H.; Wibberenz, G. Latitudinal transport effects on the modulation of a few-MeV cosmic ray electrons from solar minimum to solar maximum. *J. Geophys. Res.* **2004**, *109*, A02115. [CrossRef]
170. Strong, A.W.; Orlando, E.; Jaffe, T.R. The interstellar cosmic-ray electron spectrum from synchrotron radiation and direct measurements. *Astron. Astrophys.* **2011**, *534*, A54. [CrossRef]
171. Orlando, E. Imprints of cosmic rays in multifrequency observations of the interstellar emission. *Mon. Not. R. Astron. Soc.* **2018**, *475*, 2724. [CrossRef]
172. Phan, V.H.M.; Schulze, F.; Mertsch, P.; Recchia, S.; Gabici, S. Stochastic fluctuations of low-energy cosmic rays and the interpretation of Voyager data. *Phys. Rev. Lett.* **2021**, *127*, 141101. [CrossRef]
173. Bieber, J.W.; Burger, R.A.; Engel, R.; Gaisser, T.K.; Roesler, S.; Stanev, T. Antiprotons at solar maximum. *Phys. Rev. Lett.* **1999**, *83*, 674. [CrossRef]

 *physics*

Article

Dissipative Ion-Acoustic Solitary Waves in Magnetized κ-Distributed Non-Maxwellian Plasmas

Sharmin Sultana [1,*] and Ioannis Kourakis [2]

[1] Department of Physics, Jahangirnagar University, Savar, Dhaka 1342, Bangladesh
[2] Department of Mathematics, College of Science and Engineering, Khalifa University of Science and Technology, Abu Dhabi 127788, United Arab Emirates; ioanniskourakissci@gmail.com
* Correspondence: ssultana@juniv.edu

Abstract: The propagation of dissipative electrostatic (ion-acoustic) solitary waves in a magnetized plasma with trapped electrons is considered via the Schamel formalism. The direction of propagation is assumed to be arbitrary, i.e., oblique with respect to the magnetic field, for generality. A non-Maxwellian (nonthermal) two-component plasma is considered, consisting of an inertial ion fluid, assumed to be cold for simplicity, and electrons. A (kappa) κ-type distribution is adopted for the electron population, in addition to particle trapping taken into account in phase space. A damped version of the Schamel-type equation is derived for the electrostatic potential, and its analytical solution, representing a damped solitary wave, is used to examine the nonlinear features of dissipative ion-acoustic solitary waves in the presence of trapped electrons. The influence of relevant plasma configuration parameters, namely the percentage of trapped electrons, the electron superthermality (spectral) index, and the direction of propagation on the solitary wave characteristics is investigated.

Keywords: dissipative solitary waves; magnetized plasma; superthermal trapped electrons; kappa distribution; Schamel equation; oblique propagation of electrostatic plasma waves; suprathermals

Citation: Sultana, S.; Kourakis, I. Dissipative Ion-Acoustic Solitary Waves in Magnetized κ-Distributed Non-Maxwellian Plasmas. *Physics* **2022**, *4*, 68–79. https://doi.org/10.3390/physics4010007

Received: 21 October 2021
Accepted: 3 January 2022
Published: 20 January 2022

Publisher's Note: MDPI stays neutral with regard to jurisdictional claims in published maps and institutional affiliations.

1. Introduction

The occurrence of highly energetic particles is a ubiquitous feature in space plasmas (e.g., in the ionosphere, the auroral zone, solar wind, and at the mesosphere, etc.) and in laboratory plasmas [1–12]. The velocity distribution in such plasmas may deviate from the usual thermal Maxwellian distribution, developing a long-tail for high-velocity arguments due to an excess in the fast (superthermal) part of the population; such a behavior is effectively modeled by a (kappa) κ-type distribution function [1,4,12–19]. The kappa distribution function was initially postulated by Vasyliunas [1] in an effort to reproduce the observed power-law dependence at high velocities [17,20,21]. Since then, a large number of studies adopted kappa distributions, combining theoretical [18,19,22,23], computational [24], and even experimental [25,26] approaches to study the effect of superthermal particle populations on wave dynamics in different plasma environments.

Particle "trapping", i.e., the fact that a portion of, for example, the electron population remains confined in a finite region—thus generating vortices—in phase space, is an intrinsic characteristic of plasma dynamics, often overlooked in studies based on basic fluid theory. Phase-space structures, known as "electron-holes" are thus formed due to particles trapped in the wave potential. This mechanism, initially predicted via kinetic theory [27–29], was later observed in space and in the laboratory [30–36], and it was also shown to occur spontaneously in computer simulations [37]. Of particular relevance to current study is the fact that Simpson et al. [38] reported the presence of trapped electrons in the Saturnian magnetic field, an environment characterized by the existence of κ-distributed electrons with values of $\kappa \simeq 2$–4, as confirmed by Schippers et al. [34]. It is therefore important to consider the effect of particle nonthermality and trapping effect simultaneously to explore the properties of different electrostatic modes.

As regards the theoretical modeling of particle trapping, Schamel's original papers [27,28] showed that trapped particles led to a vortex-like electron distribution, and the kinetic model was shown to be associated with a modified version of the known integrable Korteweg–de Vries (KdV) partial differential equation. The "Schamel equation" [28,39–42] describes the evolution of nonlinear electrostatic waves under the influence of a fractional nonlinearity (in contrast with the standard KdV theory where quadratic nonlinearity is dominant). A number of theoretical studies followed [43–49] in an effort to investigate the properties of nonlinear waves (solitary waves, shocks) in the presence of trapped particles, using first principles.

The combined effect of electron superthermality and phase-space trapping was first considered by Williams et al. [40], who adopted the Schamel equation approach to model and characterize ion-acoustic solitary waves in an unmagnetized electron-ion plasma with κ-distributed electron populations subject to trapping. Following that study, the combined effect of electron superthermality and trapping was considered by Sultana and coworkers [41,42] (on ion-acoustic modes in collisionless plasmas) and by Hassan et al. [50], who investigated the nonlinear features of electron-acoustic waves in a magnetized plasma and considered the combined effect of electron trapping and electron superthermality. The study led by Hassan et al. [50] focused on electron-acoustic waves, a mode known to occur exclusively in the simultaneous presence of two distinct electron populations (usually referred to as the 'cold' and the 'hot' electrons), as it relies in fact on the inertia being provided by the cold electron component and the restoring force being provided by their hot counterpart. The associated (electronic) dynamical frequency scales are clearly distinct from the (slower, ionic) scales that are typical of the study presented here.

To our best knowledge, there is no rigorous and systematic study of the nonlinear propagation of the ion-acoustic waves in a magnetized *collisional* plasma in the presence of trapped κ-nonthermal electrons. The investigation at hand is therefore an attempt to fill in this gap by presenting a rigorous and systematic study of the characteristics of ion-acoustic waves propagating in a magnetized κ-nonthermal plasma [41], taking into account the combined impact of electron trapping and of a suprathermal electron distribution, in account of the inherent plasma collisionality. The main focus here is to investigate the influence of particle trapping on the dynamics of dissipative solitary waves, and also to analyze the effect of the ambient magnetic field and its interplay with wave damping and how these affect the characteristics of obliquely propagating ion-acoustic solitary excitations.

This article is organized as follows. The basic formalism is presented in the following Section 2. A dissipative version of the Schamel equation is derived via a multiscale perturbative approach, and the detail about the nonlinear, dispersion, and dissipative term, is discussed in Section 3. The propagation nature (basic features) of dissipative ion-acoustic waves for different relevant plasma (configuration) parameters is studied numerically in Section 4. Finally, the results obtained are summarized in the concluding Section 5.

2. Basic Plasma-Fluid-Dynamic Formalism

An electron-ion plasma is considered here being embedded in a uniform magnetic field directed along the z-axis, i.e., $\mathbf{B}_0 = B_0 \hat{z}$. Due to their large mass (relative to the electrons), inertial ions are modeled as a cold fluid, i.e., their thermal pressure is neglected for simplicity. At the ionic scale of interest, the electron inertia may be neglected: the electrons are assumed to deviate from thermal equilibrium, and hence, a κ-type distribution will be explicitly adopted to model their distribution. For the purpose of this analysis, the combined effect of electron trapping and superthermality is considered, following the steps outlined in Ref. [40].

Charge neutrality at equilibrium imposes: $z_i n_{i0} - n_{e0} = 0$, where n_{i0} and n_{e0} denote the unperturbed ion and electron number densities, respectively, while z_i is the charge state of the ion component (e.g., 1, 2,...; the value of z_i is left arbitrary here, for generality).

We are interested in modeling the dynamics of ion-acoustic excitations, whose phase speed v_{ph} may well exceed the ion thermal speed, but is far smaller than the electron thermal speed. The following fluid evolution equations are considered:

$$\frac{\partial n_i}{\partial T} + \nabla.(n_i \mathbf{u}_i) = 0, \tag{1}$$

$$\frac{\partial \mathbf{u}_i}{\partial T} + (\mathbf{u}_i.\nabla)\mathbf{u}_i = -\frac{z_i e}{m_i}\nabla\Phi$$

$$+\frac{z_i e B_0}{m_i}(\mathbf{u}_i \times \hat{z}) - v_i \mathbf{u}_i, \tag{2}$$

$$\nabla^2\Phi = 4\pi e(n_e - z_i n_i), \tag{3}$$

where n_i (n_e) denotes the number density of the ion (electron) species, $\mathbf{u}_i$ is the ion fluid speed, Φ is the electrostatic wave potential (all these quantities are dynamic functions of space and time), and e is electron charge. An ad hoc damping term is introduced in the fluid equation of motion (momentum conservation equation) to account for, e.g., ion-neutral collisions; the collision frequency v_i was defined to this effect.

The combined effect of electron trapping and deviation from Maxwell-type equilibrium was studied analytically in Ref. [40]; the tedious algebraic procedure need not be reproduced here, but the main steps are summarized. A modified κ-distribution function, effectively taking into account the trapped part of the electron population (i.e., for electrons trapped in the wave potential if their energy $E_e < 0$) is given by [40]

$$f(v,\phi) = \frac{\Gamma(\kappa)}{\sqrt{2\pi}(\kappa - 3/2)^{1/2}\Gamma(\kappa - 1/2)}$$

$$\times \left[1 + \beta\left(\frac{v^2/2 - \phi}{\kappa - 3/2}\right)\right]^{-\kappa} \text{ for } E_e \leq 0. \tag{4}$$

Here, v is the velocity, ϕ is the electrostatic potential, and κ is the superthermality index (measures deviation from the Maxwell–Boltzmann distribution), while β (<1) quantifies the efficiency of electron trapping. The known vortex-type distribution for trapped Maxwellian electrons is recovered in the limit $\kappa \to \infty$. The number density of the electrons is obtained by integration as [40]

$$n_e(\phi) = \int_{-\infty}^{-\sqrt{2\phi}} f_e^\kappa(v,\phi)\,dv + \int_{-\sqrt{2\phi}}^{\sqrt{2\phi}} f(v,\phi)\,dv$$

$$+ \int_{\sqrt{2\phi}}^{\infty} f_e^\kappa(v,\phi)\,dv, \tag{5}$$

where $f_e^\kappa(v,\phi)$ is the κ-distribution function for the free electrons; details can be found in Ref. [17].

By normalizing all variables, one obtains the following system of (dimensionless) equations:

$$\frac{\partial n}{\partial t} + \tilde{\nabla}.(n\mathbf{u}) = 0, \tag{6}$$

$$\frac{\partial \mathbf{u}}{\partial t} + (\mathbf{u}.\tilde{\nabla})\mathbf{u} = -\tilde{\nabla}\phi + \Omega_c(\mathbf{u} \times \hat{z}) - v\mathbf{u}, \tag{7}$$

$$\tilde{\nabla}^2\phi \simeq 1 - n + a_1\phi + a_2\phi^{3/2} + a_3\phi^2, \tag{8}$$

where $n = n_i/n_{i0}$, $u = [m_i u_i^2/(z_i T_e)]^{1/2}$ with m_i being the mass of ion, T_e being the electron temperature (Boltzmann's constant k_B is omitted where obvious), the electrostatic potential $\phi = e\Phi/T_e$, $\lambda_D = (T_e/4\pi e^2 z_i n_{i0})^{1/2}$, $t = \omega_{p,i}T$ (where $\omega_{p,i} = (4\pi e^2 z_i^2 n_{i0}/m_i)^{1/2}$ is the ion plasma frequency, and T is the inverse of the ion plasma frequency), $\Omega_c = \omega_{c,i}/\omega_{p,i}$

(where $\omega_{c,i} = z_i e B_0 / m_i$) is the reduced cyclotron frequency, and $\nu = \nu_i / \omega_{p,i}$ (note that all frequencies were scaled by the ion plasma frequency for convenience). The information related to electron trapping, for κ-distributed electrons, is "hidden" in the coefficients a_1, a_2, a_3 entering the normalized expression of Poisson Equation (8), which are given by [40]

$$a_1 = \frac{2\kappa - 1}{2\kappa - 3}, \quad a_2 = \frac{8\sqrt{2/\pi}(\beta - 1)\,\kappa\,\Gamma(\kappa)}{3(2\kappa - 3)^{3/2}\,\Gamma(\kappa - 1/2)},$$

$$a_3 = \frac{4\kappa^2 - 1}{2(2\kappa - 3)^2}. \tag{9}$$

Once a solution for the electrostatic potential ϕ is formally obtained, the trapped electron population's density obtained from Equation (5) can be expressed as [40]

$$n_e \simeq 1 + a_1\phi + a_2\phi^{3/2} + a_3\phi^2 \cdots, \tag{10}$$

which to be substituted into Poisson Equation (8). The known analogous expression for the trapped electron number density in the case of Maxwellian plasma [28] is readily recovered here, upon considering the limit $\kappa \to \infty$ in the latter relation. On the other hand, the limit of Equation (10) as $\beta \to 1$ leads to the classical expression for κ-distributed electrons; see e.g., Ref. [51] and elsewhere. Finally, the Maxwellian limit for free electrons, viz. $a_1 = 2a_3 = 1$, is recovered by considering $\beta \to 1$ and $\kappa \to \infty$, reducing the electron density dependence to $e^\phi \simeq 1 + \phi + \phi^2/2$, as expected.

The closed system of Equations (6)–(8), describes the evolution of the plasma (fluid) state variables, and forms the basis of the analysis here.

3. A Schamel Equation for Damped Ion-Acoustic Waves (IAWs)

To model small-amplitude ion-acoustic excitations (dissipative solitary waves) within the model under consideration, one needs to proceed by defining a set of stretched coordinates as

$$\xi = \epsilon^{1/4}\left(l_x\hat{x} + l_y\hat{y} + l_z\hat{z} - v_p t\right), \quad \tau = \epsilon^{3/4} t, \tag{11}$$

where ϵ ($\ll 1$) is a small parameter that measures the strength of the nonlinearity, v_p is a constant to be determined (in fact, representing the phase speed, scaled by the ion sound speed, $c_0 = (z_i T_e / m_i)^{1/2}$), and l_x, l_y and l_z, are directional cosines of the wave vector **k** along the x, y and z axes, respectively (for instance, $l_z = (\mathbf{k} \cdot \hat{z})/k$), hence $l_x^2 + l_y^2 + l_z^2 = 1$. Let us recall that the position variables x, y and z are all normalized by λ_D, while τ is normalized by the ion plasma period $\omega_{p,i}^{-1}$. The above *Ansatz*, which was first introduced in Ref. [28] and then later adopted by various authors (e.g., [40,52]), essentially describes a Galilean transformation into a slowly varying moving frame, wherein the time variation of the structure is even slower in time.

The dependent variables n, **u** and ϕ may now be expanded near the equilibrium states as power series of ϵ as follows:

$$\left.\begin{aligned}
n &= 1 + \epsilon n_1 + \epsilon^{3/2} n_2 + \dots, \\
u_x &= \epsilon^{5/4} u_{1,x} + \epsilon^{3/2} u_{2,x} + \dots, \\
u_y &= \epsilon^{5/4} u_{1,y} + \epsilon^{3/2} u_{2,y} + \dots, \\
u_z &= \epsilon\, u_{1,z} + \epsilon^{3/2} u_{2,z} + \dots, \\
\phi &= \epsilon\phi_1 + \epsilon^{3/2}\phi_2 + \dots
\end{aligned}\right\} \tag{12}$$

To close the series expansion of the variables, a weak dissipation [53–55] to be considered due to ion-neutral collisions by assuming that the damping coefficient scales as $\nu = \epsilon^{3/4}\nu_0$.

Let us now proceed by substituting expansions (11) and (12) into the considered fluid plasma model Equations (6)–(8) and collecting various terms arising in each order in ϵ.

The phase speed v_p is obtained as a compatibility constraint upon considering the lowest order contributions in ϵ from each of the equations; the resulting expression reads:

$$v_p = l_z / \sqrt{a_1}. \tag{13}$$

This expression for the phase speed v_p depends on the angle θ (via $l_z = \cos\theta$) and on the electron superthermality index κ. Considering parallel propagation ($l_z = 1$), the known expression for the ion sound speed in nonmagnetized plasma [22,40] is recovered as expected. Recovering dimensions for a minute, for physical transparency, Equation (13) leads to

$$V_p = \frac{\omega}{k} = \lambda_D \, \omega_{p,i} \, \frac{l_z}{\sqrt{a_1}} = \left(\frac{z_i T_e}{m_i}\right)^{1/2} \left(\frac{2\kappa - 3}{2\kappa - 1}\right)^{1/2} l_z, \tag{14}$$

where ω, k and V_p here denote the wave (angular) frequency, the wavenumber and the phase speed (in the dimensional form), respectively. The acoustic ("sound") speed is thus recovered for infinitely large κ, while a lower value (i.e., predicting slower solitary waves) is predicted for small κ, in agreement with earlier theoretical predictions [22] and with space observations [21].

The variation of the ion-acoustic phase speed, v_p, versus the electron's superthermality index, κ, is depicted in Figure 1, suggesting a slower phase speed in a plasma with significant portion of the electrons in the superthermal region (i.e., lower values of κ), when compared with the case of thermal (Maxwellian) electron. The phase speed is higher for parallel propagation than for oblique propagation, as shown in Figure 1. As expected, the curve tends to unity, asymptotically ($v_p \to 1$) for infinite kappa and for parallel propagation, prescribing the acoustic speed (in electron-ion plasma) as the phase speed in the Maxwellian limit.

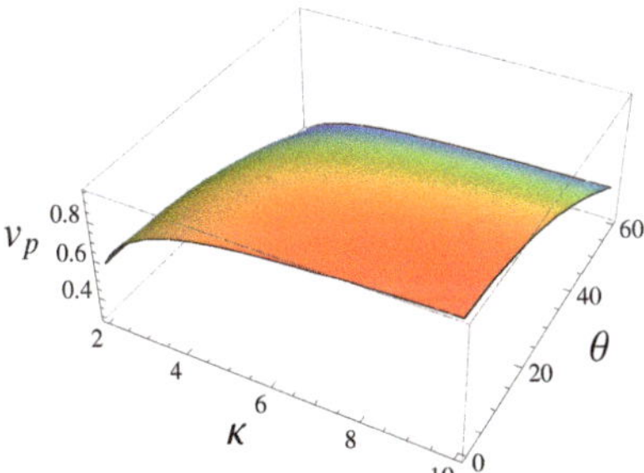

Figure 1. Phase speed, v_p, versus electron superthermality index, κ, and obliqueness angle, $\theta = \cos^{-1} l_z$, where l_z is a directional cosine of the wave vector $\mathbf{k}$ along z-axis.

The perpendicular (x and y) components of the electric field related drift of the ion fluid, in terms of the electric potential ϕ_1, can be obtained by separating the y and x-components of the momentum equation, respectively, as

$$u_{1,x} = -\frac{l_y}{\Omega_c} \frac{\partial \phi_1}{\partial \xi}, \tag{15}$$

and

$$u_{1,y} = \frac{l_x}{\Omega_c} \frac{\partial \phi_1}{\partial \xi}. \tag{16}$$

Following an analogous procedure, the parallel (z) component of the ion fluid velocity is obtained as

$$u_{1,z} = \frac{l_z}{v_p} \phi_1, \tag{17}$$

Finally, in leading order, the perturbation of the ion density $n \simeq 1 + \epsilon n_1 + \mathcal{O}(\epsilon^{3/2})$ is obtained as

$$n_1 = \left(\frac{l_z}{v_p}\right)^2 \phi_1 .\tag{18}$$

The next order in ϵ (obtained upon separating $\epsilon^{3/2}$ from the momentum equation) leads to the x and y-components of the second order drift velocity of the ion fluid in the form

$$u_{2,x} = \frac{l_x v_p}{\Omega_c^2} \frac{\partial^2 \phi_1}{\partial \xi^2} ,\tag{19}$$

$$u_{2,y} = \frac{l_y v_p}{\Omega_c^2} \frac{\partial^2 \phi_1}{\partial \xi^2} ,\tag{20}$$

Following the same procedure (i.e., separating coefficient of $\epsilon^{7/4}$ from the continuity and the z-component of the momentum equations, and then $\epsilon^{3/2}$ from the Poisson equation) and eventually eliminating n_2, $u_{2,z}$, and ϕ_2, one is led to a nonlinear partial differential equation (PDE) in the form

$$\frac{\partial \psi}{\partial \tau} + A \psi^{1/2} \frac{\partial \psi}{\partial \xi} + B \frac{\partial^3 \psi}{\partial \xi^3} + C\psi = 0 ,\tag{21}$$

where, for brevity, the leading contribution of the electrostatic potential is denoted by $\psi = \phi_1$.

Equation (21), which bears the structure of the original Schamel equation [28], with the addition of the last term (arising due to collisions being taken into account), represents an evolution equation for an electrostatic potential disturbance, $\phi \simeq \epsilon \psi + \mathcal{O}(\epsilon^3/2)$, in a region where the trapped electrons are present. The algebraic scheme implied is obvious: once ψ is obtained from Equation (21), the leading contributions for the ion density and for the ion fluid speed (three) components can be obtained from (four) Equations (15)–(18).

The nonlinearity coefficient A, which is responsible for wave steepening, is given by

$$A = -\frac{3}{4} \frac{v_p^3}{l_z^2} a_2 = -\frac{3}{4} \frac{a_2}{a_1^{3/2}} l_z .\tag{22}$$

The nonlinearity dependence enters via both θ and κ, as expected: this is seen in Figure 2a.

On the other hand, the dispersion coefficient B—which is responsible for wave broadening—is given by

$$B = \frac{v_p^3}{2l_z^2} \left(1 + \frac{1 - l_z^2}{\Omega_c^2}\right) .\tag{23}$$

The expression for the coefficient B can be simplified upon setting $v_p^3/2l_z^2 = l_z/2a_1^{3/2}$, showcasing the dependence of B on the propagation angle (via l_z) and on κ (via a_1), as shown in Figure 2b. The influence of the magnetic field (via Ω_c) disappears in the case of parallel propagation ($l_z = 1$), thus recovering a one-dimensional damped Schamel equation for unmagnetized plasma (this was intuitively expected, since the Larmor force has no component in the direction of the magnetic field, and thus does not affect parallel wave propagation).

Finally, the dissipative term C is given by

$$C = \frac{v_0}{2} ,\tag{24}$$

as imposed by compatibility requirements (i.e., balancing various terms occurring in the same order in ϵ).

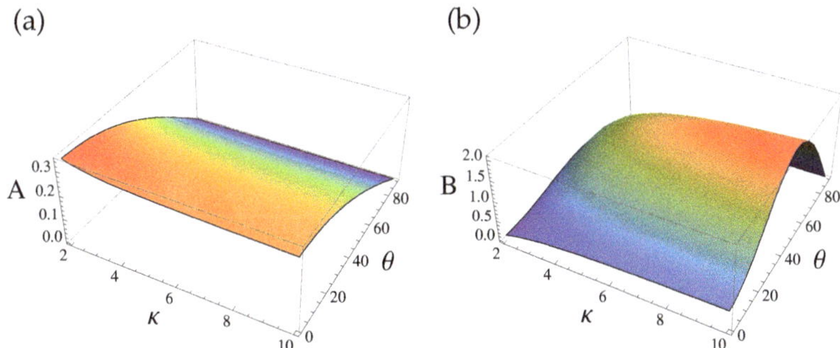

Figure 2. Nonlinearity term A (**a**) and dispersion term B (**b**) versus κ and obliqueness angle θ for $\beta = 0.5$ and $\Omega_c = 0.3$, where β denotes the efficiency of electron trapping and Ω_c is the ratio of the the the reduced cyclotron frequency to the ion plasma frequency.

Interestingly, both A and B vanish for perpendicular propagation (i.e., for $l_z = \cos(\pi/2) = 0$), as seen in Figure 2. On the other hand, considering parallel propagation ($l_z = \cos(0) = 1$) and the Maxwellian limit (infinite κ), one finds $v_p = 1$ (acoustic speed), while the two coefficients become $A = (1 - \beta)/\sqrt{\pi}$ and $B = 1/2$, thus recovering exactly the analytical form of the original Schamel equation [28]. Figure 2 examines the influence of superthermality index κ and the obliqueness θ on the nonlinear term A and the dispersion term B. One can see that the (value of the) nonlinearity term increases, while the dispersive term decreases, if one assumes stronger deviation from the Maxwellian equilibrium, i.e., for small value of the κ parameter. On the other hand, for fixed κ, the nonlinearity term A attains its highest value for parallel propagation ($\theta = 0$), as shown in Figure 2.

The dispersive term shows slightly more perplex behavior by increasing with growing θ, reaching a maximum, and then going to zero—as said above—for $\theta = \pi/2$. It is evident in Equation (22) and in Figure 3a that $a_2 \to 0$, and, hence, $A \to 0$ in the limit $\beta \to 1$; therefore, the nonlinear Equation (21) is not valid in the absence of trapped electrons. On the other hand, the dispersive term decreases with stronger magnetic field, as seen in Figure 3b.

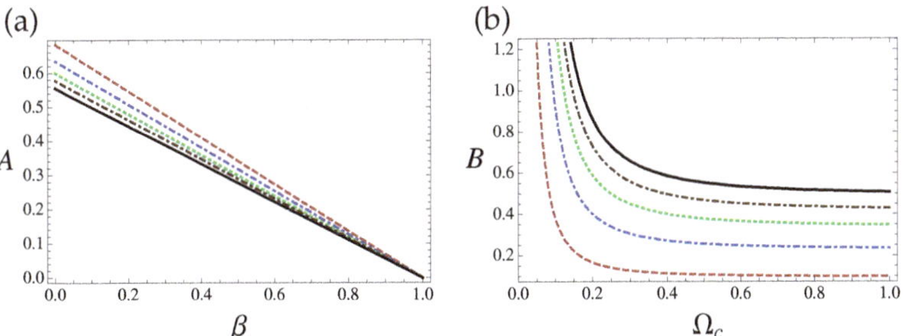

Figure 3. (**a**) Nonlinearity term A versus β. A does not depend on the magnetic field. (**b**) Dispersion term B versus Ω_c. B does not depend on β. $\theta = 10°$ is assumed and $\kappa = 2, 3, 5, 10$, and ∞ (top to bottom in (**a**) and bottom to top in (**b**)).

4. Parametric Analysis

In this Section, we are interested in tracing the influence of different plasma configuration parameters, such as the superthermality (spectral) index, κ, the electron trapping parameter, β, the collisional term, ν, the obliqueness angle, θ, and the ambient magnetic field (strength), B_0, on the propagation characteristics of ion-acoustic solitary waves within the model under consideration. To see how these plasma configuration parameters affect the dynamical properties of solitary waves, first, dissipative effect is assumed to be negligible: $\nu \to 0$. The damped Schamel Equation (21) then reduces to a κ-dependent form of the Schamel-type equation [28], which possesses a solitary wave solution in the form [28,40]

$$\psi(\xi, \tau) = \psi_0 \, \text{sech}^4\left(\frac{\xi - u_0\tau}{\delta}\right), \tag{25}$$

representing a pulse-shaped excitation with amplitude $\psi_0 = (15u_0/8A)^2$, width $\delta = \sqrt{16B/u_0}$ and velocity u_0 in the moving reference frame (note that the actual speed in the laboratory frame is $v_p + \epsilon u_0$, so the pulse structure is superacoustic; recall that v_p is essentially the sound speed). The product $\psi_0\delta^4 = (30B/A)^2$ is constant for a given (fixed) set of plasma parameters, that is in fact independent of u_0. The electric field $E \, (=-\nabla\psi)$ which is associated with the solitary potential in Equation (25) is of the form:

$$E = -E_0 \, \text{sech}^4\left(\frac{\xi - u_0\tau}{\delta}\right) \tanh\left(\frac{\xi - u_0\tau}{\delta}\right), \tag{26}$$

where $E_0 = 225 \, u_0^2/(16A^2\delta)$ (this is actually the norm of the vector; the respective components are regulated by the direction cosines $l_{x,y,z}$; recall that $l_x^2 + l_y^2 + l_z^2 = 1$). The pulse form for the potential is shown in Figure 4, while the associated bipolar electric field structures are shown in Figure 5b,d.

To trace the dynamical evolution of the solitary wave solution and to elucidate the role of different plasma configuration parameters on the properties of (nonlinear) solitary waves, the nonlinear damped Schamel Equation (21) was solved numerically by using the Wolfram MATHEMATICA^{TM} software package, adopting the solitary wave solution (25) as initial condition.

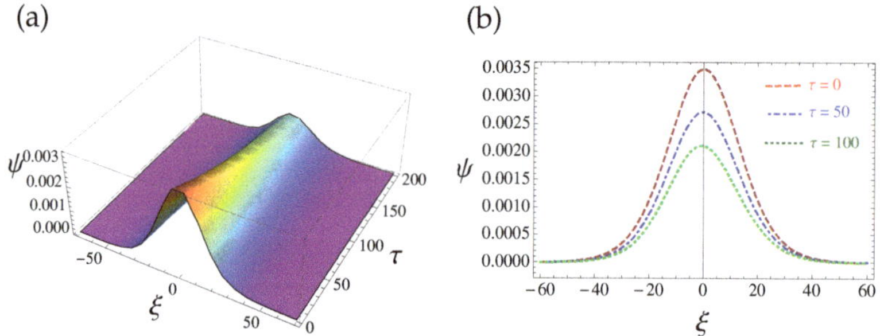

Figure 4. (**a**) The ion-acoustic solitary potential pulse ψ versus the space coordinate ξ and the time τ. (**b**) Potential pulse versus ξ at different time instants. Here, $\beta = 0.5$, $\kappa = 3$, $\Omega_c = 0.2$, $\nu = 0.01$, $\theta = 10°$, and $u_0 = 0.01$ are used. See text for details.

The time evolution of dissipative ion-acoustic solitary potential waveforms (pulses) is shown in Figure 4. The analytical solution (25) was adopted at an initial condition, and then Equation (21) was solved numerically for $\nu \neq 0$. As expected, the pulse amplitude decreases in time due to the damping, as illustrated in Figure 4.

The influence of the trapping parameter β and also of the superthermality (spectral) index κ was investigated numerically; a snapshot at $\tau = 50$ (dimensionless units) is shown in Figure 5. Here, $\Omega_c = 0.2$, $\nu = 0.01$, $u_0 = 0.01$, $\tau = 50$, and $\theta = 10°$ are used. One can see that both the height and the width of the solitary wave are affected by the trapping parameter β (Figure 5a) and by the value of κ (Figure 5c). As β increases, the waves become taller in amplitude, but the width remains unchanged; see Figure 5a. An increase in plasma superthermality (that is, a smaller value of κ) results in shorter and narrower solitary waves, as seen in Figure 5c. These results recover the theoretical predictions of Ref. [41] for the collision-free case, i.e., for $\nu = 0$.

A similar investigation is shown in Figure 6, where the solitary wave solution (25) was obtained numerically for $\kappa = 3$, $\beta = 0.5$, $u_0 = 0.01$, $\tau = 50$, and $\theta = 10°$. The role of the magnetic field (strength) B_0 (via Ω_c) is shown in Figure 6a. As B_0 increases, the width of the solitary wave decreases, while the amplitude is unaffected.

Finally, in Figure 6b, various values of the collisional parameter ν are considered (keeping all other values fixed). As expected, the pulse amplitude decreases with higher dissipation.

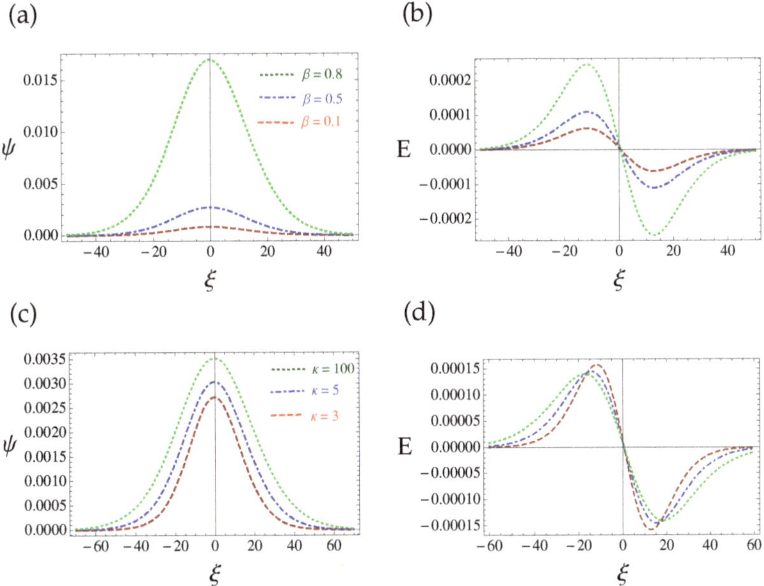

Figure 5. (a) Effect of trapping parameter β on electrostatic solitary wave (pulse) profile and (b) associated electric field structures for $\kappa = 3$. (c) Effect of superthermality index κ on solitary pulse and (d) associated electric field for $\beta = 0.5$. Here, $\Omega_c = 0.2$, $\nu = 0.01$, $u_0 = 0.01$, $\tau = 50$, and $\theta = 10°$ are used.

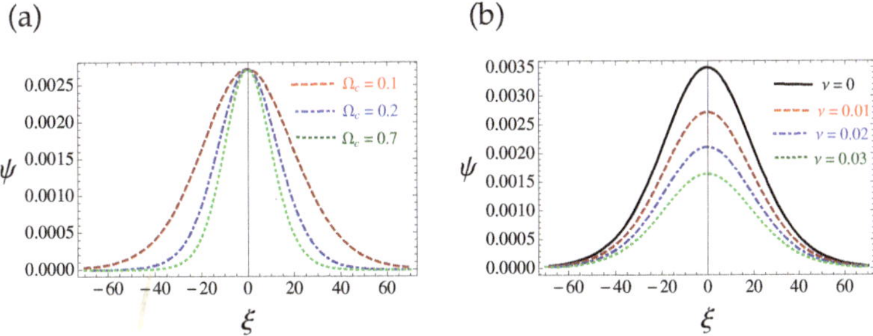

Figure 6. Effect of (a) external magnetic field Ω_c (for $\nu = 0.01$) and of (b) collisionality parameter ν (for $\Omega_c = 0.1$) on electrostatic solitary waves (pulses) for $\kappa = 3$, $\beta = 0.5$, $u_0 = 0.01$, $\tau = 50$, and $\theta = 10°$.

5. Conclusions

In this paper, the basic features of damped ion-acoustic solitary waves were investigated in the presence of trapped superthermal electrons described by a κ-type (non-Maxwellian) distribution [40]. The effect of ion-neutral collisions was also taken into account, leading to wave damping as expected.

The reductive perturbation approach were adopted to derive a nonlinear Schamel-type partial differential equation featuring an additional damping term. The solitary wave solution of the standard (nondissipative) Schamel equation was used to solve the damped Schamel equation numerically and to analyze the basic features of dissipative ion-acoustic solitary waves. The amplitude of solitary waves was found to decrease, while their width becomes narrower with an increase in superthermality (i.e., for a stronger deviation from Maxwellian equilibrium). The proportion of trapped electrons also affects solitary waves, since their amplitude increases in the presence of a larger proportion of trapped electrons in the plasma; on the other hand, rather counter-intuitively, their width remains the same. The above behavior was also observed via numerical integration of the dissipative Schamel equation.

While the nonlinear term is independent of the external magnetic field, the dispersive term depends strongly on the external magnetic field and in fact decreases for strong magnetic field (strength) values. Therefore, a steeper solitary wave with same maximum amplitude will be expected to occur in the presence of a stronger magnetic field, as confirmed by the numerical simulation here.

The current study focused on the 'simplest' version of a fluid model for magnetized plasma, i.e., assuming a uniform magnetic field and neglecting drift forces. A drift-kinetic approach would require the electrons to create a more complete picture to correctly account for the non-negligible $E \times B$ drift, for example. These aspects, as investigated, e.g., by Jovanoviç et al. [56], and summarized by Eliasson and Shukla [57], lie beyond the scope of the present study (weak $\sim \epsilon$ excitations were considered here which, in addition to the absence of an ambient electric field (bias), prescribe a negligible $E \times B$ drift).

The fundamental trapping scenario was considered in this paper, (the so-called β-trapping effect). It may occur, however, that the filamentation process in the final state of pattern formation in the electron phase space results in multiple electron transfer taking place through the separatrix; in turn, leading to additional trapping scenarios. In that case, the electric wave potential may not be expressed in a closed algebraic form, and new types of nonlinear structures may arise, as recently pointed out by Schamel [58,59]. As argued there, the existing wave theory for phase space holes, based on the linear Landau–van Kampen approach, overlooks these trapping and coherence aspects in pattern formation, and thus fails to account for a plethora of nonlinear phenomena, which are nonetheless predicted by this new approach [58,59]. Covering these aspects may form the focus of future studies.

The results, obtained here, aim to contribute to the understanding of the salient features of nonlinear electrostatic perturbations in non-Maxwellian plasmas, in account of electron trapping in phase space.

Author Contributions: Conceptualization, S.S.; formal analysis, S.S.; supervision, I.K.; writing—original draft, S.S. All authors have read and agreed to the published version of the manuscript.

Funding: Financial support by the project FSU-2021-012/8474000352 *"Modeling of Nonlinear Waves and Shocks in Space and Laboratory Plasmas"* (PI: Ioannis Kourakis), funded by Khalifa University of Science and Technology (Abu Dhabi, United Arab Emirates), is acknowledged, with thanks. Support from ADEK (Abu Dhabi Department of Education and Knowledge, United Arab Emirates), currently ASPIRE, in the form of an AARE-2018 (ADEK Award for Research Excellence 2018) grant (ADEK/HE/157/18—AARE-179) is also gratefully acknowledged.

Data Availability Statement: Not applicable.

Acknowledgments: Both authors had the pleasure and honor to have collaborated with Reinhard Schlickeiser on various projects in the past. This work is therefore dedicated to Reinhard Schlickeiser on the occasion of his 70th birthday. As a renowned plasma physicist and as an efficient supervisor, collaborator, and colleague, Reinhard Schlickeiser was a source of inspiration for both of us.

Conflicts of Interest: The authors declare no conflict of interest.

References

1. Vasyliunas, V.M. A survey of low-energy electrons in the evening sector of the magnetosphere with OGO 1 and OGO 3. *J. Geophys. Res.* **1968**, *73*, 2839–2884. [CrossRef]
2. De Feiter, L.D.; De Jager, C. Superthermal plasma nodules and their relation to solar flares. *Sol. Phys.* **1973**, *28*, 183–186. [CrossRef]
3. Scudder, J.D.; Sittler, E.C.; Bridge, H.S. A survey of the plasma electron environment of Jupiter: A view from Voyager. *J. Geophys. Res.* **1981**, *86*, 8157–8179. [CrossRef]
4. Leubner, M.P. On Jupiter's whistler emission. *J. Geophys. Res.* **1982**, *87*, 6335–6338. [CrossRef]
5. Christon, S.P.; Williams, D.J.; Mitchell, D.G.; Frank, L.A.; Huang, C.Y. Spectral characteristics of plasma sheet ion and electron populations during undisturbed geomagnetic conditions. *J. Geophys. Res.* **1989**, *94*, 13409–13424. [CrossRef]
6. Collier, M.R. On generating kappa-like distribution functions using velocity space Lévy flights. *Geophys. Res. Lett.* **1993**, *20*, 1531–1534. [CrossRef]
7. Decker, D.T.; Basu, B.; Jasperse, J.R.; Strikland, D.J.; Sharber, J.R.; Winningham, J.D. Upgoing electrons produced in an electron-proton-hydrogen atom aurora. *J. Geophys. Res.* **1995**, *100*, 21409–21420. [CrossRef]
8. Codrescu, M.V.; Fuller-Rowell, T.J.; Robble, R.G.; Evans, D.S. Medium energy particle precipitation influences on the mesosphere and lower thermosphere. *J. Geophys. Res.* **1997**, *102*, 19977–19987. [CrossRef]
9. Maksimovic, M.; Gary, S.P.; Skoug, R.M. Solar wind electron suprathermal strength and temperature gradients: Ulysses observations. *J. Geophys. Res.* **2000**, *105*, 18337–18350. [CrossRef]
10. Antonova, E.E.; Stepanova, M.V.; Teltzov, M.V.; Tverskoy, B.A. Multiple inverted-V structures and hot plasma pressure gradient mechanism of plasma stratification. *J. Geophys. Res.* **1998**, *103*, 9317–9332. [CrossRef]
11. Mori, H.; Ishii, M.; Murayama, Y.; Kubota, M.; Sakanoi, K.; Yamamoto, M.Y.; Monzen, Y.; Lummerzheim, D.; Watkins, B.J. Energy distribution of precipitating electrons estimated from optical and cosmic noise absorption measurements. *Ann. Geophys.* **2004**, *22*, 1613–1622. [CrossRef]
12. Livadiotis, G. (Ed.) *Kappa Distributions. Theory and Applications in Plasmas*; Elsevier: Amsterdam, The Netherlands, 2017.
13. Armstrong, T.P.; Paonessa, M.T.; Bell, E.V.; Krimigis, S.M. Voyager observations of Saturnian ion and electron phase space densities. *J. Geophys. Res.* **1983**, *88*, 8893–8904. [CrossRef]
14. Hasegawa, A.; Mima, K.; Duong-van, M. Plasma Distribution function in a superthermal radiation field. *Phys. Rev. Lett.* **1985**, *54*, 2608–2611. [CrossRef]
15. Hellberg, M.A.; Mace, R.L.; Armstrong, R.J.; Karlstad, G. Electron-acoustic waves in the laboratory: An experiment revisited. *J. Plasma Phys.* **2000**, *64*, 433–443. [CrossRef]
16. Baluku, T.K.; Hellberg, M.A. Dust acoustic solitons in plasmas with kappa-distributed electrons and/or ions. *Phys. Plasmas* **2008**, *15*, 123705. [CrossRef]
17. Hellberg, M.A.; Mace, R.L.; Baluku, T.K.; Kourakis, I.; Saini, N.S. Comment on "Mathematical and physical aspects of Kappa velocity distribution" [Phys. Plasmas 14, 110702 (2007)]. *Phys. Plasmas* **2009**, *16*, 094701. [CrossRef]
18. Sultana, S.; Kourakis, I.; Saini, N.S.; Hellberg, M.A. Oblique electrostatic excitations in a magnetized plasma in the presence of excess superthermal electrons. *Phys. Plasmas* **2010**, *17*, 032310. [CrossRef]
19. Sultana, S.; Kourakis, I. Electrostatic solitary waves in the presence of excess super-thermal electrons: Modulational instability and envelope soliton modes. *Plasma Phys. Control. Fusion* **2011**, *53*, 045003. [CrossRef]
20. Mauk, B.H.; Mitchell, D.G.; McEntire, R.W.; Paranicas, C.P.; Roelof, E.C.; Williams, D.J.; Krimigis, S.M.; Lagg, A. Energetic ion characteristics and neutral gas interactions in Jupiter's mag-netosphere. *J. Geophys. Res.* **2004**, *109*, A09S12.
21. Hapgood, M.; Perry, C.; Davies, J.; Denton, M. The role of suprathermal particle meas-urements in CrossScale studies of collisionless plasma processes. *Planet. Space Sci.* **2011**, *59*, 618–629. [CrossRef]
22. Kourakis, I.; Sultana, S.; Hellberg, M.A. Dynamical characteristics of solitary waves, shocks and envelope modes in kappa-distributed non-thermal plasmas: An overview. *Plasma Phys. Control. Fusion* **2012**, *54*, 124001. [CrossRef]
23. Atteya, A.; Sultana, S.; Schlickeiser, R. Dust-ion-acoustic solitary waves in magnetized plasmas with positive and negative ions: The role of electrons superthermality. *Chinese J. Phys.* **2018**, *56*, 1931–1939. [CrossRef]
24. Lotekar, A.; Kakad, A.; Kakad, B. Generation of ion acoustic solitary waves through wave breaking in superthermal plasmas. *Phys. Plasma* **2017**, *24*, 102127. [CrossRef]
25. Goldman, M.V.; Newman, D.L.; Mangeney, A. Theory of weak bipolar fields and electron holes with applications to space plasmas. *Phys. Rev. Lett.* **2007**, *99*, 145002. [CrossRef]
26. Sarri, G. Observation and characterization of laser-driven phase space electron holes. *Phys. Plasmas* **2010**, *17*, 010701. [CrossRef]
27. Schamel, H. Stationary solitary, snoidal and sinusoidal ion acoustic waves. *Plasma Phys.* **1972**, *14*, 905–924. [CrossRef]
28. Schamel, H. A modified Korteweg-de Vries equation for ion acoustic wavess due to resonant electrons. *J. Plasma Phys.* **1973**, *9*, 377–387. [CrossRef]
29. Schamel, H. Cnoidal electron hole propagation: Trapping, the forgotten nonlinearity in plasma and fluid dynamics. *Phys. Plasma* **2012**, *19*, 020501. [CrossRef]
30. Ergun, R.E.; Carlson, C.W.; McFadden, J.P.; Mozer, E.S.; Delory, G.T.; Peria, W.; Chaston, C.C.; Temerin, M.; Roth, I.; Muschietti, L.; et al. FAST satellite observations of large-amplitude solitary structures. *Geophys. Res. Lett.* **1998**, *25*, 2041–2044. [CrossRef]
31. Ergun, R.E.; Andersson, L.; Main, D.S.; Su, Y.J.; Carlson, C.W.; McFadden, J.P.; Mozer, F.S. Parallel electric fields in the upward current region of the aurora: Numerical solutions. *Phys. Plasmas* **2002**, *9*, 3685–3694. [CrossRef]

32. Andersson, L.; Ergun, R.E.; Newman, D.L.; McFadden, J.P.; Carlson, C.W.; Su, Y.J. Characteristics of parallel electric fields in the downward current region of the aurora. *Phys. Plasmas* **2002**, *9*, 3600–3609. [CrossRef]
33. Cattell, C.; Neiman, C.; Dombeck, J.; Crumley, J.; Wygant, J.; Kletzing, C.A.; Peterson, W.K.; Mozer, F.S.; André, M. Large amplitude solitary waves in and near the Earth's magnetosphere, magnetopause and bow shock: Polar and cluster observations. *Nonlinear Proc. Geophys.* **2003**, *10*, 13–26. [CrossRef]
34. Schippers, P.; Blanc, M.; André, N.; Dandouras, I.; Lewis, G.R.; Gilbert, L.K.; Persoon, A.M.; Krupp, N.; Gurnett, D.A.; Coates, A.J.; et al. Multi-instrument analysis of electron populations in Saturn's magnetosphere. *J. Geophys. Res.* **2008**, *113*, A07208. [CrossRef]
35. Lynov, J.P.; Michelsen, P.; Pecseli, H.L.; Rasmussen, J.J.; Saeki, K.; Turikov, V.A. Observations of solitary structures in a magnetized plasma loaded waveguide. *Phys. Scr.* **1979**, *20*, 328–335. [CrossRef]
36. Goldman, M.V.; Newman, D.L.; Ergun, R.E. Phase-space holes due to electron and ion beams accelerated by a current-driven potential ramp. *Nonlinear Proc. Geophys.* **2003**, *10*, 37–44. [CrossRef]
37. Jenab, S.M.H.; Brodin, G.; Juno, J.; Kourakis, I. Ultrafast electron holes in plasma phase space dynamics. *Sci. Rep.* **2021**, *11*, 16358. [CrossRef]
38. Simpson, J.A.; Bastian, T.S.; Chenette, D.L.; Lentz, G.A.; McKibben, R.B.; Pyle, K.R.; Tuzzolino, A.J. Saturnian trapped radiation and its absorption by satellites and rings: The first results from Pioneer 11. *Science* **1980**, *207*, 411–415. [CrossRef]
39. Verheest, F.; Hereman, W. Conservations laws and solitary wave solutions for gen-eralized Schamel equations. *Phys. Scr.* **1994**, *50*, 611–614. [CrossRef]
40. Williams, G.; Verheest, F.; Hellberg, M.A.; Anowar, A.G.M.; Kourakis, I. A Schamel equation for ion acoustic waves in superthermal plasmas. *Phys. Plasmas* **2014**, *21*, 092103. [CrossRef]
41. Sultana, S.; Islam, S.; Mamun, A.A.; Schlickeiser, R. Oblique propagation of ion-acoustic solitary waves in a magnetized plasma with electrons following a generalized distribution function. *Phys. Plasmas* **2019**, *26*, 012107. [CrossRef]
42. Sultana, S.; Mannan, A.; Schlickeiser, R. Obliquely propagating electron-acoustic solitary waves in magnetized plasmas: The role of trapped superthermal electrons. *Eur. Phys. J. D* **2019**, *73*, 220. [CrossRef]
43. Mamun, A.A.; Cairns, R.A.; Shukla, P.K. Effects of vortex-like and non-thermal ion distributions on non-linear dust-acoustic waves. *Phys. Plasmas* **1996**, *3*, 2610–2614. [CrossRef]
44. Nejoh, Y.-N. The dust charging effect on electrostatic ion waves in a dusty plasma with trapped electrons. *Phys. Plasmas* **1997**, *4*, 2813–2819. [CrossRef]
45. Mamun, A.A. Nonlinear propagation of ion-acoustic waves in a hot magnetized plasma with vortexlike electron distribution. *Phys. Plasmas* **1998**, *5*, 322–324. [CrossRef]
46. Mamun, A.A.; Shukla, P.K. Electron-acoustic solitary waves via vortex electron distribution. *J. Geophys. Res. Space Phys.* **2002**, *107*, 1135. [CrossRef]
47. Mamun, A.A.; Shukla, P.K.; Stenflo, L. Obliquely propagating electron-acoustic solitary waves. *Phys. Plasmas* **2002**, *9*, 1474–1477. [CrossRef]
48. Tribeche, M.; Djebarni, L.; Schamel, H. Solitary ion-acoustic wave propagation in the presence of electron trapping and background nonextensivity. *Phys. Lett. A* **2012**, *376*, 3164–3174. [CrossRef]
49. Hafez, M.G.; Roy, N.C.; Talukder, M.R.; Ali, M.H.; Hafez, M.G.; Roy, N.C.; Talukder, M.R.; Ali, M.H. Effects of trapped electrons on the oblique propagation of ion acoustic solitary waves in elec-tron-positron-ion plasmas. *Phys. Plasmas* **2016**, *23*, 082904. [CrossRef]
50. Hassan, M.R.; Rajib, T.I.; Sultana, S. Electron-acoustic solitons in magnetized collisional nonthermal lasmas. *arXiv* **2019**, arXiv:1912.04756.
51. Sultana, S.; Kourakis, I. Electron-scale electrostatic solitary waves and shocks: The role of superthermal electrons. *Eur. Phys. J. D* **2012**, *66*, 100. [CrossRef]
52. Ferdousi, M.; Sultana, S.; Mamun, A.A. Oblique propagation of ion-acoustic solitary waves in a magnetized electron-positron-ion plasma. *Phys. Plasmas* **2015**, *22*, 032117. [CrossRef]
53. Sultana, S.; Kourakis, I. Electron-scale dissipative electrostatic solitons in mul-ti-species plasmas. *Phys. Plasmas* **2015**, *22*, 102302. [CrossRef]
54. Sultana, S. Ion acoustic solitons in magnetized collisional non-thermal dusty plasmas. *Phys. Lett. A* **2018**, *382*, 1368–1373. [CrossRef]
55. Sultana, S.; Schlickeiser, R.; Elkamash, I.S.; Kourakis, I. Dissipative high-frequency envelope soliton modes in nonthermal plasmas. *Phys. Rev. E* **2018**, *98*, 033207. [CrossRef]
56. Jovanović, D.; Shukla, P.K.; Stenflo, L.; Pegoraro, F. Nonlinear model for electron phase-space holes in magnetized space plasmas. *J. Geophys. Res. Space Phys.* **2002**, *107*, 1110. [CrossRef]
57. Eliasson, B.; Shukla, P.K. Formation and dynamics of coherent structures involving phase-space vortices in plasmas. *Phys. Rep.* **2006**, *422*, 225–290. [CrossRef]
58. Schamel, H. Two-parametric, mathematically undisclosed solitary electron holes and their evolution equation. *Plasma* **2020**, *3*, 166–179. [CrossRef]
59. Schamel, H. Pattern formation in Vlasov-Poisson plasmas beyond Landau, as caused by the continuous spectra of electron and ion hole equilibria. *arXiv* **2021**, arXiv:2110.01433.

 *physics*

Article

Determining Pitch-Angle Diffusion Coefficients for Electrons in Whistler Turbulence

Felix Spanier [1,*], Cedric Schreiner [2,3] and Reinhard Schlickeiser [4,5]

1   Institut für Theoretische Astrophysik, Universität Heidelberg, Albert-Ueberle-Strasse 2, 69120 Heidelberg, Germany
2   Centre for Space Research, Northwest-University, Potchefstroom 2520, South Africa; mail@cschreiner.de
3   Max-Planck-Institute for Solar System Research, Justus-von-Liebig-Weg 3, 37077 Göttingen, Germany
4   Institut für Theoretische Physik IV, Ruhr-Universität Bochum, Universitätsstrasse 150, 44801 Bochum, Germany; rsch@tp4.rub.de
5   Institut für Theoretische Physik und Astrophysik, Christian-Albrechts-Universität zu Kiel, Leibnizstr. 15, 24118 Kiel, Germany
*   Correspondence: felix.spanier@uni-heidelberg.de

Abstract: Transport of energetic electrons in the heliosphere is governed by resonant interaction with plasma waves, for electrons with sub-GeV kinetic energies specifically with dispersive modes in the whistler regime. In this paper, particle-in-cell simulations of kinetic turbulence with test-particle electrons are performed. The pitch-angle diffusion coefficients of these test particles are analyzed and compared to an analytical model for left-handed and right-handed polarized wave modes.

Keywords: cosmic-ray transport; turbulence; plasma waves; particle-in-cell simulations

Citation: Spanier, F.; Schreiner, C.; Schlickeiser, R. Determining Pitch-Angle Diffusion Coefficients for Electrons in Whistler Turbulence. *Physics* **2022**, *4*, 80–103. https://doi.org/10.3390/physics4010008

Received: 30 August 2021
Accepted: 7 December 2021
Published: 20 January 2022

Publisher's Note: MDPI stays neutral with regard to jurisdictional claims in published maps and institutional affiliations.

1. Introduction

The solar wind and most phases of the interstellar medium are strongly turbulent media with high magnetic Reynolds numbers of 10^{14} [1]. Turbulence manifests itself in a spectrum of plasma waves at various length scales and frequencies. The energy distribution as a function of the frequency follows a characteristic power law. The current understanding of the turbulent processes is such that energy is injected at large scales, i.e., at small wave numbers and frequencies, and then cascades to smaller spatial scales.

The energy spectrum can be divided into several regimes, each may span several orders of magnitude in wave number or frequency. At the largest scales, the injection range is found, which then transitions into the inertial range. The inertial range can be described by magnetohydrodynamic (MHD) theory, and turbulence is dominated by the interaction of Alfvén waves. At smaller scales, kinetic effects of the particles come into play.

This high wave number regime is often referred to as the kinetic, dispersive, or dissipation range of the spectrum, since the waves become dispersive and dissipation starts to set in. While the spectrum extends to even smaller scales, damping eventually becomes dominant and leads to an exponential cutoff of the energy spectrum.

Power-law distributions of the fluctuating magnetic energy are expected in the injection, inertial, and dissipation range of the spectrum. However, the spectral indices may differ among the individual regimes. Goldreich and Sridhar [2,3] presented a detailed model of the turbulent energy cascade in the inertial regime. Their model predicts a spectral index of $-5/3$, which actually seems to be realized in the solar wind [4]. Subsequent models by Galtier et al. [5,6] give rise to a spectral index $k_\perp^{-2}$.

Kinetic turbulence in the dissipation range is an active field of research [7]. In particular, the composition of the wave spectrum is subject to discussion, because a transition from non-dispersive Alfvén waves to dispersive wave modes is expected. Possible candidates for the waves in the dissipation range are the so-called kinetic Alfvén waves (KAWs) and whistler waves [8].

The impact of kinetic turbulence on the transport of energetic particles is another major topic. The transport of energetic protons is well described by models of Alfvénic turbulence since the protons mainly interact with these waves at low frequencies. The theoretical framework of quasi-linear theory (QLT) can be used to describe particle transport by a series of resonant interactions with the magnetic fields of Alfvén waves, which leads to scattering of the particles [9–11]. This theory describes changes of the particles' pitch angles (the angle of the velocity vector relative to a background magnetic field), momenta or positions as diffusion processes and allows to predict diffusion coefficients and other quantities, such as the mean free path, which can be compared to observations.

Dispersive waves are more difficult to handle in (analytical) theory. Nonetheless, QLT can also yield predictions for particle transport in a medium containing dispersive waves [12,13]. The introduction of dispersive waves can even solve some of the problems that are encountered if a purely Alfvénic spectrum of waves is assumed [14]. Still, the model remains an approximation, and computer simulations are used to clarify some of the details that are not included in the analytical theory. Different kinds of simulations are used to study different physical regimes and processes—from the acceleration of particles [15,16] to the transport of energetic particles—considering both non-dispersive [17] and dispersive waves [18,19].

The key problem that has been chosen for the subject of this study is the process of electron transport in dispersive turbulence. The transport of electrons at sub-GeV energies has been of high interest for quite some time [20]. As mentioned above, particle acceleration in Alfvénic turbulence in the inertial range is well understood. However, turbulence on kinetic scales still poses problems for both observations and modeling.

2. Theory

2.1. Turbulence Theory

From the observations, it is not entirely clear which types of plasma waves constitute the spectrum of turbulent waves in the dispersive and dissipative regime. KAWs and whistler waves are both possible candidates [8]. While KAWs represent the continuation of the Alfvén mode (Equation (13) below) in the dispersive regime with very large perpendicular wave numbers (perpendicular wavelength in the range of proton gyroradius), whistler waves (Equation (16) below) are right-handed polarized modes at large wave numbers. It is reasonable to assume that non-dispersive Alfvén waves simply transition to KAWs. However, observations reveal that whistler waves are also present in various regions of the heliosphere, such as in the interplanetary medium [21], close to interplanetary shocks [22,23] or planetary bow shocks [24], and also in the Earth's ionosphere and foreshock region [25,26].

Whereas left-handed polarized Alfvén waves are damped by protons and cannot cascade to frequencies above the proton cyclotron frequency, a spectrum of whistler waves may extend to frequencies beyond the ion-cyclotron frequencies. Whistler waves primarily interact with electrons and are also damped by electrons at higher frequencies (close to the electron cyclotron frequency). This is an interesting aspect of kinetic turbulence, since a population of whistler waves can heat the electrons in the solar wind or even accelerate particles in the high-energy tail of the thermal spectrum.

Kinetic turbulence includes the smallest length scales, where the interaction of waves and particles becomes important. Although the wave-particle interactions are not explicitly included in the theory, their effect has to be considered by allowing dispersive waves and damping. This regime is generally more complicated and less understood than MHD turbulence.

The general picture associated with turbulence is as follows. Energy is injected into the system at a large outer scale, small wave numbers k. The energy is transported via the interaction of waves to smaller spatial scales (larger wave numbers), and the (magnetic) energy spectrum, $E_B(k)$, follows a power-law distribution. This is the inertial range. The spectrum steepens as the kinetic regime or dissipation range is reached, but energy is still transported to smaller scales. First, only ion effects will start influencing the plasma dynamics, but at even larger wave numbers, the electrons can also interact with the plasma waves. This is where the energy spectrum is cut off. One aspect that has been discussed in greater detail since the seminal paper of Goldreich and Sridhar [3] is the possible anisotropy with respect to the background magnetic field: the turbulent spectrum may behave differently depending on wave numbers parallel ($k_\parallel$) or perpendicular ($k_\perp$) to the background magnetic field.

The special case of whistler turbulence has been discussed in greater detail. The properties of this whistler turbulence have been analyzed using kinetic simulations in two [27,28] and three [29–32] dimensions. These studies suggest a steeper energy spectrum than for Alfvénic turbulence, with a spectral index s in the range between -3.7 and -5.5 and a possible break in the energy spectrum [32]. Results by Chang [30] also suggest that the wave vector anisotropy depends on the choice of the plasma beta. A relatively isotropic spectrum is obtained for a plasma beta $\beta \sim 1$, whereas $\beta < 1$ yields an anisotropic cascade which favors the transport of energy to larger $k_\perp$. The plasma beta is the ratio of thermal to magnetic energy. The anisotropy additionally depends on the energy deployed to the electromagnetic fields of the turbulent whistler waves [29].

2.2. Subluminal Parallel Waves in Cold Plasmas

In warm thermal plasmas with low plasma betas, the real part of the dispersion relation agrees well with the cold plasma dispersion relation, so, the latter is used here. In addition, the resonance broadening effects, caused by a finite imaginary part of the dispersion relation in warm plasmas implying a finite weak-damping rate; those effects were considered by Schlickeiser and Achatz [33].

Using the convention of frequencies $\omega > 0$, where ω is the real part of the generally complex frequency, but $k_\parallel \in [-\infty, \infty]$ (here, the case $k_\perp = 0$, also known as slab case, is treated), the dispersion relation of right-handed ("R") and left-handed ("L") polarized undamped low-frequency ($\omega \leq \Omega_{e,0}$, with $\Omega_{e,0}$ being the electron gyrofrequency) parallel Alfvén wave in a cold electron-proton background plasma reads [34]:

$$n_L^2 = 1 - \frac{\omega_{pi}^2}{\omega(\omega - \Omega_i)} - \frac{\omega_{pe}^2}{\omega(\omega + \Omega_e)}, \tag{1}$$

$$n_R^2 = 1 - \frac{\omega_{pi}^2}{\omega(\omega + \Omega_i)} - \frac{\omega_{pe}^2}{\omega(\omega - \Omega_e)}, \tag{2}$$

$$\frac{k_\parallel^2 c^2}{\omega_{L,R}^2} = 1 - \frac{\omega_{pi}^2}{\omega(\omega \mp \Omega_i)} - \frac{\omega_{pe}^2}{\omega(\omega \pm \Omega_e)}, \tag{3}$$

$$\frac{k_\parallel^2 c^2}{\omega_{L,R}^2} - 1 = -\frac{c^2 \Omega_i^2}{v_A^2} \frac{M+1}{(\omega \mp \Omega_i)(\omega \mp M\Omega_i)} \tag{4}$$

with the proton-to-electron mass ratio, $M = m_p/m_e = 1836$, and the Alfvén speed, $v_A = \beta_A c = 2.18 \times 10^{11} B[\mathrm{G}] n_i^{-1/2}[\mathrm{cm}^{-3}]$ Here, ω_{pi} is the ion plasma frequency, ω_{pe} is the electron plasma frequency, Ω_i is the ion gyrofrequency, Ω_e is the electron gyrofrequency, B is the magnetic field, β is the plasma beta, and c is the speed of light. For subluminal wave phase speeds,

$$\left| V_{\mathrm{phase}} \right| = \left| \frac{\omega_{L,R}}{k_\parallel} \right| \ll c, \tag{5}$$

which has to be checked a posteriori, the dispersion relation (4) simplifies to

$$\frac{k_\parallel^2 c^2}{\omega_{L,R}^2} \simeq -\frac{(M+1)\Omega_i^2 \omega_{L,R}^2}{v_A^2(\omega \mp \Omega_i)(\omega \mp M\Omega_i)}. \tag{6}$$

It is convenient to introduce dimensionless frequencies and parallel wave numbers by defining

$$y_{L,R} = \frac{\omega_{L,R}}{\Omega_i} > 0 \quad \text{and} \quad k = \frac{k_\parallel}{k_c}, \tag{7}$$

respectively, with

$$k_c = \frac{\Omega_i}{v_A}, \tag{8}$$

so that the subluminal dispersion relation becomes

$$k^2 = -\frac{(M+1)y_{L,R}^2}{(y_{L,R} \mp 1)(y_{L,R} \pm M)} \tag{9}$$

with the two solutions,

$$k_{1,2} = -\frac{\sqrt{(M+1)}\, y_{L,R}}{\sqrt{(y_{L,R} \mp 1)(y_{L,R} \pm M)}}. \tag{10}$$

As $y_{L,R}$ is always positive, the solution $k_1 > 0$ describes forward-moving waves with positive phase speed, whereas the negative solution $k_2 = -k_1 < 0$ describes backward-moving waves.

2.2.1. Left-Handed Modes

Equation (9) indicates that no left-handed polarized solution with $y_L > 1$ exists, so, a further simplification of Equation (9) for left-handed polarized waves is possible, assuming that $M \gg 1$:

$$k^2 \simeq \frac{y_L^2}{1 - y_L} \tag{11}$$

with the solutions,

$$y_L(k) = \frac{k^2}{2} \pm \sqrt{\frac{k}{2}}\sqrt{k^2 - 4} \simeq \begin{cases} |k| & \text{for } k \ll 1, \\ 1 - 1/k^2 & \text{for } k \gg 1, \end{cases} \tag{12}$$

corresponding to

$$\omega_L \simeq \begin{cases} v_A |k_\parallel| & \text{for } k_\parallel \ll k_c, \\ \Omega_i\left(1 - k_c^2/k_\parallel^2\right) & \text{for } k_\parallel \gg k_c. \end{cases} \tag{13}$$

2.2.2. Right-Handed Modes

The right-handed solutions of Equation (9),

$$k^2 = -\frac{(M+1)y_R^2}{(y_R + 1)(y_R + M)}, \tag{14}$$

can be approximated under the assumption that $M \gg 1$:

$$y_R = \frac{(M+1)k^2}{2(1 + k^2 + M)}\left(1 - \frac{2}{M+1} + \sqrt{1 + \frac{4M}{k^2(M+1)}}\right). \tag{15}$$

Depending on k, different regimes can be identified:

$$\omega_R = \begin{cases} v_A |k_\parallel| & \text{for } |k_\parallel| < k_c, \\ \Omega_i + \dfrac{v_A^2 k_\parallel^2}{\Omega_i} & \text{for } k_c \leq |k_\parallel| \leq M^{1/2}k_c, \\ \Omega_e\left(1 - \dfrac{Mk_c^2}{k_\parallel^2}\right) & \text{for } |k_\parallel| > M^{1/2}k_c. \end{cases} \tag{16}$$

The first range describes the linear dispersion regime, the second range is the whistler regime, and the last range is the electron–cyclotron range. While these approximate solutions provide good estimates to the real solution, there is a major problem: the solutions do not provide continuous coverage. An alternative approximation is

$$y_R(k) \simeq |k|(1+|k|). \tag{17}$$

In the following, the particle scattering by parallel waves at electron or ion cyclotron frequencies are ignored as soon as these are highly damped in a realistic warm thermal plasma, so that the resonant interaction does not apply; see [33] for a discussion of wave-particle interactions with damped waves. For left-handed and right-handed polarized waves, this restricts the normalized wave numbers to values of $k \leq 1$ and $k \leq M$, respectively.

2.3. Particle Transport

Any charged particle of given velocity v, Lorentz factor $\gamma = (1 - (v^2/c^2))^{-1/2}$, pitch-angle cosine $\mu = v_\parallel / v$, mass m, charge $q_i = e|Z_i|Q$ with $Q = \mathrm{sgn}(Z_i)$, Z_i being the ion charge number, and relativistic gyrofrequency $\Omega' = Q\Omega/\gamma$ with positive $\Omega = |q|B_0/(mc)$ with Ω being the particle's non-relativistic gyrofrequency, q being the electric charge, and B_0 being the magnetic background field, interacts with parallel right-handed and left-handed polarized plasma waves whose wave number, k, and real frequency, $\omega_{R,L}$, obey the resonance condition,

$$v\mu k_\parallel - \omega_{R,L}(k_\parallel) \mp \frac{Q\Omega}{\gamma} = 0. \tag{18}$$

Introducing

$$x = \frac{p}{m_e c}, \quad \epsilon = \frac{v_A}{v} = \beta_A \frac{(1+x^2)^{1/2}}{x}, \tag{19}$$

with the particle momentum p and the dimensionless frequency and wave number (7), the resonance condition (18) reads:

$$\Omega\left(\frac{\mu k}{\epsilon} - y_{R,L}(k) \mp S_i\right) = 0 \tag{20}$$

with

$$S_i = \frac{Q|Z_i|m_p}{m\gamma} = \frac{1}{\gamma}\begin{cases} 1 & \text{for protons,} \\ -M & \text{for electrons.} \end{cases} \tag{21}$$

The quasilinear Fokker–Planck coefficient for the pitch-angle cosine, μ, is given by

$$\begin{aligned}
D_{\mu\mu}(\mu) &= \frac{\pi^2 \Omega^2 (1-\mu^2)}{B_0^2} \int_{-\infty}^{\infty} dk_\parallel \\
&\quad \times \left[g_R(k_\parallel)\,\delta(vk_\parallel - \omega_R - \Omega')\left(1 - \frac{\mu\omega_R}{k_\parallel v}\right)^2 \right. \\
&\quad \left. + \; g_L(k_\parallel)\,\delta(vk_\parallel - \omega_L + \Omega')\left(1 - \frac{\mu\omega_L}{k_\parallel v}\right)^2 \right]
\end{aligned} \tag{22}$$

with the magnetic fluctuation spectra of right/left-handed polarized waves $g_{R,L}(k_\parallel)$, where the total magnetic field fluctuations are determined as in [35]:

$$(\delta B)^2 = 2\pi \int_{-\infty}^{\infty} dk_\parallel [g_L(k_\parallel) + g_R(k_\parallel)]. \tag{23}$$

In Equation (22), the frequencies $\omega_{R,L}(k_\parallel)$ are determined by the solutions of the dispersion relations, discussed in Section 2.2 above. The function $\delta(x)$ is the Dirac delta function.

In terms of the normalized wave number, $k_\parallel = k_c k$, and frequency, $\omega_{R,L} = \Omega_i y_{R,L}$, the Fokker–Planck coefficient (22) reads:

$$
\begin{aligned}
D_{\mu\mu}(\mu) \;=\; & \frac{\pi^2 \Omega^2 k_c (1-\mu^2)}{\Omega_i^2 B_0^2} \int_{-\infty}^{\infty} dk_\parallel \\[2mm]
& \times \left[g_R(k_\parallel)\, \delta(k\mu/\epsilon - y_R(k) - S_i)\left(1 - \frac{\mu\omega_R}{k_\parallel v}\right)^2 \right. \\[2mm]
& \left. + \; g_L(k_\parallel)\, \delta(k\mu/\epsilon - y_L + S_i)\left(1 - \frac{\mu\omega_L}{k_\parallel v}\right)^2 \right].
\end{aligned}
\tag{24}
$$

The calculation of the Fokker–Planck coefficient requires a knowledge of the magnetic fluctuation spectra. The correct theoretical description is complicated, as described in Section 2.1 above, but results from numerical calculations can be inferred.

Deriving the Fokker–Planck coefficients in general is quite an involved task but one can derive some limiting cases. It is helpful to account for the relative abundance of forward-moving and backward-moving waves and the corresponding polarization states. Let us introduce the cross helicities, $H_{L,R} \in [-1,1]$ for left/right-handed polarized waves to express the spectra (23) of backward-propagating ("b") and forward-propagating ("f") waves

$$
g^b_{L,R} = \frac{1 - H_{L,R}}{2} g_{L,R}(k)\, \Theta(-k),
\tag{25}
$$

$$
g^f_{L,R} = \frac{1 + H_{L,R}}{2} g_{L,R}(k)\, \Theta(k).
\tag{26}
$$

The step function $\Theta(\pm k)$ appears because backward-moving and forward-moving waves only occur at negative and positive wave numbers, respectively. In general, these cross helicities can depend on the wave number, but throughout this article isospectral turbulence is adopted, where $H_{L,R}$ are constants (independent of k). The magnetic helicity $\sigma(k) \in [-1,1]$ characterizes the relative abundances of left-handed and right-handed polarized waves:

$$
g_L(k) = \frac{1 + \sigma(k)}{2} g_{\text{tot}}(k),
\tag{27}
$$

$$
g_R(k) = \frac{1 - \sigma(k)}{2} g_{\text{tot}}(k)
\tag{28}
$$

where $g_{\text{tot}}(k)$ is the total wave abundance at a specific wave number. For parallel plasma waves, $\sigma(y > 1) = -1$ as soon as no left-handed polarized waves exist at these normalized frequencies.

Using the helicities introduced, the Fokker–Planck coefficient (24) reads:

$$
\begin{aligned}
D_{\mu\mu}(\mu) \;=\; & \frac{\pi^2 \Omega^2 k_c (1-\mu^2)}{\Omega_i^2 B_0^2} \int_{-\infty}^{\infty} dk_\parallel g_{\text{tot}}(k) \\[2mm]
& \times \left[(1-H_R)(1-\sigma(k))\, \delta(k\mu/\epsilon + y_R + S_i)\left(1 + \frac{\epsilon\mu y_R(k)}{k}\right)^2 \right. \\[2mm]
& + \; (1-H_L)(1+\sigma(k))\, \delta(k\mu/\epsilon + y_R(k) - S_i)\left(1 + \frac{\epsilon\mu y_L(k)}{k}\right)^2 \\[2mm]
& + \; (1+H_R)(1+\sigma(k))\, \delta(-k\mu/\epsilon + y_R + S_i)\left(1 - \frac{\epsilon\mu y_R(k)}{k}\right)^2 \\[2mm]
& \left. + \; (1+H_L)(1-\sigma(k))\, \delta(-k\mu/\epsilon + y_R(k) - S_i)\left(1 - \frac{\epsilon\mu y_L(k)}{k}\right)^2 \right].
\end{aligned}
\tag{29}
$$

2.3.1. Interactions in the Whistler Regime

For frequencies above the ion cyclotron frequency, only right-handed waves exist obeying the dispersive whistler mode dispersion relation. As discussed above, the turbulent spectrum typically has a much softer spectral index than 2 (theoretical values are in the range of 3 to 6 [7,36]) in this case.

We consider the case $H_R = -1$ and $\sigma(k) = -1$, only backward-moving right-handed polarized waves, which reduces the Fokker–Planck coefficient (29) to

$$
D_{\mu\mu}(\mu) = \frac{\pi^2 \Omega^2 k_c (1 - \mu^2)}{\Omega_i^2 B_0^2} \int_{k_0}^{\infty} dk_\| g_{\text{tot}}(k)
$$
$$
\times \left[\delta(k\mu/\epsilon + y_R + S_i) \left(1 + \frac{\epsilon \mu y_R(k)}{k} \right)^2 \right].
$$
(30)

The result for forward-moving waves is similar:

$$
D_{\mu\mu}(\mu) = \frac{\pi^2 \Omega^2 k_c (1 - \mu^2)}{\Omega_i^2 B_0^2} \int_{k_0}^{\infty} dk_\| g_{\text{tot}}(k)
$$
$$
\times \left[\delta(-k\mu/\epsilon + y_R + S_i) \left(1 - \frac{\epsilon \mu y_R(k)}{k} \right)^2 \right].
$$
(31)

With Equation (17), the resonance condition with positive values of k reads:

$$
0 = S_i + k\frac{\mu}{\epsilon} + \begin{cases} |k| & \text{for } k \leq 1, \\ k^2, & \text{for } 1 \leq k \leq M^{1/2} = 43. \end{cases}
$$
(32)

It can be shown that this equation for protons and electrons has only one solution $k_r > 0$. In the Alfvénic wave number range ($k \leq 1$), this is trivial: $k_r = -S_i/(1 + (\mu/\epsilon))$, which can be positive depending on the signs of S_i and μ.

In the whistler wave number range ($0 \leq k \leq 43$), the proof is a bit more involved. Here, Equation (32) has two solutions:

$$
k_1 = \frac{1}{2} \left(\sqrt{\frac{\mu^2}{\epsilon^2} - 4S_i} - \frac{\mu}{\epsilon} \right),
$$
(33)

$$
k_2 = -\frac{1}{2} \left(\sqrt{\frac{\mu^2}{\epsilon^2} - 4S_i} + \frac{\mu}{\epsilon} \right).
$$
(34)

To obtain real-valued solutions (33) and (34), the condition $\mu^2 \geq 4S_i\epsilon^2$ must be fulfilled. Assuming that this is fulfilled, one notes that for non-negative values of $\mu \geq 0$, the solution $k_2(\mu \geq 0) < 0$ is always negative, leaving only one solution for $k_r = k_1(\mu \geq 0) > 0$. Alternatively, for negative values of $\mu = -|\mu|$ the solutions (33) and (34) become

$$
k_1(\mu < 0) = \frac{1}{2} \left(\sqrt{\frac{\mu^2}{\epsilon^2} - 4S_i} - \frac{|\mu|}{\epsilon} \right),
$$
(35)

$$
k_2(\mu < 0) = \frac{1}{2} \left(\frac{|\mu|}{\epsilon} - \sqrt{\frac{\mu^2}{\epsilon^2} - 4S_i} \right).
$$
(36)

To further evaluate the solution, it is necessary to distinguish between positive (for protons) and negative values (for electrons) of S_i. For electrons the solution $k_2(\mu < 0) < 0$ is again always negative. For protons, both solutions (35) and (36) are positive, but the second one is

$$
k_2(\mu < 0, S_i > 0) \leq S_i^{1/2} = \frac{1}{2\gamma} < 1
$$
(37)

always (for protons $S_i = 1/\sqrt{2\gamma}$). This solution is positive but is in contradiction with the above-made assumption of the modes in the whistler regime with $k > 1$. This leaves us with only one solution for the resonant wave number $k_r = k_1(\mu < 0, S_i > 0)$ in the whistler wave number range.

One then obtains:

$$
\begin{aligned}
D_{\mu\mu}(\mu) &= \frac{\pi^2 \Omega^2 k_c (1-\mu^2)}{\Omega_i^2 B_0^2} \Theta(k_r - k_0)\,\Theta(M^{1/2} - k_r) \\
&\times \frac{g_{\text{tot}}(k_r)}{\left|\frac{dy_R}{dk} + \frac{\mu}{\epsilon}\right|_{k=k_r}} \left(1 + \frac{\epsilon \mu y_R(k_r)}{k_r}\right)^2.
\end{aligned}
\tag{38}
$$

The case of forward-moving right-handed polarized waves is similar to the backward-moving ones. The main difference is the resonant wave number,

$$
k_r = \frac{1}{2}\left(\sqrt{\frac{\mu^2}{\epsilon^2} - 4S_i} + \frac{\mu}{\epsilon}\right).
\tag{39}
$$

2.3.2. Alfvén and Whistler Contributions

The total Fokker–Planck coefficient can be written as a sum of Alfvén and whistler contributions:

$$
D_{\mu\mu}(\mu) = D_{\mu\mu}^{\text{A}} + D_{\mu\mu}^{\text{W}}.
\tag{40}
$$

For the Alfvén wave Fokker–Planck coefficient, inserting the asymptotic expansions $y_{R,L}(k \le 1) \simeq k$ of Equations (12) and (15) one obtains:

$$
\begin{aligned}
D_{\mu\mu}^{\text{A}}(\mu) &= \frac{\pi^2 \Omega^2 k_c (1-\mu^2)}{\Omega_i^2 B_0^2} \int_{k_0}^{1} dk_{\parallel} g_{\text{tot}}(k) \\
&\times \{(1+\epsilon\mu)^2[(1-H_R)(1-\sigma)\,\delta(k(1+\mu/\epsilon) + S_i) \\
&+ (1-H_L)(1+\sigma)\,\delta(k(1+\mu/\epsilon) - S_i)] \\
&+ (1-\epsilon\mu)^2[(1+H_R)(1-\sigma)\,\delta(k(1-\mu/\epsilon) + S_i) \\
&+ (1+H_L)(1+\sigma)\,\delta(k(1-\mu/\epsilon) - S_i)]\,\}.
\end{aligned}
\tag{41}
$$

The whistler contribution is calculated above, while for completeness is given here in the same form:

$$
\begin{aligned}
D_{\mu\mu}^{\text{W}}(\mu) &= \frac{\pi^2 \Omega^2 k_c (1-\mu^2)}{\Omega_i^2 B_0^2} \int_{1}^{M^{1/2}} dk_{\parallel} g_{\text{tot}}(k) \\
&\times \left[(1+\epsilon\mu)^2(1-H_R)(1-\sigma)\,\delta(k^2 + k\mu/\epsilon + S_i)\right. \\
&+ \left.(1-\epsilon\mu)^2(1+H_R)(1-\sigma)\,\delta(k^2 - k\mu/\epsilon + S_i)\right].
\end{aligned}
\tag{42}
$$

2.3.3. Electrons

The Fokker–Planck coefficient derived in the previous section hold for any particle. In general, the integration is not complicated as the delta function of the resonance condition helps to simplify the calculations. However, a specific turbulent spectrum has to be defined. We refrain from performing the integral here but point out which interaction will take place. There is a clear difference between electrons and protons, and the discussion in this Section is limited to the electron case.

For Alfvén waves, one can distinguish the interaction of electrons with forward/backward-moving left/right-handed modes. Defining

$$\mu_R(x) = \frac{\beta_A(M - \sqrt{1 + x^2})}{x}, \tag{43}$$

$$\mu_L(x) = \frac{\beta_A(M + \sqrt{1 + x^2})}{x}, \tag{44}$$

one can constrain the waves for which the resonant interaction with electrons is possible:

1. backward-moving right-handed polarized Alfvén waves for all pitch-angle cosines with $\mu \geq \mu_R(x)$ and $\mu \geq -\epsilon = -\beta_A\sqrt{1 + x^2}/x$;
2. backward-moving left-handed polarized Alfvén waves for all pitch-angle cosines with $\mu \leq \mu_L(x)$; and $\mu \leq -\epsilon = -\beta_A\sqrt{1 + x^2}/x$
3. forward-moving right-handed polarized Alfvén waves for all pitch-angle cosines with $\mu \leq \mu_R(x)$ and $\mu \geq \epsilon = \beta_A\sqrt{1 + x^2}/x$;
4. forward-moving left-handed polarized Alfvén waves for all pitch-angle cosines with $\mu \geq \mu_L(x)$ and $\mu \geq \epsilon = \beta_A\sqrt{1 + x^2}/x$.

For whistler waves, additionally

$$\mu_0(x) = \frac{\beta_A M^{1/2}(\sqrt{1 + x^2} - 1)}{x} \tag{45}$$

is defined. The resonant interaction takes place within the following range:

$$-\mu_0(x) \leq \mu \leq \mu_R(x), \tag{46}$$

The resulting total Fokker–Planck scattering coefficient is shown in Figure 1.

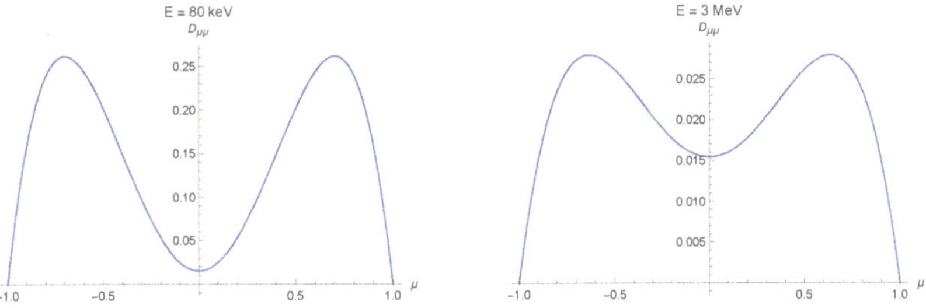

Figure 1. The pitch-angle dependence of the Fokker–Planck scattering coefficient model calculation for electrons at extreme ends. A power-law spectrum with with spectral index $s = 3$ in the wave number range $0.01 < k < M^{1/2}$, where M is the proton-to-electron mass ratio, is assumed. Electrons at 80 keV and 3 MeV are considered. The proton gyrofrequency $\Omega_p = 5 \times 10^4$ Hz, and the Alfvén speed $v_A = 0.001c$ in fractions of the speed of light c.

3. Numerical Methods

3.1. Particle-in-Cell Simulations

To be able to model dispersive waves of different kinds and to obtain a self-consistent description of electromagnetic fields and charged particles in the plasma, a fully kinetic particle-in-cell (PiC) approach [37] is employed here. In particular, the explicit second-order PiC code ACRONYM [38] is used which is fully relativistic, parallelized and three-dimensional. Although the PiC method might not be the most efficient numerical technique when dealing with proton effects, we still favor this approach because of its versatility. A more detailed discussion of advantages and drawbacks as well as a direct comparison of PiC and MHD approaches with the specific problem of the interaction of protons and left-handed polarized waves can be found in Sections 3 and 6 of Ref. [39]. However, the PiC approach is well-suited for the study of electron scattering as soon as the time and length scales of electron interactions are closer to the scales of time step lengths and cell sizes in PiC simulations, thus reducing computing time compared to simulations, in which proton interactions are studied.

The details of the PiC code are not discussed here. The simulation technique used here does not differ from standard techniques. Two points, however, to be mentioned: the initialization of turbulence and the tracking of test particles. Turbulence is discussed in Section 3.1.1 just below, because the numerical treatment is inherently connected to the physical processes of turbulence. The treatment of energetic particles is divided into to two parts: the injection of particles and the analysis of the particle data.

3.1.1. Setup of Turbulence Simulations

Here, a simulation setup that was inspired by Gary et al. [27] is used. Waves are initialized at low k, $k < \Omega_i/V_A$. The layout of the initial waves in the wave number space is explained below and is drawn in Figure 2 for the two-dimensional setup. However, in the PiC simulation, the velocity space and the electromagnetic fields are considered three-dimensional. As shown below, the two-dimensional simulations give an energy cascade similar to that of the three-dimensional simulations. As highlighted by Gary et al. [29], the three-dimensional case differs mainly in the anisotropy and a break at $k_\perp c/\Omega_e$. The question whether particle transport is different in two and three dimensions may be understood from the fact that particle motion is still three-dimensional. The theoretical description is assuming gyrotropy anyway, so the different perpendicular directions are therefore averaged out.

In the simulations, the natural mass ratio, $m_p/m_e = 1836$, is used. Waves are initialized with equal amplitudes and a random phase angle. The total magnetic energy density of the initial waves is set to 10% of the energy density of the background magnetic field. This can be expressed by $\delta B^2/B_0^2 = 0.1$, where $\delta B^2 = \sum_j \delta B_j^2$ and j denotes individual waves.

To analyze the simulations, the spectra of the magnetic energy density, $E_B = |\vec{B}^2(\vec{k})|/(8\pi)$ in wave number space are considered. A two-dimensional energy spectrum, $E_B(k_\parallel, k_\perp)$, can be obtained by Fourier transforming the field data to obtain the perpendicular coordinate $k_\perp$. The parallel direction is equivalent to the z-direction of the simulation, whereas the perpendicular direction is represented by the x-direction in a two-dimensional simulation or by the x-y-plane in a three-dimensional simulation. A one-dimensional spectrum $E_B(k)$ can be obtained by integrating over the angle θ in the $k_\parallel$-$k_\perp$-plane. Additional one-dimensional spectra, $E_B(k_\parallel)$ and $E_B(k_\perp)$, are obtained by integrating $E_B(k_\parallel, k_\perp)$ over $k_\perp$ and $k_\parallel$, respectively.

Electron transport is studied in two simulations, S1 and S2. The aim is to resolve several wave numbers in both the undamped and damped regimes of the whistler mode. This should allow us to see differences in the spectral slope or the anisotropy in both regimes. To resolve the relatively large spatial scales of the undamped regime, large simulation boxes are required. Thus, the decision was taken to restrict the investigations to two-dimensional setups.

Simulations S1 and S2 are characterized by the physical and numerical parameters listed in Tables 1 and 2, respectively.

The setups are aimed to simulate decaying turbulence with a set of 42 initially excited whistler waves according to Figure 2.

Table 1. Physical parameters for simulations S1 and S2: electron plasma frequency, $\omega_{p,e}$ electron cyclotron frequency, Ω_e, and thermal speed, $v_{\text{th},e}$ of electrons, sum δB^2 of the squares of the magnetic field amplitudes of the individual waves, and plasma beta β.

Simulation	$\omega_{p,e}$ (rad/s)	$|\Omega_e|$ $(\omega_{p,e})$	$v_{\text{th},e}$	$\delta B^2/B_0^2$	β
S1	1.966×10^8	0.447	$0.10\,c$	0.10	0.20
S2	1.966×10^8	0.447	$0.05\,c$	0.10	0.05

Table 2. Numerical parameters for the two-dimensional simulations S1 and S2: number of cells, $N_\parallel$ and $N_\perp$, in the directions parallel and perpendicular to the background magnetic field, respectively; number of time steps, N_t, grid spacing, Δx, time step length Δt, and the number of particles (electrons and protons combined) per cell (ppc).

Simulation	$N_\parallel$ (Δx)	$N_\perp$ (Δx)	N_t (Δt)	Δx $(c\,\omega_{p,e}^{-1})$	Δt $(\omega_{p,e}^{-1})$	ppc
S1	2048	2048	1.0×10^5	7.0×10^{-2}	4.1×10^{-2}	256
S2	2048	2048	1.0×10^5	3.5×10^{-2}	2.0×10^{-2}	256

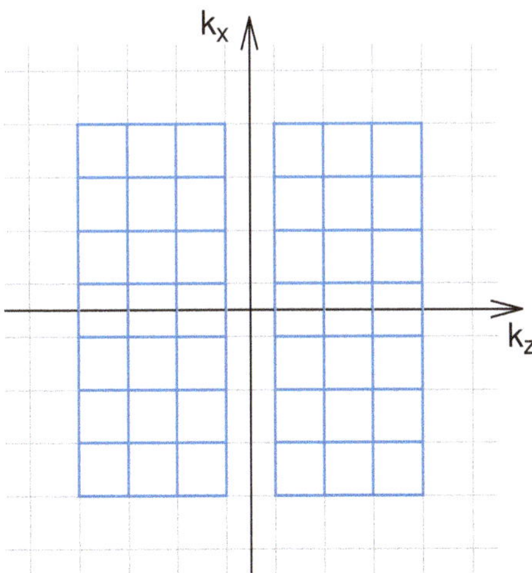

Figure 2. Schematic representation of two-dimensional wave number space. The discretized wave vectors are represented by the gray boxes, and the axes mark the directions parallel (k_z) and perpendicular (k_x) to the background magnetic field $\vec{B}_0$ in the case of a two-dimensional simulation. For the simulation of decaying turbulence in two dimensions, a set of 42 initial waves is excited, where each wave occupies one position on the grid. These positions are indicated by the blue boxes in accordance with the setup specified by Gary [27].

3.1.2. Turbulence Spectra

Here, the simulations S1 and S2 with the parameters from Tables 1 and 2 are briefly discussed.

Figure 3 shows the perpendicular spectra $E_B(k_\perp)$ of the magnetic field energy from simulations S1 (Figure 3a) and S2 (Figure 3b). At small, perpendicular wave numbers, the magnetic energy distribution follows a power-law with spectral index $s_\perp = -3.1$ and -3.0 for S1 and S2, respectively. After the break, the spectra steepen, and significant differences between both simulations become obvious in the different spectral indices.

The numerical noise level in simulation S2 is about one order of magnitude lower than in S1, which allows an energy cascade to higher wave numbers. This can be explained by the lower plasma temperature in S2, leading to less kinetic energy of the particles and therefore less fluctuations in the electromagnetic fields. The flatter spectrum in S2 (after the break; compared to S1) agrees with results from Chang [30], who reported a more efficient perpendicular energy transport with decreasing plasma beta β.

Chang [30] also observed stronger anisotropy in simulations with lower β. However, this is not supported by the data from simulations S1 and S2. The parallel spectra $E_B(k_\parallel)$ are depicted in Figure 4. In both cases, the parallel spectra do not contain a break and are steeper than the perpendicular spectra at small wave numbers. For S1, the parallel spectrum reaches the noise level approximately at the position where cyclotron damping is assumed to set in. Figure 4a,b, however, shows that the parallel spectrum in simulation S2 extends to wave numbers in the damped regime. The slope does not change at the transition into the dissipation range and is flatter than the slope in the perpendicular spectrum at corresponding $k_\perp$. Thus, the parallel energy transport is assumed to dominate at large wave numbers. Unfortunately, the turbulent cascade reaches the numerical noise level prior to that.

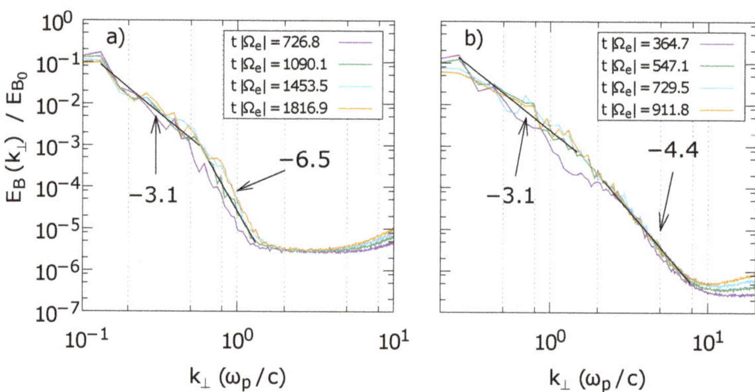

Figure 3. Normalized magnetic field energy distribution $E_B(k_\perp)/E_{B_0}$ over the perpendicular wave number, $k_\perp$ normalized to c/ω_p, for simulations S1 (**a**) and S2 (**b**). Here, E_{B_0} is the magnetic field energy of the background, ω_p is the plasma frequency, and c is the speed of light. The data are obtained at four times $t|\Omega_e|$ as indicated, where Ω_e is the electron gyrofrquency. At the earliest time steps, the spectra reach their steady states. Later in the simulations, the shapes of the spectra do not change significantly. Power-law fits to the data are indicated by the black lines at times $t|\Omega_e| = 1090.1$ (**a**) and 547.1 (**b**). The corresponding spectral indices are highlighted by the arrows.

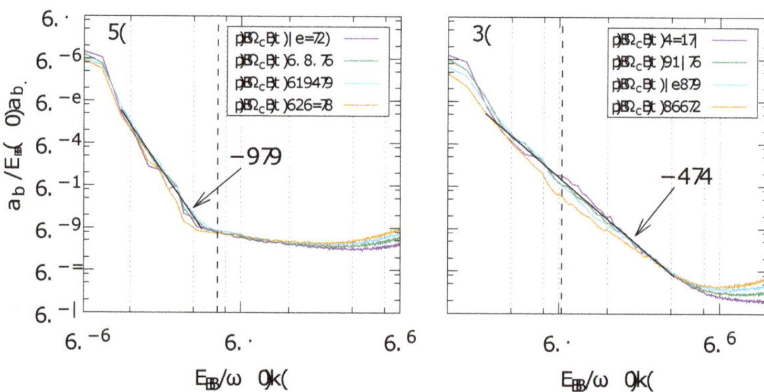

Figure 4. Normalized magnetic field energy distribution $E_B(k_\parallel)/E_{B_0}$ over the parallel wave number, $k_\parallel$ normalized to c/ω_p, for simulations S1 (**a**) and S2 (**b**). The data are obtained at four times $t|\Omega_e|$ as indicated. The spectra reach their steady state at the earliest time steps. Power-law fits to the data are indicated by the black lines at times $t|\Omega_e| = 1090.1$ (**a**) and 547.1 (**b**). The spectral indices are highlighted by the arrows. The dashed vertical lines indicatethe expected onset of cyclotron damping for purely parallel propagating waves.

The energy distribution $E_B(k_\parallel, k_\perp)$ in two-dimensional wave number space supports the claim that parallel energy transport becomes important in simulation S2, as Figure 5 shows. Figure 5b shows the distribution of magnetic field energy in simulation S2. Although hardly any (quasi-)parallel waves are produced above $k_\parallel c/\omega_p \approx 1$ (where cyclotron damping sets in), this critical parallel wave number can be passed at higher $k_\perp$. At small wave numbers, however, the perpendicular cascade clearly dominates. In simulation S1, the situation is different, as Figure 5a shows. The perpendicular cascade at small wave numbers is similar to S2, as expected, but at larger $k_\perp$, there is hardly any energy transport to higher parallel wave numbers, in agreement with the spectra given in Figures 3 and 4.

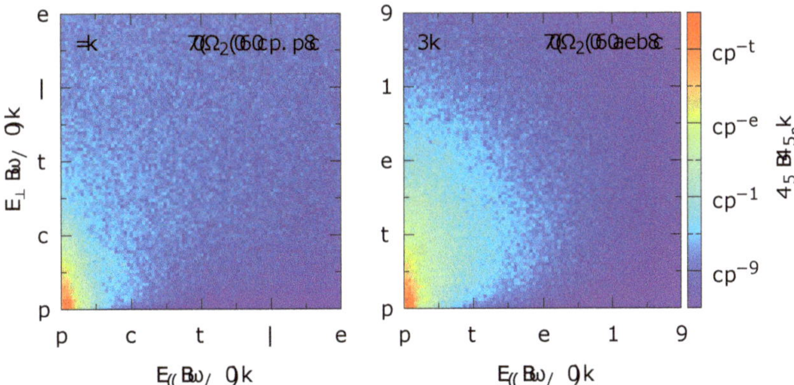

Figure 5. Two-dimensional magnetic field energy distribution in wave number space for simulations S1 (**a**) and S2 (**b**). Note different scales on the axes of (**a**) vs. (**b**).

3.2. Simulation of Energetic Particles

In order to study wave-particle scattering, a specific initialization of a test particle population is prepared. The ACRONYM code allows for different particle species (typically protons and electrons, but also positrons or heavier ions) and different particle populations (a background plasma and, for example, additional jet populations, non-thermal particles, etc.).

The simulations S1 and S2, discussed here, employ a thermal background plasma (see Table 1) and an additional population of non-thermal test particles to study the transport of energetic electrons. It is found that the test particles have no noticeable influence on the background plasma, even if the ratio $N_\mathrm{t}/N_\mathrm{bg}$ of numerical particles in the test to the background particle population is of the order of unity.

3.2.1. Initialization and Analysis

Test particles are initialized as a mono-energetic population; i.e., the particles have the same absolute speed, but their direction of motion is chosen randomly. The speed is calculated from the resonance condition for waves in the plasma. Solving Equation (18) for the speed of a particle of species α yields:

$$v_\alpha = \left| \frac{k_\parallel \, \omega \, |\mu_\mathrm{res}| \pm |\Omega_\alpha| \sqrt{k_\parallel^2 \, \mu_\mathrm{res}^2 + (\Omega_\alpha^2 - \omega^2)/c^2}}{k_\parallel^2 \, \mu_\mathrm{res}^2 + \Omega_\alpha^2/c^2} \right|, \tag{47}$$

where μ_res is the desired resonant pitch-angle cosine. The sign in the numerator changes depending on the polarization of the wave, its direction of propagation, and the particle species.

The directions of motion of the bulk of the test particles are chosen at random, using the speed calculated from Equation (47), a random polar angle cosine μ, and a random azimuth angle ϕ. This yields an isotropic distribution of the velocity vectors in μ-ϕ space. It is convenient to choose an isotropic distribution in $\mu = \cos\theta$ (instead of θ), because the analysis of pitch-angle scattering relies on the pitch-angle cosine and not on the pitch-angle itself.

A fraction of the test particle population is not initialized as described above, but instead uses a parabolic distribution of polar angle cosines. This is done by assigning

$$\mu = A \left(R + B\right)^{1/3} - C \tag{48}$$

to the particles, where R is a random number between zero and one, and A, B, and C are parameters describing the shape of the parabola. The parabolic distribution is required for the analysis of pitch-angle scattering using the method of Ivascenko et al. [40] Ivascenko et al. suggest the use of a half-parabola, i.e., $A = 2$, $B = 0$, $C = 1$, but other distributions are also possible (see Figure 6). The resulting angular distribution of the entire test particle population is, therefore, not entirely isotropic.

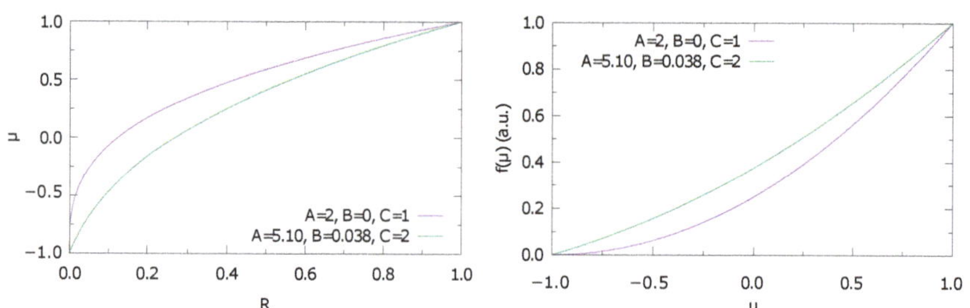

Figure 6. A fraction of the test particle population has pitch-angle cosines assigned according to a parabolic distribution. **Left panel:** the assigned pitch-angle cosine μ as a function of the random number $R \in [0, 1]$, which is used to generate the distribution. The curves follow Equation (48) and employ two sets of parameters A, B, and C, as indicated. **Right panel:** the resulting particle distribution $f(\mu)$ as a function of μ. The purple lines employ the parameters suggested by Ivascenko et al. [40], whereas the green curves show an improved implementation that is used in the ACRONYM code. Note that the derivative $df(\mu)/d\mu \neq 0$ over the whole range of pitch-angle cosines in the latter case, whereas it becomes zero at $\mu = -1$ when the parameters of Ivascenko et al. [40] are used.

The technique described above to create a population of energetic test particles for the study of wave-particle scattering was designed for a single plasma wave in the simulation [41]. However, it can also be applied to simulations with several plasma waves.

The test particle population is not injected at the start of the simulations S1 and S2 but at a later time t_{inj} for the following reason: it is expected that turbulence develops from the initial conditions of the simulation, i.e., from a small set of seed waves that interact and start the turbulent cascade. This process takes time, and it may be desired to wait until a turbulent cascade is established before the transport of energetic test particles can be studied. Therefore, an optional deployment of test particles at later times in the simulation is favored. The particles are created at a a pre-defined time step, and the initialization is carried out as described earlier. Those particles can then be tracked for the rest of the simulation.

To evaluate particle transport in the turbulent plasma, the test particle data can be analyzed after the simulation to obtain the diffusion coefficient $D_{\mu\mu}$. The diffusion coefficient is calculated from a simplified Fokker–Planck equation, Equation (22), where pitch-angle diffusion is assumed to be the only relevant diffusion process:

$$\frac{\partial f_\alpha}{\partial t} - \frac{\partial}{\partial \mu} D_{\mu\mu} \frac{\partial f_\alpha}{\partial \mu} = 0. \tag{49}$$

This equation can be rewritten to yield

$$\frac{\partial f_\alpha(\mu, t)}{\partial t} = \left(\frac{dD_{\mu\mu}(\mu)}{d\mu}\right) \frac{\partial f_\alpha(\mu, t)}{\partial \mu} + D_{\mu\mu}(\mu) \frac{\partial^2 f_\alpha(\mu, t)}{\partial \mu^2}. \tag{50}$$

The method described by Ivascenko et al. [40] is based on integrating Equation (49) over μ, which yields the pitch-angle current j_μ:

$$\int_{-1}^{\mu} \mu \, \frac{\partial f_\alpha(\mu, t)}{\partial t} = D_{\mu\mu}(\mu) \, \frac{\partial f_\alpha(\mu, t)}{\partial \mu} = -j_\mu. \tag{51}$$

The diffusion coefficient is then obtained by dividing j_μ by $\partial f_\alpha / \partial \mu$.

3.2.2. Physical Parameters

Using the setups of simulations S1 and S2, the transport of energetic electrons in kinetic turbulence is studied. In the following, the exact parameters for the test particle energy distribution are presented.

In simulations S1 and S2, decaying whistler turbulence is simulated, as was shown in the Section 3.2. As can be seen in the magnetic energy spectra presented in Figures 3 and 4, a steady state in terms of the power-law slope of the spectral energy distribution is established after a given time in each of the two simulations. As soon as this stage of the simulation is reached, a population of energetic test electrons can be injected as described in Section 3.2.1.

The time step for the checkpoint and subsequent restart is chosen to be $t\,|\Omega_e| = 726.8$ for S1 and $t\,|\Omega_e| = 364.7$ for S2. For each of these two setups, six test electron configurations are prepared. The simulations are labeled according to the physical setup (S1 or S2) followed by a letter referring to the test particle configuration ("a" through "f"). The parameters of the test particles can be found in Table 3 and describe the test electron speed v_e and kinetic energy E_e.

Table 3. Test electron characteristics for the simulations of particle transport: test electron speed v_e and corresponding kinetic energy $E_{\mathrm{kin},e}$. The individual simulations (letters "a" through "f") are based on the simulations of kinetic turbulence Sj with $j = 1$ and 2, which are described in Section 3.1.2 (Tables 1 and 2). Note that simulations Sje and Sjf employ the same test electron energies. However, they differ in the way the test electron distribution is initialized (see text).

Simulation	Sja	Sjb	Sjc	Sjd	Sje	Sjf
v_e (c)	0.546	0.862	0.941	0.979	0.999	0.999
$E_{\mathrm{kin},e}$ (eV)	1.0×10^5	5.0×10^5	1.0×10^6	2.0×10^6	1.0×10^7	1.0×10^7

The test electron energy is increased from simulation S1a (S2a) to S1e (S2e). Simulation S1f (S2f) uses the same particle energy as S1e (S2e), but a different parabolic angular distribution of the particles. Here the particle density $f(\mu)$ increases with increasing μ, while in the other simulations it decreases with increasing pitch-angle cosine. This change in the pitch-angle distribution allows to check for systematic errors in the particle data.

4. Results

Pitch-Angle Diffusion Coefficients

The test particle simulations are analyzed as described in Section 3.2.1. The energetic electrons are tracked for several electron cyclotron time scales, and the resulting pitch-angle diffusion coefficients $D_{\mu\mu}$ are presented in Figures 7 and 8 for data based on the setup of S1 and S2, respectively. Time is measured as the interval Δt from the time of the injection of the particles to the current time step.

The results of both sets of simulations, one based on S1 and the other based on S2, do not differ qualitatively, as would be expected from the two setups. The only difference between the physical parameters for S1 and S2 is the plasma temperature, which has no direct influence on the test electrons. Although the magnetic energy spectrum differs at high wave numbers (see Figure 3), the distribution of magnetic energy at small wave numbers is very similar. As it is explained below, this low-wave-number regime represents the dominant influence on particle transport. Thus, the two sets of simulations are discussed simultaneously in what follows.

Although particle data can be obtained for an arbitrary number of time steps, the interval that can be used for the analysis is still limited. The method of Ivascenko et al. [40] critically depends on the particle distribution $f(\mu)$ in pitch-angle space. In order for the method to work, the initial distribution must be slightly disturbed, but the perturbations must not be too strong. This leaves only a brief period of time for the optimal efficiency of the method.

Figures 7a and 8a show the typical behavior of the derived $D_{\mu\mu}$ over time. Shortly after the injection of the test electrons, the perturbations of $f(\mu)$ are small, resulting in a low amplitude of $D_{\mu\mu}$ (purple lines). With increasing time, the amplitude grows and reaches a maximum (green and blue lines). At later times, the amplitude decreases again, as the perturbations become too strong and the method becomes unreliable (orange lines). The other panels of the two figures show the time evolution until the maximum amplitude of $D_{\mu\mu}$ is reached for other test electron energies.

Although the turbulent cascade is assumed to be symmetric about $\mu = 0$, the panels of Figures 7 and 8 show an obvious asymmetry in the pitch angle diffusion coefficients derived from the test electron data. The amplitude of $D_{\mu\mu}$ is generally larger for $\mu < 0$. While the energy spectrum itself is isotropic in μ, one could argue that the polarization of the waves' magnetic fields relative to the direction of the background magnetic field B_0 is different (i.e., the plasma physics definition of the polarization).

The magnetic helicity of the plasma waves is one of the possible causes of this anisotropy. Another reason for the asymmetry found in Figures 7 and 8 is that the parabolic distribution $f(\mu)$ of the test particles implies that there are more test electrons at negative pitch-angle cosines (except for simulations S1f and S2f). Therefore, the particle statistics is more reliable for negative μ, and the method of Ivascenko et al. [40] produces more accurate diffusion coefficients. While $D_{\mu\mu}$ can also be calculated for $\mu > 0$, it is more prone to errors, and statistical fluctuations play a more important role, as Figure 9 indicates. However, small-scale statistical fluctuations can be suppressed by use of a Savitzky-Golay filter, as suggested by Ivascenko et al. [40].

Especially for early time steps, it can be seen that $D_{\mu\mu}$ is found to diverge at $\mu = 1$. This is, of course, not a physical effect. At $\mu = 1$, the derivative of the initial parabolic distribution $f(\mu)$ becomes (almost) zero. In this case, the method of Ivascenko et al. [40] becomes numerically unstable.

Another numerical effect causes $D_{\mu\mu}$ to become negative. This can be seen in Figure 8d and Figure 8e for early times. Negative solutions are most likely related to statistical fluctuations in the particle distribution, which drown the signal at early times, when the physically motivated perturbations of $f(\mu)$ are still developing.

Besides these flaws, the derived pitch-angle diffusion coefficients appear reasonable. They develop a (more or less) symmetric shape about $\mu = 0$, indicating that neither direction is preferred. This is expected from the setup of the turbulence simulations S1 and S2, which employ a symmetric layout of initial waves and therefore should produce turbulent cascades that are symmetric in μ. This, however, cannot be proven by the plots of the energy distribution in wave number space, since the information about the direction of propagation of the waves is lost.

An interesting observation is that the pitch-angle diffusion coefficients grow in amplitude with the particle energy increasing from 100 keV to 2 MeV. At the highest test electron energy, however, the amplitude of $D_{\mu\mu}$ is significantly lower than in all other cases. Both Figures 7e and 8e also show that $D_{\mu\mu}$ forms a single peak close to $\mu = 0$ in the case of the highest electron energy, whereas all other simulations produce a double peak structure. The reason for these differences is not clear. However, it is assumed that the different behavior of the 10 MeV electrons is related to their scattering characteristics. These high energy particles resonate with all of the initially excited waves in the simulations (see Figure 2), which is not the case in the simulations of less energetic electrons. Since the initial waves contain the most energy, they are also assumed to significantly influence particle transport, especially if wave–particle resonances may occur.

In fact, the $D_{\mu\mu}$ in Figures 7e and 8e exhibit distinct peaks at early times (purple and green curves). Similar behavior is also found in simulations S1f and S2f, which are not included in Figures 7 and 8. For the example of one time step in simulation S1f, the peak structures in $D_{\mu\mu}$ are related to wave–particle resonances calculated according to Equation (18). The result is shown in Figure 10, where the colored vertical lines mark the expected positions of resonances. It can be seen that the resonances coincide with the positions of the peaks in $D_{\mu\mu}$. The region around $\mu = 0$ is most densely populated by resonances, which might explain the single peak in $D_{\mu\mu}$ at later times as seen in Figures 7e and 8e.

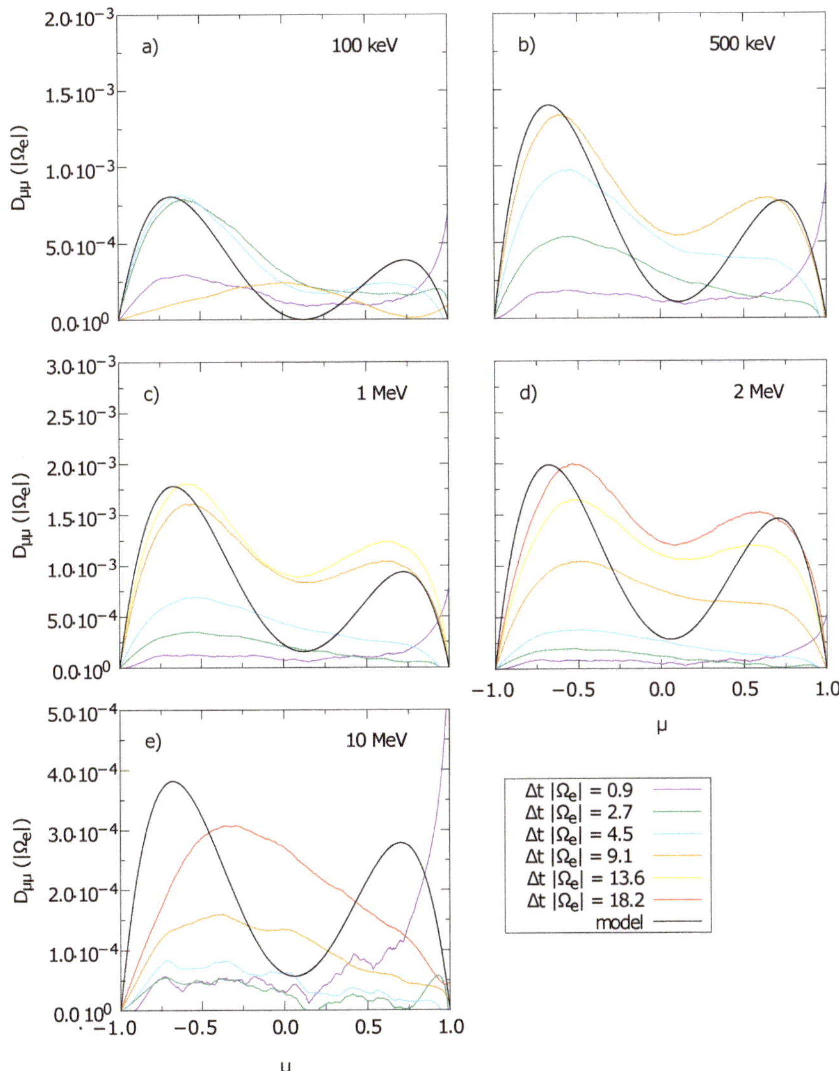

Figure 7. Pitch-angle diffusion coefficients $D_{\mu\mu}$ for test electrons with different energies for simulations S1a to S1e (**a**–**e**). The colored lines denote the diffusion coefficients derived from the simulation data at various times. The black lines represent the model predictions derived in Section 2.3.3.

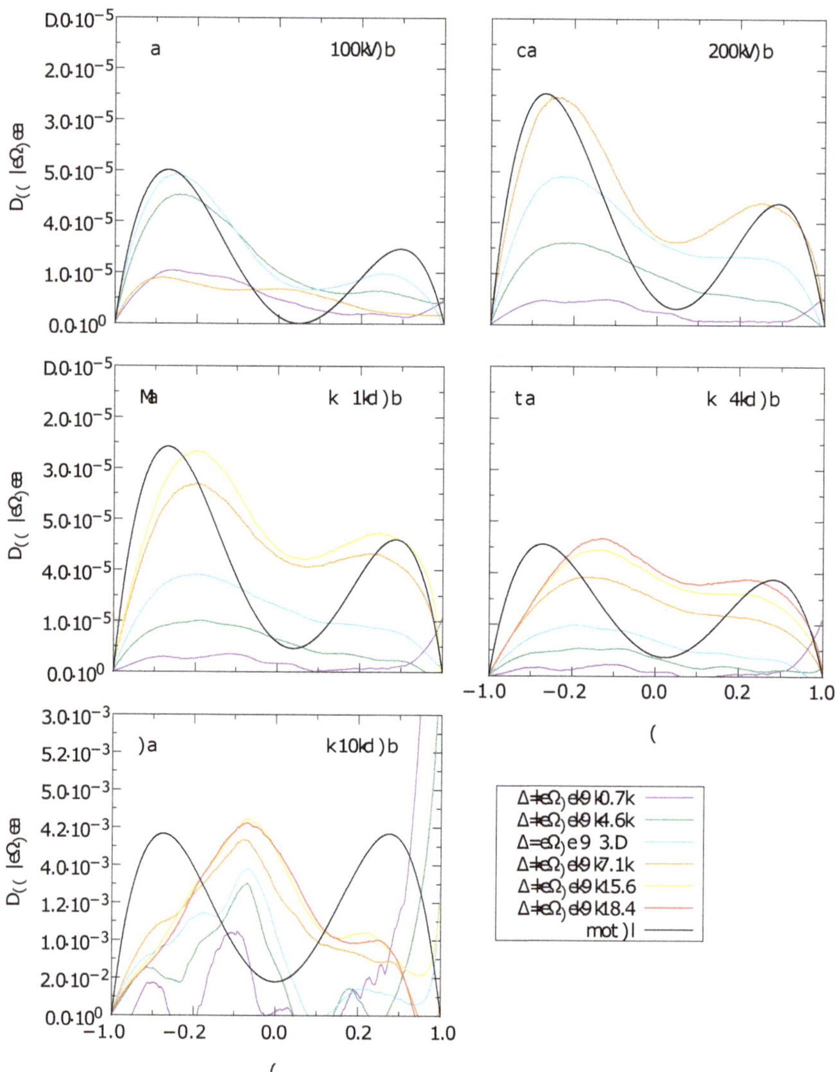

Figure 8. Pitch-angle diffusion coefficients $D_{\mu\mu}$ for test electrons with different energies for simulations S2a to S2e (**a**–**e**). The colored lines denote the diffusion coefficients derived from the simulation data at various times. The black lines follow the model predictions derived in Section 2.3.3.

Finally, Figures 7 and 8 also include the model predictions from Equations (41) and (42). Some of the parameters required can be directly obtained from the setup of the simulations: the ratio $\delta B^2/B_0^2$ is listed in Table 1, and the test electron speed speed v_e and the electron cyclotron frequency Ω_e are found in Table 3. However, the minimum wave number $k_{\min}$, the spectral index s, and the cross helicity and magnetic helicity are not as trivial to find.

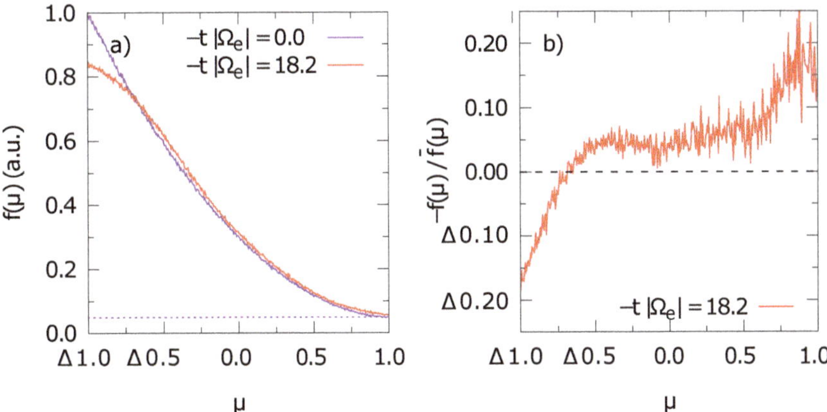

Figure 9. Test electron distribution in pitch-angle space: (**a**) the initial distribution $f(\mu)$ at the time of the injection of the test electrons ($\Delta t = 0$) and at a later time in simulation S1c; (**b**) the relative deviation $\Delta f / \bar{f}$ of the distributions at these two time steps. The deviation is defined as the difference of the two distributions over their mean value. Statistical fluctuations are visible to become more significant for larger μ values.

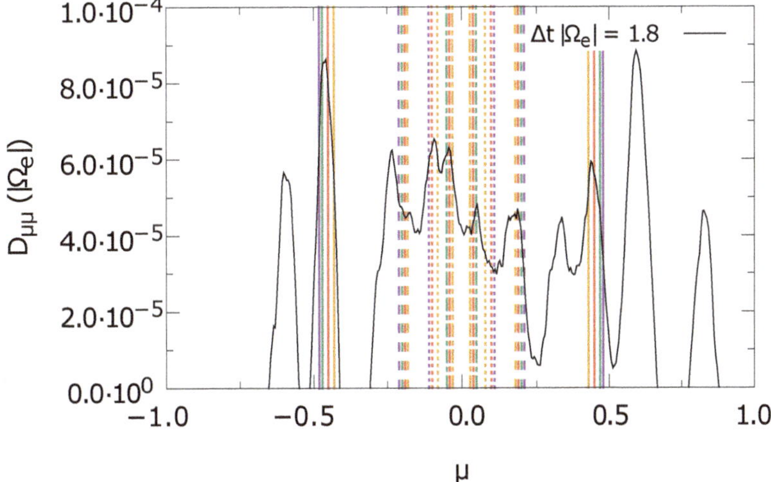

Figure 10. Pitch-angle diffusion coefficient $D_{\mu\mu}$ at one point in time as derived from the data of simulation S1f (black line). A noticeable number of peaks in $D_{\mu\mu}$ coincide with the positions of wave–particle resonances predicted by the resonance condition (18), which are marked by the colored, vertical lines. The colors denote the parallel wave numbers $k_\parallel$ (in numerical units) from one to four: purple, green, red, and orange. The line style refers to perpendicular wave numbers $k_\perp$ (also in numerical units) from zero to three: solid, dashed, dotted, dot-dashed. For example, the red dashed lines represent the resonance with a wave at ($k_\parallel = \pm 3$, $k_\perp = 1$). Only resonances of the first order, i.e., $\mathcal{N} = \pm 1$ in Equation (18), are shown. Note that $k_\perp$ does not enter the resonance condition explicitly but is required to calculate the frequency $\omega(k_\parallel, k_\perp)$ according to the cold plasma dispersion relation.

For the minimum wave number, the magnetic energy spectra in Figures 3 and 4 have been considered. The second smallest resolved wave number $k_{\min} = 2 \Delta k$ has been chosen, where Δk is the grid spacing in wave number space. In a square simulation box, where the numbers of grid cells $N_\parallel$ and $N_\perp$ in the parallel and perpendicular directions are equal, the grid spacing is given by $\Delta k = 2\pi/(N_\parallel \Delta x) = 2\pi/(N_\perp \Delta x)$. The minimum wave number $k_{\min}$ marks the beginning of the downward slope of the energy spectrum. Waves at small wave numbers are assumed to dominate the interaction with the particles due to their high energy content and the steep spectral slope. Therefore, the spectral index $s = |s_\perp| = 3.1$ was chosen, in accordance with the index of the perpendicular spectrum in Figure 3a. This spectral index corresponds to simulation S1, but since the index in S2 is similar, $s = 3.1$ was used in both cases.

Finally, the magnetic helicity σ was chosen to be 0. The effect is in fact rather small since electrons mostly resonate with right-handed polarized modes.

From this starting point, the three parameters $k_{\min}$, s, and H_R were fitted according to the numerical data from each simulation. The resulting parameters, which are used in the plots in Figures 7 and 8, are listed in Table 4. It can be seen that most simulations can be described with the initial choices for $k_{\min}$ and s explained above.

Table 4. Parameters assumed for the model: spectral index s, cross-helicity H_R, and minimum wave number $k_{\min}$. The latter is given in units of the grid spacing $\Delta k = \{4.4,\, 8.7\} \times 10^{-2}\, \omega_p/c$ in wave number space in simulations S1 and S2, respectively.

Simulation	S1a	S1b	S1c	S1d	S1e	S1f	S2a	S2b	S2c	S2d	S2e	S2f
s	3.1	3.1	3.1	3.1	3.1	3.1	3.1	3.1	3.1	3.1	3.1	3.1
H_R	1.00	0.55	0.55	0.26	0.26	0.25	0.99	0.59	0.42	0.26	0	0
$k_{\min}\,(\Delta k)$	2	2	2	2	1	1	2	2	2	2	1	1

In general, the model describes the data surprisingly well. Position and amplitude of the maxima and the inclination of the flanks are in good agreement. The contributions at $\mu = 0$ are in disagreement; this is however not unexpected as for quasi-linear theory. Still, a non-zero contribution at medium energies is found, which is different from a non-dispersive quasi-linear approach. The agreement of the model and the simulation results also supports the claim that the waves at small wave numbers dominate the interactions with the particles. Otherwise, the spectral index s would have to be changed according to the particle energy. The low energy particles, e.g., 100 keV, resonate with plasma waves in the high wave number regime, where the spectrum is steeper. Thus, according to Figures 3 and 4, the effective spectral index s should increase for these particles, if the resonant interactions with high-k waves were important. However, this seems not to be the case. But as the results in Figures 7 and 8 show, a change of the model equations for $D_{\mu\mu}$ is not necessary.

The model only fails for the simulations of 10 MeV-electrons (S1e, S1f, S2e, S2f). This might already be expected from the considerations discussed above: the high energy electrons are able to resonate with the initially excited waves. These waves contain the most energy and thus dominate the interaction of the particles with the turbulent spectrum. However, the initial waves cannot be considered to be part of the power-law spectrum itself. As Figures 3 and 4 show, the energy distribution forms a plateau at smallest wave numbers, where the initial waves are located. The initial waves are also only few in number, thus not forming a continuous spectrum but a population of distinct, individual waves. Representing a continuous spectrum on a discretized grid is always doomed to fail, but at larger wave numbers, the higher number of individual waves at least creates a rudimentary approximation of a continuum. Thus, the whole model assumption, i.e., a continuous power-law spectrum, is invalid. As Figure 10 shows, the pitch-angle diffusion coefficient derived from the simulation data can be described reasonably well by individual resonances with a number of waves.

Finally, it is worth taking a look at simulations S1f and S2f, which have not been discussed so far. These simulations, which employ the same test electron energies as S1e and S2e, were carried out to test whether the initial particle distribution $f(\mu)$ has an influence on the resulting $D_{\mu\mu}$. It was already discussed above that the statistical fluctuations tend to become more noticeable at those μ where fewer particles are located. Thus, reversing the slope of the initial parabola should shift the dominant influence of statistical fluctuations from positive μ to negative.

Figure 11 depicts the pitch-angle diffusion coefficients derived from simulations S2e and S2f in panels a and b, respectively. This example is chosen because physical results for $D_{\mu\mu}$ are only obtained at negative μ at early times in S2e. Should this also be the case in S2f, this would mean that some physical process prefers the interaction of waves and energetic particles that propagate opposite to the background magnetic field. However, as Figure 11 shows, this is not the case. The pitch-angle diffusion coefficient derived from S2f appears to be more symmetric about $\mu = 0$ at early times.

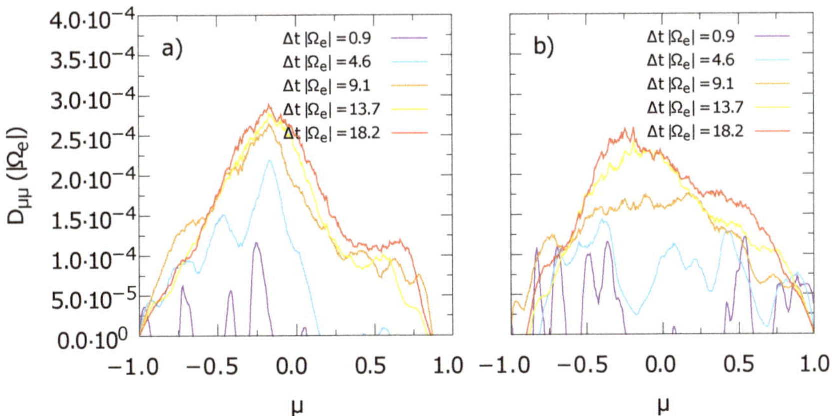

Figure 11. Comparison of the pitch-angle diffusion coefficients $D_{\mu\mu}$ derived from the test electron data of simulations S2e (**a**) and S2f (**b**). The two simulations differ by the slope of the parabolic particle distribution $f(\mu)$ used to initialize the test electrons. The asymmetry of $D_{\mu\mu}$ in S2e at early times is not reproduced by S2f, which suggests a numerical or statistical reason for the asymmetry. At late times, the $D_{\mu\mu}$ become similar, with a single peak near $\mu = 0$ in both simulations.

At late times both simulations produce a single peak in $D_{\mu\mu}$, which is located near $\mu = 0$. The peak is slightly shifted to negative μ in both S2e and S2f. This may hint at a physical process leading to the peak not being centered exactly around $\mu = 0$. Such an asymmetry is sometimes predicted in theoretical models, e.g., Schlickeiser [11]. However, considering the results of simulations of energetic particles and their interaction with individual waves, an asymmetry is not expected here.

Thus, the results of simulations S2e and S2f depicted in Figure 11 do not entirely agree with the expectations. It might be worthwhile to investigate the behavior of the pitch-angle diffusion coefficient in more detail in a future project. Changing the initial particle distribution $f(\mu)$ once more (e.g., by altering the parameters A, B, and C in Equation (48)) or reversing the direction of the integration over μ in the method of [40] might help to distinguish between a physically motivated asymmetry and numerical artifacts.

5. Conclusions

In this paper, a set of pitch-angle diffusion coefficients for dispersive whistler waves are derived. Using a particle-in-cell code turbulence in the dispersive regime was simulated. Test particle electrons were injected into the simulated turbulence and their transport parameters were derived.

The conducted turbulence simulations yield power-law spectra of the magnetic field energy in wave number space. The measured spectral indices are in agreement with the findings of Refs. [27,29]. Numerical noise limits the energy spectra at high wave numbers, thus hindering the production of an energy cascade in the dissipation range.

While the theory is limited to parallel waves, simulations were performed in two-dimensional wave-number space. The theoretical description of oblique, dispersive waves is not practically doable, while one-dimensional turbulence simulations are not producing an energy cascade. The approximation of a parallel spectrum makes this difference between dimensionalities reasonable.

The simulations of energetic particle transport in kinetic turbulence show that the steep energy spectrum leads to wave–particle interactions primarily in the low wave number regime. While low-energy particles, in principle, resonate with waves in the dispersive or dissipative regime of the turbulent cascade, these interactions are subordinate to interactions with non-resonant waves at lower wave numbers. The reason for this is that the energy content of dispersive waves decreases rapidly with increasing wave number due to the steep power-law spectrum. Thus, the waves at low wave numbers dominate the spectrum as far as particle transport is concerned.

This can be seen when comparing simulation data to the theoretical model. The test electron data from the simulations allows us to derive pitch-angle diffusion coefficients $D_{\mu\mu}$ using the method of [40]. The presented model for $D_{\mu\mu}$ in plasma turbulence with dispersive waves allows for the prediction of pitch-angle diffusion coefficient for Alfvén and whistler turbulence.

Simulation data and model match rather well for low-energy electrons. Contributions at $\mu = 0$ are not modeled correctly as is expected for a quasi-linear model. The cross-helicity assumed in model parameters may not necessarily represent the cross-helicity of the plasma, but may be to some degree a numerical artifact. At higher electron energies, particles interact with the small number of excited plasma waves, which are used as a seed population for the generation of kinetic turbulence. The resulting $D_{\mu\mu}$ does not match the prediction for the interaction with the (continuous) turbulent spectrum but can be explained by resonant scattering with several waves at discrete wave numbers.

In general simulations, dispersive whistler turbulence and the corresponding particle transport are possible but are also still too expensive in terms of computing resources.

Author Contributions: Conceptualization, F.S., C.S. and R.S.; methodology, F.S. and R.S.; software, C.S.; writing—original draft preparation, F.S., C.S. and R.S.; writing—review and editing, F.S., C.S. and R.S.; visualization, F.S. and C.S.; funding acquisition, F.S. All authors have read and agreed to the published version of the manuscript.

Funding: This research was funded by Deutsche Forschungsgemeinschaft (DFG) through Grant No. SP 1124/9.

Institutional Review Board Statement: Not applicable.

Informed Consent Statement: Not applicable.

Data Availability Statement: Not applicable.

Acknowledgments: F.S. would like to thank the Deutsche Forschungsgemeinschaft (DFG) for support through Grant No. SP 1124/9. The authors gratefully acknowledge the data storage service SDS@hd supported by the Ministry of Science, Research and the Arts Baden-Württemberg (MWK) and the German Research Foundation (DFG) through grant INST 35/1314-1 FUGG and INST 35/1503-1 FUGG. The authors gratefully acknowledge the Gauss Centre for Supercomputing e.V. (www.gauss-centre.eu, accessed on 30 November 2021) for funding this project by providing computing time on the GCS Supercomputer SuperMUC at Leibniz Supercomputing Centre (www.lrz.de, accessed on 30 November 2021) through grant pr84ti. We acknowledge the use of the *ACRONYM* code and would like to thank the developers (Verein zur Förderung kinetischer Plasmasimulationen e.V.) for their support.

Conflicts of Interest: The authors declare no conflict of interest.

References

1.   Borovsky, J.E.; Funsten, H.O. Role of solar wind turbulence in the coupling of the solar wind to the Earth's magnetosphere. *J. Geophys. Res. Space Phys.* **2003**, *108*, 1246. [CrossRef]

2. Sridhar, S.; Goldreich, P. Toward a theory of interstellar turbulence. I: Weak Alfvénic turbulence. *Astropart. Phys.* **1994**, *432*, 612–621. [CrossRef]
3. Goldreich, P.; Sridhar, S. Toward a theory of interstellar turbulence. II: Strong Alfvénic turbulence. *Astropart. Phys.* **1995**, *438*, 763–775. [CrossRef]
4. Sahraoui, F.; Goldstein, M.L.; Belmont, G.; Canu, P.; Rezeau, L. Three Dimensional Anisotropic k Spectra of Turbulence at Subproton Scales in the Solar Wind. *Phys. Rev. Lett.* **2010**, *105*, 131101. [CrossRef] [PubMed]
5. Galtier, S.; Nazarenko, S.V.; Newell, A.C.; Pouquet, A. A weak turbulence theory for incompressible magnetohydrodynamics. *J. Plasma Phys.* **2000**, *63*, 447–488. [CrossRef]
6. Galtier, S.; Nazarenko, S.V.; Newell, A.C.; Pouquet, A. Anisotropic Turbulence of Shear-Alfvén Waves. *Astrophys. J.* **2002**, *564*, L49–L52. [CrossRef]
7. Howes, G.G. Kinetic turbulence. In *Magnetic Fields in Diffuse Media*; Lazarian, A., de Gouveia Dal Pino, E., Melioli, C., Eds.; Springer: Berlin/Hedelberg, Germany, 2015; pp. 123–152. [CrossRef]
8. Gary, S.P.; Smith, C. Short-wavelength turbulence in the solar wind: Linear theory of whistler and kinetic Alfvén fluctuations. *J. Geophys. Res.* **2009**, *114*, A12105. [CrossRef]
9. Jokipii, J.R. Cosmic-Ray Propagation. I. Charged Particles in a Random Magnetic Field. *Astrophys. J.* **1966**, *146*, 480–487. [CrossRef]
10. Lee, M.A.; Lerche, I. Waves and Irregularities in the Solar Wind. *Rev. Geophys. Space Phys.* **1974**, *12*, 671–687. [CrossRef]
11. Schlickeiser, R. Cosmic-ray transport and acceleration. I-Derivation of the kinetic equation and application to cosmic rays in static cold media. II-Cosmic rays in moving cold media with application to diffusive shock wave acceleration. *Astrophys. J.* **1989**, *336*, 243–293. [CrossRef]
12. Steinacker, J.; Miller, J.A. Stochastic gyroresonant electron acceleration in a low-beta plasma. I—Interaction with parallel transverse cold plasma waves. *Astrophys. J.* **1992**, *393*, 764–781. [CrossRef]
13. Vainio, R. Charged-Particle Resonance Conditions and Transport Coefficients in Slab-Mode Waves. *Astrophys. J. Suppl. Ser.* **2000**, *131*, 519. [CrossRef]
14. Achatz, U.; Dröge, W.; Schlickeiser, R.; Wibberenz, G. Interplanetary transport of solar electrons and protons: Effect of dissipative processes in the magnetic field power spectrum. *J. Geophys. Res.* **1993**, *98*, 13261–13280. [CrossRef]
15. Ng, C.K.; Reames, D.V. Focused interplanetary transport of approximately 1 MeV solar energetic protons through self-generated Alfvén waves. *Astrophys. J.* **1994**, *424*, 1032–1048. [CrossRef]
16. Vainio, R.; Kocharov, L. Proton transport through self-generated waves in impulsive flares. *Astron. Astrophys.* **2001**, *375*, 251–259. [CrossRef]
17. Lange, S.; Spanier, F.; Battarbee, M.; Vainio, R.; Laitinen, T. Particle scattering in turbulent plasmas with amplified wave modes. *Astron. Astrophys.* **2013**, *553*, A129. [CrossRef]
18. Gary, S.P.; Saito, S. Particle-in-cell simulations of Alfvén-cyclotron wave scattering: Proton velocity distributions. *J. Geophys. Res.* **2003**, *108*, 1194. [CrossRef]
19. Camporeale, E. Resonant and nonresonant whistlers-particle interaction in the radiation belts. *Geophys. Res. Lett.* **2015**, *42*, 3114–3121. [CrossRef]
20. Ma, C.Y.; Summers, D. Formation of power-law energy spectra in space plasmas by stochastic acceleration due to whistler-mode waves. *Geophys. Res. Lett.* **1998**, *25*, 4099–4102. [CrossRef]
21. Dröge, W. Particle Scattering by Magnetic Fields. In *Cosmic Rays and Earth: Proceedings of the an ISSI Workshop, Bern, Switzerland, 21– 26 March 1999*; Bieber, J.W., Eroshenko, E., Evenson, P., Flückiger, E.O., Kallenbach, R., Eds.; Springer: Dordrecht, The Netherlands, 2000; pp. 121–151. [CrossRef]
22. Coroniti, F.V.; Kennel, C.F.; Scarf, F.L. Whistler mode turbulence in the disturbed solar wind. *J. Geophys. Res.* **1982**, *87*, 6029–6044. [CrossRef]
23. Aguilar-Rodriguez, E.; Blanco-Cano, X.; Russel, C.T.; Luhmann, J.G.; Jian, L.K.; Ramírez Vélez, J.C. Dual observations of the interplanetary shocks associated with stream interaction regions. *J. Geophys. Res.* **2011**, *116*, A12109. [CrossRef]
24. Fairfield, D.H. Whistler waves observed upstream from collisionless shocks. *J. Geophys. Res.* **1974**, *79*, 1368–1378. [CrossRef]
25. Kennel, C.; Petschek, H. Limit on stably trapped particle fluxes. *J. Geophys. Res.* **1966**, *71*, 1–28. [CrossRef]
26. Palmroth, M.; Archer, M.; Vainio, R.; Hietala, H.; Pfau-Kempf, Y.; Hoilijoki, S.; Hannuksela, O.; Ganse, U.; Sandroos, A.; von Alfthan, S.; et al. ULF foreshock under radial IMF: THEMIS observations and global kinetic simulation Vlasiator results compared. *J. Geophys. Res. Space Phys.* **2015**, *120*, 8782–8798. [CrossRef]
27. Gary, S.P.; Saito, S.; Li, H. Cascade of whistler turbulence: Particle-in-cell simulations. *Geophys. Res. Lett.* **2008**, *35*. [CrossRef]
28. Che, H.; Goldstein, M.L.; Viñas, A.F. Bidirectional energy cascades and the origin of kinetic Alfvénic and whistler turbulence in the solar wind. *Phys. Rev. Lett.* **2014**, *112*, 061101. [CrossRef] [PubMed]
29. Gary, S.P.; Chang, O.; Wang, J. Forward cascade of whistler turbulence: Three-dimensional particle-in-cell simulations. *Astrophys. J.* **2012**, *755*, 142. [CrossRef]
30. Chang, O.; Gary, S.P.; Wang, J. Whistler turbulence at variable electron beta: Three-dimensional particle-in-cell simulations. *J. Geophys. Res. Space Phys.* **2013**, *118*, 2824–2833. [CrossRef]
31. Gary, S.P.; Hughes, R.S.; Wang, J.; Chang, O. Whistler anisotropy instability: Spectral transfer in a three-dimensional particle-in-cell simulation. *J. Geophys. Res. Space Phys.* **2014**, *119*, 1429–1434. [CrossRef]

32. Chang, O.; Gary, S.P.; Wang, J. Whistler Turbulence Forward Cascade versus Inverse Cascade: Three-dimensional Particle-in-cell Simulations. *Astrophys. J.* **2015**, *800*, 87. [CrossRef]
33. Schlickeiser, R.; Achatz, U. Cosmic-ray particle transport in weakly turbulent plasmas. Part 1. Theory. *J. Plasma Phys.* **1993**, *49*, 63–77. [CrossRef]
34. Swanson, D.G. *Plasma Waves*; Academic Press: San Diego, CA, USA, 1989. [CrossRef]
35. Schlickeiser, R. *Cosmic Ray Astrophysics*; Springer: Berlin/Heidelberg, Germany, 2002. [CrossRef]
36. Terry, P.W.; Almagri, A.F.; Fiksel, G.; Forest, C.B.; Hatch, D.R.; Jenko, F.; Nornberg, M.D.; Prager, S.C.; Rahbarnia, K.; Ren, Y.; et al. Dissipation range turbulent cascades in plasmasa. *Phys. Plasmas* **2012**, *19*, 055906. [CrossRef]
37. Hockney, R.W.; Eastwood, J.W. *Computer Simulation Using Particles*; CRC Press/Taylor & Francis Group: Boca Raton, FL, USA, 1988. [CrossRef]
38. Kilian, P.; Burkart, T.; Spanier, F. The influence of the mass ratio on particle acceleration by the filamentation instability. In *High Performance Computing in Science and Engineering'11*; Nagel, W.E., Kröner, D.B., Resch, M., Eds.; Springer: Berlin/Heidelberg, Germany, 2012; pp. 5–13. [CrossRef]
39. Schreiner, C.; Spanier, F. Wave-particle-interaction in kinetic plasmas. *Comput. Phys. Commun.* **2014**, *185*, 1981–1986. [CrossRef]
40. Ivascenko, A.; Lange, S.; Spanier, F.; Vainio, R. Determining pitch-angle diffusion coefficients from test particle simulation. *Astrophys. J.* **2016**, *833*, 223–231. [CrossRef]
41. Schreiner, C.; Kilian, P.; Spanier, F. Particle scattering off of right-handed dispersive waves. *Astrophys. J.* **2017**, *834*, 161–179. [CrossRef]

physics

Review

A Review on Scene Prediction for Automated Driving

Anne Stockem Novo [1,2,*], Martin Krüger [3,4], Marco Stolpe [4] and Torsten Bertram [3]

[1] Institute of Computer Science, Hochschule Ruhr West, 45479 Mülheim an der Ruhr, Germany
[2] Division Electronics and ADAS, ZF Automotive Germany GmbH, 45881 Gelsenkirchen, Germany
[3] Institute of Control Theory and Systems Engineering, TU Dortmund University, 44227 Dortmund, Germany; martin2.krueger@tu-dortmund.de (M.K.); Torsten.Bertram@tu-dortmund.de (T.B.)
[4] Automated Driving & Integral Cognitive Safety, ZF Automotive GmbH, 40547 Düsseldorf, Germany; marco.stolpe@zf.com
* Correspondence: anne.stockem-novo@hs-ruhrwest.de

Abstract: Towards the aim of mastering level 5, a fully automated vehicle needs to be equipped with sensors for a 360° surround perception of the environment. In addition to this, it is required to anticipate plausible evolutions of the traffic scene such that it is possible to act in time, not just to react in case of emergencies. This way, a safe and smooth driving experience can be guaranteed. The complex spatio-temporal dependencies and high dynamics are some of the biggest challenges for scene prediction. The subtle indications of other drivers' intentions, which are often intuitively clear to the human driver, require data-driven models such as deep learning techniques. When dealing with uncertainties and making decisions based on noisy or sparse data, deep learning models also show a very robust performance. In this survey, a detailed overview of scene prediction models is presented with a historical approach. A quantitative comparison of the model results reveals the dominance of deep learning methods in current state-of-the-art research in this area, leading to a competition on the cm scale. Moreover, it also shows the problem of inter-model comparison, as many publications do not use standardized test sets. However, it is questionable if such improvements on the cm scale are actually necessary. More effort should be spent in trying to understand varying model performances, identifying if the difference is in the datasets (many simple situations versus many corner cases) or actually an issue of the model itself.

Keywords: automated driving; data-driven modeling; deep learning; scene prediction; trajectory prediction

Citation: Stockem Novo, A.; Krüger, M.; Stolpe, M.; Bertram, T. A Review on Scene Prediction for Automated Driving. *Physics* **2022**, *4*, 132–159. https://doi.org/10.3390/physics4010011

Received: 1 November 2021
Accepted: 7 January 2022
Published: 1 February 2022

1. Introduction

In the last five years, huge progress has been made in the technology of self-driving cars. Advanced Driver Assistance Systems (ADAS) have become very mature and are part of any new vehicle nowadays. First, functions beyond the ADAS level, e.g., lane keeping or lane change assistance, were introduced into the series market of conventional Original Equipment Manufactures (OEMs). Even more compelling were the advances in higher automated driving levels made by Tesla with the autopilot and full self-driving capability functions [1], even though, other than the feature's names would suggest, the driver is still responsible and needs to monitor the driving process at all times.

The Society of Automotive Engineers (SAE) defines [2] six levels of Automated Driving (AD) (see Figure 1). ADAS systems are limited to levels 1 and 2, where the human driver is still the one operating the vehicle. Although the system can take full control already for specific use cases at level 2, the human driver needs to be attentive all the time, meaning eyes on the traffic at any time. It allows for temporary hands-off but the system reminds the driver to take back control after a short time period.

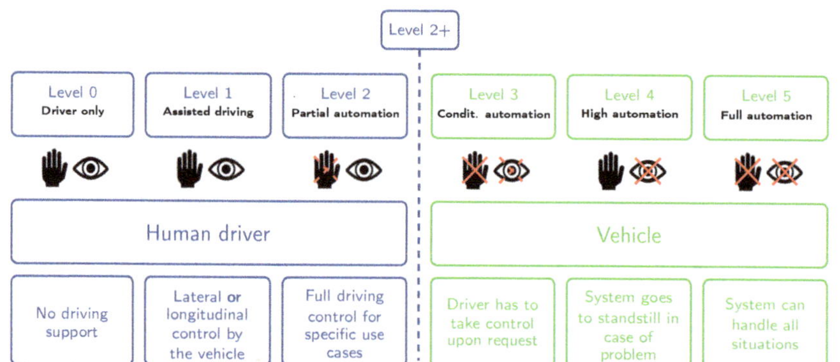

Figure 1. Automated driving (AD) levels according to the definition of the Society of Automotive Engineers.

The actual AD use case starts from level 3. The challenging task in designing a level 3 system is that a human driver who can go hands- and eyes-off must be notified on time to take back control in a period of around 10 s. Bridging the time span of 10 s in case of a complex situation is a great challenge such that some OEMs are discussing the strategy of skipping this level, going straight to high automation level 4. Here, in contrast to the reaction times of human drivers, the vehicle can handle problems on its own within just fractions of a second, or go to a fail-safe condition if no other solutions can be found. The final level 5 can handle all kinds of situations and use cases. A steering wheel might not be present anymore.

In 2019, the first autonomous taxi fleets were announced for the year 2020 [3]. However, this was delayed partially due to legislation and partially due to technical issues. Only slowly, the first test cases have been introduced. While the use case automated highway driving or level 4 driving on well-defined urban test fields can be handled quite well already, the big challenge is handling fully AD level 5 for all possible scenarios and corner cases. This includes strongly populated cities where it is not possible to anticipate the traffic scene for more than 1–2 s reliably or situations that occur very seldom.

The advantages of AD are numerous and quite obvious: an increase in comfort for the passengers, along with more flexibility for young, old or disabled persons. Furthermore, AD will reduce the number of accidents. A simulation study of fatal crashes predicted a reduction of collisions by 82% [4]. Naturally, also machines will fail but the amount and severeness of accidents is expected to be less compared to human drivers. This aspect is followed by a number of ethical questions, e.g., how to decide which action to take in case of an unavoidable crash? This is still a question under discussion. Moreover, human failures are better accepted than accidents caused by a machine. A further advantage to mention is that driverless systems can be optimized in economic aspects. Car sharing concepts are being designed which will reduce the number of required vehicles by efficient usage and fewer parking spaces will be needed. Traffic jams often occur due to inattentive drivers and unnecessary braking maneuvers [5]. This can easily be avoided in driverless traffic with the help of car-to-car communication [6].

While trying to reach the goal of fully automated vehicles, it is important having in mind the demands of the industry. First, AD vehicles need to be equipped with multiple sensors. However, the costs need to be low enough that either individual rides with a robo-taxi service or the purchase of an actual automated vehicle is affordable, which obviously addresses people of different income classes. Second, current systems are often requiring high-performing computer hardware. The aim is to go towards embedded systems which are drastically limited in run-time and memory. Last but not least, an AD system has to be robust and secure, meaning that faulty sensors or adversarial attacks [7,8] need to be handled reliably. In order to get permission to drive on public roads, a company needs to undergo a formal procedure that guarantees the functional safety of the system. This latter topic is not trivial and is still a road blocker for bringing some technical applications to the market.

Despite the recent advances in AD, there still exist challenges which are some of the main reasons why fully AD cars cannot be brought onto the streets yet. One of the most challenging problems focused on in this review and being a field of increasingly active research is the prediction of future traffic participants' behavior. The intentions that lead to certain maneuvers or actions are not always obvious but rather hidden as latent factors. Being able to identify such intentions several seconds before the event, makes it possible to comfortably adjust the own driving trajectory.

A second major challenge is the handling of uncertainties. The perception of the environment is associated with uncertainties, especially when fusing information from different sources. On the one hand, multiple signals provide redundancy and thus more reliability, on the other hand, creating higher uncertainties if the individual signals do not match. Based on these signals, robust environment models need to be developed for a planning of the ego vehicle trajectory. The problem is highly complex and dynamic due to unpredictable external factors, e.g., changing weather conditions, varying environments possibly without road marking or sudden movements of humans and cyclists. Deep learning techniques are quite robust against these uncertainties.

Scene prediction is a field of active research with increasing interest as can be seen from Figure 2. In the following, this review is delimited from previous surveys. Surveys on AD often focus on perception issues and challenges related to fusion of different sources [9–11] or psychological aspects related to the interaction between the human driver and the machine [12–14]. Some earlier surveys of scene prediction give a good overview of the different approaches but do not consider yet advanced deep learning techniques [15,16]. The review by Xue et al. [17]. focuses on scene understanding, event reasoning and intention prediction, distinguishing between short-term predictions (time horizon of a few seconds) and long-term predictions (time horizon of several minutes) but is kept on a rather high level in terms of methodology.

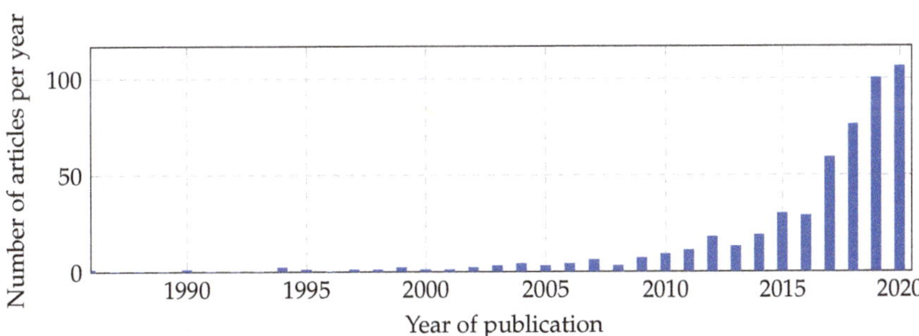

Figure 2. Number of papers per year addressing scene prediction (based on [18]).

Scene prediction often takes over new deep learning techniques from the field of natural language processing, a fast moving research area where huge progress has been made in the last three years. Naturally, those latest techniques are contained only in the most recent reviews. Refs. [19,20] highlight the importance of deep learning models but put emphasis on the visual information as input source, such as feature extraction from 3-dimensional (3D) images and videos. In the study by Yin et al. [21] and Tedjopurnomo et al. [22], deep learning techniques are presented in detail but the evaluation is done for traffic flow models. Such models have a time horizon of several minutes in contrast to the models in the paper at hand which are on the horizon of up to 10 s.

In contrast to the paper by Rasouli [23], where a detailed overview of state-of-the-art deep learning techniques is also presented, here, the approach of a historical review is chosen by analyzing the development of the methodology in this field of research. A comparison of the model results reveals the dominance of neural networks in current state-of-the-art methods. We experience the difficulty of quantitative inter-model comparison from [23] since many models are not evaluated on standardized test datasets. However, the value of a competition of performance improvements on the cm scale is questioned here. The focus should be rather on understanding the reasons for significantly varying performances on different datasets which has not been addressed in detail yet.

The review starts with a concise introduction to AD in Section 2 with the aim of placing scene prediction in the context of AD and describing the challenges. In Section 3, the standard models for scene prediction are presented. The historical context and major achievements in scene prediction are presented in Section 4, followed by a short presentation of publicly available datasets in Section 5. The model comparison is discussed in Section 6. Finally, conclusions are drawn and future challenges are addressed in Section 7.

2. The Context of Scene Prediction for Automated Driving

2.1. Sensors

The foundation of a well-performing AD system is a robust perception model. This can only be achieved with a redundant sensor setup, e.g., such as the one shown in Figure 3.

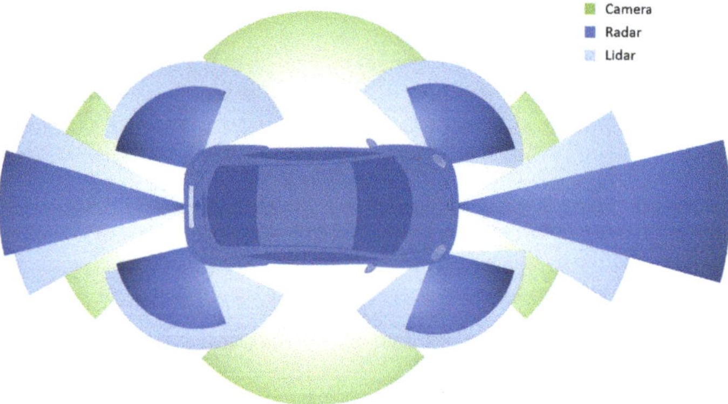

Figure 3. Example sensor setup for a self-driving vehicle.

Besides standard sensors such as an Integrated Motion Unit (IMU) for measuring acceleration and a Global Positioning System (GPS), AD vehicles are typically equipped with a camera system, radars, and Lidars, each providing a 360° surround view in the best case. The variety of sensors takes advantage of the different physical properties but also brings redundancy to the system.

The optical camera is usually good for classification tasks such as distinguishing the type of a road user, recognizing lane markers or traffic signs. While the performance on measuring distances and velocities is rather weak, this information can be retrieved well from radars. Lidars are complementary to the other two sensors, showing only a few weaknesses. Distances and velocities can be estimated with very high accuracy. The only disadvantage are the high costs of a Lidar system.

Furthermore, high-definition maps as an additional sensor are currently being integrated into level 2+ systems, providing a spatial resolution of a few centimeters. This information is especially important for the urban environment. The detailed road infrastructure, such as lane marking and shape or traffic lights and signs, is collected in huge data collection campaigns and is then abstracted with the above mentioned sensor output.

2.2. Evolutionary versus Revolutionary Approach

According to the above definition of AD levels, current series market technology has reached level 2+. This is an intermediate level between 2 and 3 which has been introduced to indicate that level 2 has technically been exceeded already. Due to some technical and legal aspects which are also discussed in this paper, the transition to level 3 is not trivial and has not been accomplished yet. There are basically two strategies for progressing to higher automation levels: conventional OEMs and Tier 1 suppliers follow a conventional bottom-up approach, which is often also called an evolutionary approach. It consists of a building block architecture as shown in Figure 4. The task *automated driving* is broken down into three main components: a perception of the environment, a decision making part with the sub-task *scene prediction* located in the module *behavior planning*, and an acting part. Each of such components is further broken down into smaller modules so that an individual module itself can be tested in terms of function safety and quality.

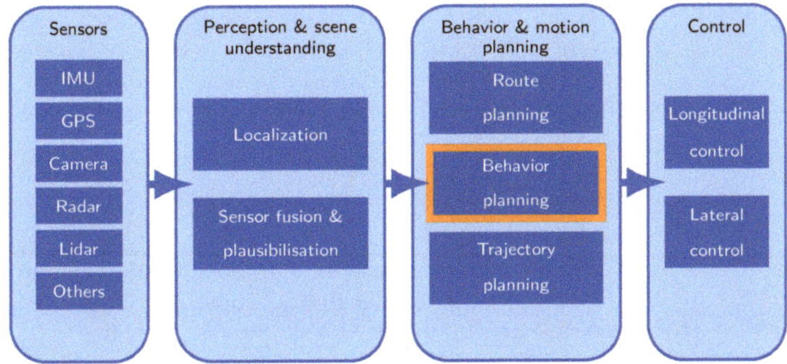

Figure 4. Simplified high-level architecture for self-driving vehicles following the conventional approach of most OEMs and Tier1s. See text for details.

Companies such as Waymo or Zoox follow a fundamentally different approach, often called the disruptive or revolutionary approach. Taking the human as a potential operator of the vehicle out of the loop allows for entirely different vehicle concepts. For instance, in a level 5 there is no need to equip the vehicle with driving control input devices such as a steering wheel or pedals. Therefore, an electric vehicle could become direction-independent, making it equally drive forward and backward. A separate scene prediction module or sub-module may not be required in this approach anymore.

2.3. Scene Prediction and Its Challenges

The goal of scene prediction is to anticipate how a traffic scene will evolve within the next seconds. All relevant agents contributing to the scene are described by their states (position, velocity and heading angle), which shall be predicted with the highest accuracy possible. A relevant agent or object is one that influences the trajectory of the ego vehicle within the considered time frame. For SAE level 3, it is aimed to reach a prediction horizon in the order of 10 s. Based on the prediction, the ego trajectory can be planned and maneuvers can be executed.

Among the biggest challenges in scene prediction are the complex spatio-temporal dependencies and high dynamics. In addition an action of a traffic participant is affecting all surrounding participants. These subtile indications are not obvious and thus not possible to describe with simple physical models. Data-driven techniques and especially deep learning models have the potential to make these predictions and to satisfy the challenges. Simple data-driven models identify the most common trajectories in historical data, deriving typical maneuver classes. More advanced and deeper models are capable of identifying typical patterns self-consistently, often generating the most likely trajectories associated with a probability. The requirement is that the scene prediction should anticipate intented actions of others in the scene with some time ahead at least as well as or better than a human driver. This also means, however, that one needs to accept the existance of unpredictable situations, such as sudden decisions, which neither a human driver, nor an automated scene prediction could ever anticipate.

The uncertainties from perception, which have already been mentioned as AD challenge, are also a determining factor for the performance of a scene prediction model. In order to understand the current situation, the system has to make the right decisions often based on noisy or sparse data. Additionally, for this aspect, deep learning models show the best performance compared to other approaches due to their ability for generalization.

3. Methods for Scene Prediction

In this section, the most widely applied methods for scene prediction are presented. These models are generic models, not specific to scene prediction or AD, and can be divided into model-driven and data-driven approaches. The first have a rather small prediction horizon which is why they only play a minor role these days but are mentioned for completeness.

3.1. Model-Driven Approaches for Scene Prediction

3.1.1. Kinematic Models

The simplest approach for extrapolating the trajectory of a traffic participant is by considering purely the kinematics of an object. Thus, the object's trajectory $\mathbf{x}(t)$, $\mathbf{x} \in \mathbb{R}^2$ and t is the time, in the plane is simply described by:

$$\mathbf{x}(t) = \mathbf{x}(0) + \mathbf{v}t + \frac{1}{2}\mathbf{a}t^2 \tag{1}$$

with $\mathbf{v}$ being the velocity and $\mathbf{a}$ the acceleration. Often, the assumption of constant velocity or constant acceleration is made, which works well for highway situations with not much traffic, but cannot take into account more dynamic situations which involve interactions among the traffic participants. A more sophisticated approach is applying a transformation into a curvi-linear coordinate system, resulting in a more consistent driving behavior.

3.1.2. Dynamic Models

More advanced models also take into account the different forces acting on an object. The models usually start from the action,

$$S(\mathbf{q}) = \int_{t_1}^{t_2} L(t, \mathbf{q}(t), \dot{\mathbf{q}}(t)) \, \mathrm{d}t, \tag{2}$$

where $L(\mathbf{q}, \dot{\mathbf{q}}, t)$ is the Lagrangian depending on the generalized coordinates $\mathbf{q}$, their t-derivatives $\dot{\mathbf{q}}$ and time t. In the context of curvy roads, the jerk $\mathbf{j}(t) = \dot{\mathbf{a}}(t)$ is especially important for vehicle dynamics.

Such models can become quite complicated, while still describing only a small time horizon, which is why they receive only minor attention in the context of scene prediction.

3.1.3. Adding Uncertainties to the Model

Uncertainties in the prediction of moving objects are commonly addressed by adding a Gaussian noise term, a normal distribution centered around the mean value. This approach is closely connected to the Kalman filter [24] which assigns a Gaussian noise profile to the sensor measurement. The uncertainty is then propagated iteratively in a step-wise calculation of the estimated rajectory,

$$\mathbf{x}(t + \Delta t) = \Phi(t + 1; t)\mathbf{x}(t) + \mathbf{u}(t), \tag{3}$$

where Φ is the transition matrix and $\mathbf{u}$ a Gaussian noise term with expectation value zero.

Monte Carlo simulations are a further method for considering uncertainties in model-driven approaches [25]. Starting from a set of input variables, the vehicle trajectories are modeled based on physical assumptions and sampling of dynamic variables. This technique also allows to introduce contraints, e.g., due to road boundaries.

3.2. Data-Driven Approaches for Scene Prediction

In data-driven development, the modelling can be done only on a large database of recorded or simulated traffic situations. The first step in model building is feature extraction, which covers the identification of relevant data and data preprocessing. These data are then used for deriving a generalized model.

3.2.1. Classic Methods

While deep learning models have become state-of-the-art for traffic modelling, let us first give a brief overview of classic data-driven approaches.

Hidden Markov Models

Markov processes are statistical models that describe the probability or likelihood of observation sequences [26]. Given a system with N distinct states, $S_1, \ldots, S_N$, at any time t, the probability of state transitions between two states is:

$$a_{ij} = P[q_t = S_j | q_{t-1} = S_i] \qquad \text{for } 1 \leq i, j \leq N, \tag{4}$$

with q_t being the observable state at time t. The transition probabilites can be extracted from a set of data.

Hidden Markov Models (HMMs) [26] are a further extension of Markov processes for non-observable, underlying states. Equation (4) applies also in this case, with an important difference of non-observable states, $S = S_1, \ldots, S_N$, and M distinct observations, $V = v_1, \ldots, v_M$. The distribution of observed states is then given by:

$$p_j(\alpha) = P[q_t = S_j], \qquad \text{for } 1 \leq j \leq N, \ 1 \leq \alpha \leq M. \tag{5}$$

The aim is to model the above parameters such that the observed sequence of states is correctly modeled. The transition probabilities as well as the relationship between the observed states and non-observable events is learned from data.

In the context of scene prediction it is the prediction of consecutive traffic maneuvers. The drawback of this method is that it considers distinct maneuvers and does not take into account interactions of traffic participants [26].

Regression Models

For a given dataset, regression models find a continuous function to describe the relationship between independent and dependent variables. Different types of regression models exist and the choice depends on the problem formulation. Polynomial regression models work well for trend forecasting. Logistic regression outputs values between 0 and 1 and is therefore good for classification tasks. Bayesian regression assumes that the data is described by a normal distribution, trying to estimate the posterior probability based on a prior distribution.

3.2.2. Neural Networks

The output of a neural network is either a classification or regression. In correspondence to biological neurons in the brain, the model consists of artifical neurons (knots) and connections (edges) which are stacked in layers. The impact of a transmitted information from one neuron to another is modeled by a weighted connection.

Feed-Forward Neural Network

In a simple Feed-Forward Neural Network (FFNN), the information is passed from one layer to the next by a simple weighted summation of the input, see Figure 5. The output at any neuron k in the higher layer is given by $y_k = \sigma[\sum_{i=0}^{N} W_{ik} X_i + B_k]$ with a non-linear activation function, σ, input, $X = (X_0, \ldots, X_N)$, weight matrix, W, of $\mathbb{R}^{N \times m}$, and bias, B, of $\mathbb{R}^m$. The compact form can be written as:

$$y = \sigma\left[W^T X + B\right] \tag{6}$$

with $y = (y_0, \ldots y_m)$ and $B = (B_0, \ldots, B_m)$. The weights and bias terms are learned in an iterative process via the backpropagation algorithm [27].

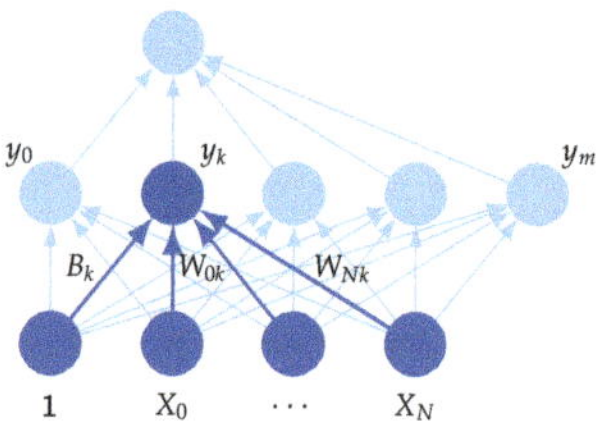

Figure 5. A simple feed-forward neural network. See text for details.

One shortcoming of feed-forward neural networks in the context of scene prediction is that they do not directly model temporal dependencies. Instead, temporal information has to be manually encoded into the input representation, of which there are many. Finding a good representation then becomes itself a complicated task.

Recurrent Neural Network

For time series prediction tasks a more sophisticated network structure is generally used. In Recurrent Neual Networks (RNN) the information from previous time steps is stored in so-called memory cells, i.e., feedback loops that memorize the information in hidden state vectors. This is shown in Figure 6 for a simple RNN model with a feedback loop.

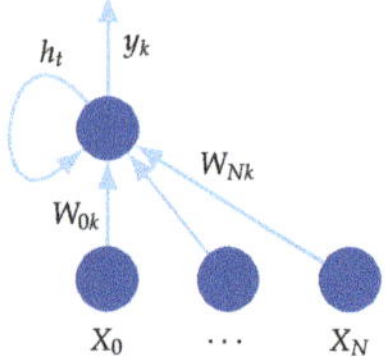

Figure 6. A simple recurrent neural network. See text for details.

The additional state vector,

$$h_t = \sigma[WX_t + Uh_{t-1} + B],\qquad(7)$$

serves for memorizing information. The matrix U is a further parameter that needs to be learned during the training process. h_t and y_k are identical in this case.

Long Short-Term Memory

Simple RNN models provide the benefit of storing past information but they show some major problems. The stored information is limited to short sequences only and they are hard to train since they suffer from vanishing and exploding gradients [28,29]. A more robust RNN architecture that almost completely circumvents these problems, is the long short-term memory neural network (LSTM) [30]. This neural network architecture is capable of memorizing hundreds of time steps. The problem of vanishing and exploding gradients is solved by the introduction of three so-called gates (input, forget, and output gate) controlling the flow of information through the LSTM layer and a truncation of the gradients in the learning algorithm.

The architecture of an LSTM is shown in Figure 7. Similar to the simple RNN, one input to a LSTM cell is the input from the layer before at current time step t, X_t. Additionally, a second input is the hidden state vector from the previous time step h_{t-1}. This state vector stores the short-term information. On the other hand, an additional cell state c_{t-1} is introduced which stores the long-term information. The output of the LSTM cell is again y_t as well as the hidden and cell state vectors, h_t and c_t. Note that only y_t is passed to the next layer, which is identical to h_t. The details of the computation are given in Appendix A.

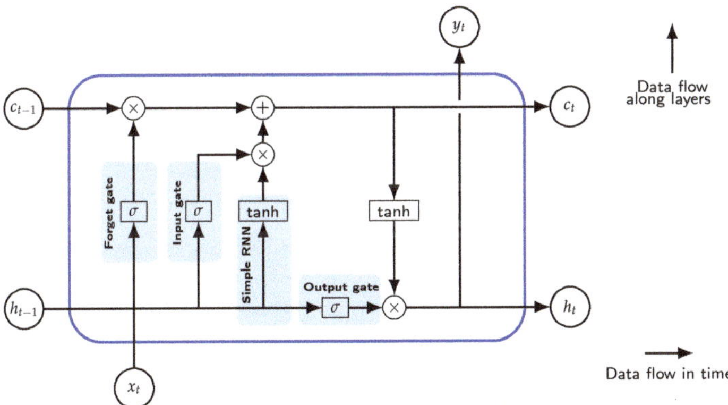

Figure 7. A long short-term memory neural network. See text for details.

There exist several variations of the LSTM architecture, e.g., the Gated Recurrent Unit (GRU), a simpler form with only two gates. The number of trainable parameters is reduced, thus it is very efficient and performs especially well on smaller data sets. The original GRU architecture [31] consists of a reset gate and an update gate. More details are given in Appendix B.

3.2.3. Encoder–Decoder Models and Attention Mechanism

Natural language processing is one of the hottest topics in machine learning these days. It was found that those models perform well also in other fields, such as scene prediction. One major improvement is the so-called attention mechanism. It is based on an encoder–decoder architecture, a sequence-to-sequence model.

A simple encoder–decoder model is shown in Figure 8, each consisting of a single recurrent unit. The encoder takes as input a time series X of length M, for which the hidden states are calculated. Only the last hidden state, h_M, is passed to the decoder, where it serves as the initial state vector. The decoder takes as input the state vector, h_M, as well as one item at a time of the labeled output sequence, such that it can predict the next item in the sequence based on the previous output, i.e., <start> $\rightarrow y_0$, $y_0 \rightarrow y_1, \ldots, y_L \rightarrow$ <end>.

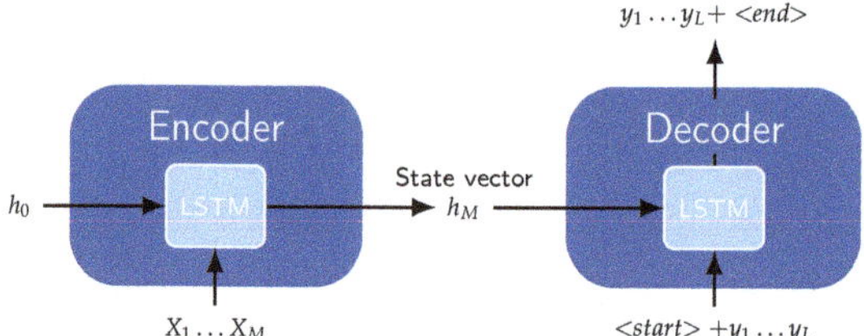

Figure 8. A simple encoder–decoder model for time series. See text for details.

Attention models extend this architecture by outputting not only the final state vector h_M but all hidden states which are then combined to a context vector [32]. The context vector is a weighted sum of the hidden states:

$$h' = \sum_{m=1}^{M} \alpha_m h_m \tag{8}$$

with the weights being normalized with a softmax function:

$$\alpha_m = \frac{\exp(b_m)}{\sum_{k=1}^{M} \exp(b_k)}. \tag{9}$$

b_k is a trainable parameter which takes the context vector and the hidden states of the decoder as input. In this approach, the model gets information from all hidden states. It is trained to adjust the weights in a way that it gives higher value—or attention—to more important parts of the input sequence.

Variational Autoencoder

An autoencoder is a system of encoder and decoder which is used in an unsupervised learning fashion. When data is fed into the encoder, it learns an abstraction into latent space by dimensionality reduction. The decoder receives the output from the encoder and is then trained to reconstruct the original data. The original data thus serves as the ground truth label during the training process, with the goal of minimizing the reconstruction error.

The problem with the simple autoencoder is that it tends to overfit the latent representation if no regularisation methods are used. Therefore, during encoding into latent space not just one data sample is used but a probability distribution with mean and standard deviation, often a Gaussian distribution, resulting in multiple different model outcomes. This extension of the simple autoencoder is called a variational autoencoder [33]. It became very popular in computer vision for generation of virtual images, especially in the context of StyleGAN2 [34]. The idea is to use the model reconstruction for generating new data samples.

A conditional variational autoencoder [33] learns a distribution of a so-called *latent variable z*. One part of the training objective, the Evidance Lower Bound (ELBO) is the Kullback–Leibler-divergence between the prior and the posterior distribution of the latent variable (the second part of the loss function is the data log likelihood). For trajectory prediction, the prior is usually only conditioned on the observation period, while the posterior is conditioned on the observation period and the ground truth future trajectory. During training, the model is supposed to learn to approximate the posterior distribution with the prior distribution, as the information about the ground truth future trajectory is not available during inference. Instead it is actually the task of the model to predict this future trajectory. While inference the prior is used to sample an actual value z from the latent variable distribution z. This sample is then used to condition the trajectory prediction on. Repetitive sampling from z allows to generate an entire probability distribution for the prediction. Therefore, such a distribution can capture all the initially addressed issues of uncertainty, ambiguity, and multi-modality due to the variety of those trajectory samples. Often the conditional variational autoencoder framework is combined with a RNN-based encoder–decoder architecture.

Convolutional Neural Network Models

Time series can also be handled quite well with 1D Convolutional Neural Networks (CNNs). As the name suggests, 1D CNNs use a 1D filter kernel in order to apply convolutions [35]. The discrete 1D convolution for a time series $X = (X_0, \ldots, X_N)$ is then given by

$$\varphi_i = (X * K)(i) = \sum_{m=0}^{M} X_{i-m*s} \, k_m \tag{10}$$

with 1D filter kernel $K = (k_0, \ldots, k_M)$ which is a trainable parameter for the network. The dimension length M of the filter kernel is specified by the user and determines the length of the time sequence that is taken into account, as well as the stride s which determines the frequency at which the input sequence is sampled. For combining spatio-temporal information, the model can be extended to higher dimensions. The biggest advantage in comparison to standard feed-forward-neural networks is that the input representation, this means which features are important and should be extracted, here can be learned.

Transformer Models

Another powerful state-of-the-art neural network model from natural language processing is the transformer architecture [36]. Such models consist of several stacked encoders and decoders, each constructed in the same way: a combination of *attention* and FFNN layers.

The input is passed sequentially along the stacked encoders and decoders, respectively, but also from each encoder to the corresponding decoder. The self attention layers serve as putting an item of a sequence in the entire context and can determine its importance for the entire sequence.

Graph Neural Network Models

An alternative representation of the data is used in Graph Neural Networks (GNN) which have become popular recently [37]. The input data is presented in a graph structure, which reflects the structure of the system. This approach provides the benefit that only existing connections are considered, in contrast to a fixed array representation in the previous approaches. Therefore, CNNs can be considered as special cases of GNNs, where the local connectivity of all nodes represented by convolutional filters. It is especially effective when dealing with sparse data.

A graph is made up of nodes N and edges E (see Figure 9):

$$G = (N, E). \tag{11}$$

The binary adjacency matrix A contains the relationships of the nodes, which can be either undirected (A is symmetric) or directed (A is not symmetric). Furthermore, each node is associated with a feature matrix X. The goal is to train a neural network that takes a representation of the graph structure as input, which is then being transformed into an embedding.

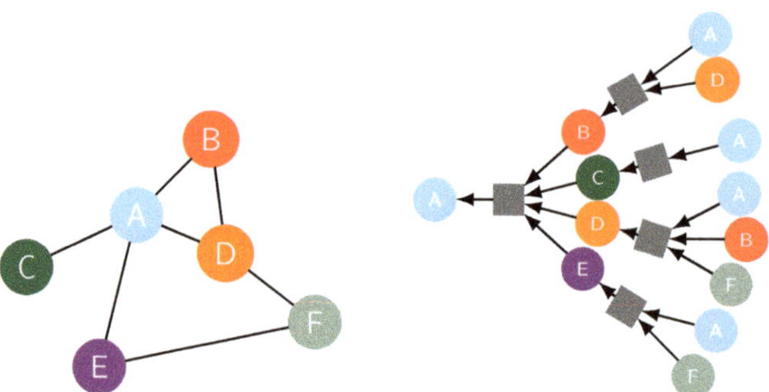

Figure 9. Left: Graph representation. **Right:** Aggregation of neighbors for node embedding. See text for details.

The key idea with GNNs is that the node embeddings are learned with neural networks. A popular approach is the GraphSAGE [38], a graph convolution network (GCN), where all neighbors of a target node are identified and their contribution to a mother node is aggregated. The squares in Figure 9 represent neural networks. The node messages are then calculated in the following way. The first embedding in the 0-th layer is just the node feature itself,

$$h_v^0 = X_v, \qquad (12)$$

with v referring to the v-th node. Embeddings in further layers are then calculated as

$$h_v^{l+1} = \sigma\left(W_{l+1} \cdot \text{CONCAT}\left[h_v^l, h_{N(v)}^{l+1}\right]\right) \qquad (13)$$

for $l \in (0, \cdots, L-1)$ with L the number of embedding layers, $N(v)$ being the neighborhood function and weight matrix W_l and CONCAT being the concatenation function [38].

4. Historical Review of Relevant Work

This Section illustrates the evolution of the field *scene prediction* from a historical point of view. Contributions with highest impact, starting from the beginnings of this field of research, and ending with the state-of-the-art research are highlighted. The presented selection is based on the impact on the community, quantified by the number of citations in the context of AD.

As highlighted here, there has been a strong increase in the amount of research that was done on trajectory prediction in the past five years. Additional expertise has been focused on vehicle trajectory prediction originating from the robotics, deep learning, and computer vision community. Beside leading to great progress on the problem itself, the gathering of researchers and research groups in this field has led to a strong parallelization of that research. So, many similar approaches were developed around the same time, leading to clusters of certain methods. Compared to less intensive investigated research fields, this also has led to a less successive development in the research history of trajectory prediction. Due to the great progress made in the last few years there emerges one import question: How accurate does trajectory prediction have to be (to enable a certain level of automation–according to the SAE definition)? Or shorter: How accurate is accurate enough? While the exact answer to this question is out of scope of this paper and seems very hard to answer too, there is one important aspect for us regarding its core. Due to the continuing focus on certain datasets for the development and the provided infrastructure and Application Programming Interfaces (APIs), trajectory prediction gradually becomes a research challenge. While the competition between single researches and research groups leads to more and more precise prediction models, improvements in the magnitude of centimeters may decide about the order on the leader board but it is an open question how much such improvements contribute a better driving experience of an automated vehicle.

Closing the loop, this is related to the challenge of a very active and dynamic research field and how to interpret and evaluate the results. Less recent approaches are characterized by an additional problem, which is related to used data. While the most recent approaches focus on only very few datasets for evaluation, older papers have worked on different datasets, making it hard to compare those approaches and their results against each other. All those thoughts motivated the structure of the upcoming chapter, the selection of the presented papers, and their evaluation and assessment.

The term "scene prediction", as used in this paper, is a fully self-consistent description for predicting all traffic participants in a certain area constituting one driving situation. Since this is a very complex task, initially only sub-tasks of scene prediction were focused on.

4.1. Recognition of Other Drivers' Intentions

First approaches towards the modeling of traffic behavior were focused on specific use cases. The first highlighted use case is the lane change prediction. Then, the use case of car-following, which focuses only on the longitudinal part of the trajectory, is briefly presented.

4.1.1. Lane Change Prediction

The goal of the lane change prediction is to identify the intention of other drivers to change the lane. It is possible to detect such an intention several seconds before the actual event, such that the automated ego vehicle can prepare to react on it.

The input to such a model is usually a feature vector $\mathbf{X}$ describing the state of a target vehicle for a history over discrete time steps. $\mathbf{X}$ contains the distance and relative velocity to the ego vehicle and surrounding vehicles in the vicinity of the target vehicle. The output is a classification of the maneuver with classes "lane change left", "lane change right" and "lane keeping".

The first step towards tackling the lane change use case was understanding the underlying decision process. In 1986, Gipps identified the three central items before making a lane change as (i) the physical possibility for changing the lane, (ii) the necessity as well as (iii) the desirability which were decided upon in a flowchart [39]. The describing terms were initially addressed by simple mathematical formulas, providing the fundamentals for a microscopic approach. Based on this study, Toledo et al. [40] refined the model by a probabilistic description of the utility terms.

Most of the following publications consider this decision process as a latent variable which is not directly observable, and focus on classifying whether a lane change is going to happen. Hunt and Lyons [41] developed a simple neural network, which is a great method for classification tasks, achieving reasonable results on a simulation data set. The encountered problems during model development were "little guidance [...] available on the selection of network architecture and the most appropriate paradigm" [41] which is the determining factor for the success of the outcome, as well as the worse performance on real data during inference.

A study [42] that received much attention, analyzed the driver's behavior during the lane change process and found that the typical sine-wave steering pattern was accompanied by a slight deceleration before the actual maneuver. Surprisingly, only 50% of the events had an activated turn indicator signal with 90% reaching into the lane change maneuver. By observing the eye movement, it was found that the driver takes off focus from the current driving lane approximately 5 s before the start of the lane change.

A typical approach for modeling the intention of the driver are HMMs. It seems that the probabilistic step-wise state progression fits well the execution of a lane change maneuver. The probabilistic modeling of the state transitions, allows to predict the maneuver approximately 1 s before it takes place [43,44]. Approaches based on Bayesian statistics which are often combined with Gaussian mixture models achieved similar performance [45–48]. Only few publications use physics-based models, e.g., describing the traffic flow as a continuous fluid [49]. More recent approaches rely on machine learning techniques such as support vector machines [50–52] or neural networks [16,53,54].

4.1.2. Car-Following

A further use case is car-following, which is similar to the adaptive cruise control function and part of lane keeping. In order to keep a dynamic safety distance to a leading vehicle, a smooth acceleration model is required to avoid immediate reaction on any sudden acceleration or braking of the leading vehicle. Interestingly, the first models were developed for city traffic, which for scene prediction is a use case considered more difficult than highway driving.

The input to car-following models is usually a feature vector **X**, where spatial coordinates and velocities are one-dimensional and aligned with the driving direction. The output of this model is an acceleration or velocity for the ego vehicle.

Due to the reduced dimensionality of the problem, this use case is more often approached with physics-based models; see, e.g., [55,56]. Here, the study by Helbing and Tilch in 1998 [57], which received more than 5000 citations, to be mentioned. The authors developed a generalized force model which uses the formalism of molecular dynamics for many particle systems. Equations of motion describe the effective acceleration and deceleration forces. The so-called social forces acting on an agent tries to reflect internal motivations. Context information such as speed limits, acceleration capabilities of vehicles, the average vehicle length, visibility and reaction time were introduced as direct parameters. The advantage of this model is a good understanding and insight into the model due to easy interpretation of the parameters.

4.2. Full Trajectory Prediction

Trajectory prediction is so complex that (almost) all approaches are data-driven. The difficulty lies in the modeling of all factors determining the internal object states and perception. The input to the model is a feature vector **X** of the observed scene. **X** can be represented by trajectory coordinates (Figure 10, left), occupied grid cells on an occupancy map (Figure 10, right), semantical features or a combination of those. These data are collected with sensors and can be raw or further processed signals. The output is a prediction of the positions and states of all traffic participants within a defined time horizon. The output can be a binary classification for an occupancy grid, or a regression model outputting floating numbers for future trajectory points of the objects.

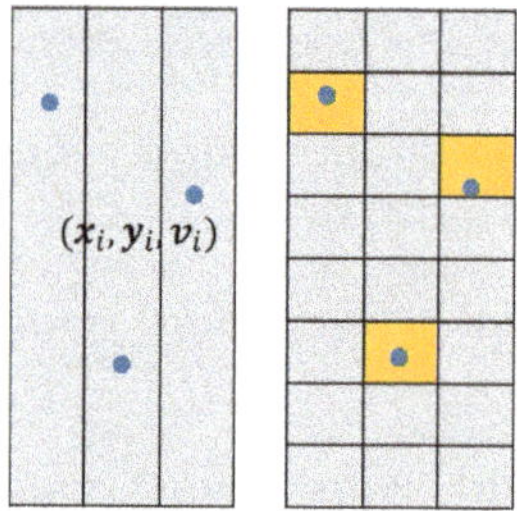

Figure 10. Left: Objects are represented by their position and velocity. **Right**: Objects are associated to a cell in an occupancy grid.

In Table 1, quantitative evaluation results of full trajectory prediction models are collected. If available, the Final Displacement Error (FDE), Average Displacement Error on the entire trajectory (ADE), Root Mean Squared Error (RMSE) or Mean Absolute Error (MAE) is given with prediction horizons in brackets. For an evaluation of multiple modes, the number of samples K is also mentioned.

Table 1. Overview of publications on scene prediction for AD. Neural network models are grouped under "L" (LSTM/GRU), "C" (CNN), "G" (GNN), and "A" (Attention). Some papers are also available at [58]. For the error it is specified the number of modes, K, over which the prediction is sampled, if applicable. See abbreviations list.

Authors	Data	Year	Method	Horizon [s]	Metrics	Error [m]
Hermes et al. [59]	real	2009	Clustering	3	RMSE	5.0 ± 0.6
Houenou et al. [60]	real	2013	CYRA	4	RMSE	0.45
Deo et al. [61]	real	2018	VGMM	5	MAE	2.18
Casas et al. [62]	real	2018	C	3	MAE	1.61
Park et al. [63]	real	2018	L	2	MAE	0.93 ($K = 5$)
Cui et al. [64]	real	2019	C	6	ADE	2.31 ($K = 3$)
Altche et al. [65]	NGSIM	2017	L	10	RMSE	0.65 [1]
Deo et al. [66]	NGSIM	2018	L+C	5	RMSE	4.37
Chandra et al. [67]	NGSIM	2019	L+C	5	ADE/FDE	5.63/9.91
Zhao et al. [68]	NGSIM	2019	L+C	5	RMSE	4.13
Tang et al. [69]	NGSIM	2019	G+A	5	RMSE	3.80 ($K = 5$)
Song et al. [70]	NGSIM	2020	L+C	5	RMSE	4.04
Chandra et al. [71]	NGSIM	2020	G+L	5	ADE/FDE	0.40/1.08
Lee et al. [72]	KITTI	2017	L+C	4	RMSE	2.06
Choi et al. [73]	KITTI	2020	G+L+C	4	ADE/FDE	0.75/1.99 ($K = 10$)
Lee et al. [72]	SDD	2017	L+C	4	RMSE	5.33
Chai et al. [74]	SDD	2019	C	5	ADE	3.50 ($K = 5$)
Mangalam et al. [75]	SDD	2020	A	5	ADE/FDE	0.18/0.29 ($K = 20$)
Tang et al. [69]	Argoverse	2019	G+A	3	ADE	1.40 ($K = 3$)
Chandra et al. [71]	Argoverse	2020	G+L	5	ADE/FDE	0.99/1.87
Park et al. [76]	Argoverse	2020	L+A	3	ADE/FDE	0.73/1.12 ($K = 6$)
Song et al. [77]	Argoverse	2021	L+C+A	3	ADE/FDE	1.22/1.56 ($K = 6$)
Zeng et al. [78]	Argoverse	2021	G+C	3	ADE/FDE	0.9/1.45 ($K = 6$)
Casas et al. [79]	nuScenes	2020	G+C	3	RMSE	1.45
Phan-Minh et al. [80]	nuScenes	2020	C	6	ADE/FDE	1.96/9.26 ($K = 5$)
Liang et al. [81]	nuScenes	2020	L+C	3	ADE/FDE	0.65/1.03
Park et al. [76]	nuScenes	2020	L+A	3	ADE/FDE	0.64/1.17 ($K = 6$)
Narayanan et al. [82]	nuScenes	2021	C+L	4	ADE/FDE	1.10/1.66 ($K = 10$)
Casas et al. [79]	ATG4D	2020	G+C	3	RMSE	0.96
Liang [81]	ATG4D	2020	L+C	3	ADE/FDE	0.68/1.04
Chandra et al. [71]	Lyft	2020	G+L	5	ADE/FDE	2.65/2.99
Chandra et al. [71]	Apolloscape	2020	G+L	3	ADE/FDE	1.12/2.05
Li et al. [83]	INTERACTION	2020	G+C+A	5	ADE/FDE	1.31/3.34
Choi et al. [73]	H3D	2020	G+L+C	4	ADE/FDE	0.42/0.96 ($K = 10$)
Song et al. [70]	HighD	2020	L+C	5	RMSE	2.63
Mohta et al. [84]	X17k	2021	C	3	FDE	0.85

[1] Only lateral position considered.

4.2.1. 1980s–2015

A very early model following the disruptive approach for an end-to-end ego path planning was suggested by Pomerleau in 1989 [85]. This neural network based model received much attention with a novel approach using camera and laser data as input, outputting the road curvature which was followed for lane centering. The network contained one hidden layer only and completed training in just half an hour. Due to the simplicity of the framework, laser data was found to play only a minor role. Pomerleau recognized the importance of this study with the concluding remark: "We certainly believe it is important to begin researching and evaluating neural networks in real world situations, and we think autonomous navigation is an interesting application for such an approach" [85]. Further work picked up on this approach, applying fully connected neural networks, but reducing the problems to lane change decisions [41,86]. These models were mainly trained on simulated data.

It was recognized that the motion of an object followed characteristic patterns. Therefore, unsupervised clustering techniques were applied for classification of these trajectory types. The advantage of this technique is that it allows for a long-term prediction of the trajectory, given that the correct maneuver class was identified.

Vasquez and Fraichard [87] followed this approach by applying a pairwise clustering algorithm based on a dissimilarity measure on simulated and real data. After calculating the mean value and standard deviation of the clusters, the likelihood was estimated that an observed trajectory belongs to a cluster. This paper considers pedestrian motion patterns. The observed problems are that an early prediction is associated with a high error, and individual aspects of the trajectories could not be taken into account. Li et al. [88] used a multi-stage approach on clustering techniques. The algorithms were optimized by a strategy of refinement, first creating general clusters which are then processed by a second clustering algorithm.

Hermes et al. [59] make use of the radial basis functions, which were used as a priori probability for trajectory prediction. Among the highlighted references, this study was found to be the first with a quantitative evaluation of experimental results on trajectory prediction. 24 test trajectories from real drives were evaluated regarding the RMSE. Although their model did not outperform the standard model on straight trajectories, it showed strong improvement for curves.

HMMs are also used for scene prediction. The input is a graph with nodes containing the object states and edges as transitions between the states. The initial state is a stochastic representation of the states, the transition probabilities are then evaluated iteratively in discrete time steps. Hidden refers to the fact that the object states are not directly observable. The process then contains two steps: determination of structure (number of nodes + edges) and parameters (state prior, transition and observation probabilities) which are learned. Often clustering techniques are used to determine the structure of the HMM. Vasquez et al. [89] use growing HMM, meaning that their model can change its structure. The data was taken on parking spots, resulting in a model error on the range of meters.

In 2013, a dynamic approach by Houenou et al. [60] received quite some attention. The Constant Yaw Rate and Acceleration (CYRA) model, using constant yaw rate and acceleration, selects the best trajectory which minimizes a cost function depending on acceleration and maneuver duration. The model achieves a very low mean displacement error, but it was evaluated on their own collected dataset only and is thus not comparable to other models.

4.2.2. 2016: The Rise of Deep Learning Techniques

In 2016 and following years, a shift in techniques is observable. The number of papers using neural networks strongly increases and a more systematic evaluation of object positions is observable.

The objects' states can be represented as sequences of discrete historical data spanning usually a few seconds, e.g., $\mathbf{X} = (X_{t-\Delta T}, X_{t-\Delta T+1}, ..., X_{-1}, X_0)$, where x_i stands for position, velocity or any other physical quantity at a given time t within the observed time frame ΔT. The goal of the model is to predict the state variables at the end of the observation period. At inference, this time should lie some seconds in the future. Time series can be described well with simple RNN or LSTM networks. Therefore, it is not surprising that many authors jumped on this trend after the first models appeared in the community. The first LSTM models were still used for classification of maneuvers [54,90] but prediction horizons up to 10 s suddenly seemed within reach [65]. The simplest LSTM models are lacking a description of mutual interactions of the traffic participants which was then approached by representing the objects on an occupancy grid [91] as visualized in Figure 10 (right).

This drawback was furthermore addressed in the following more complex models. By introducing a concatenation layer, in which the information from individual LSTMs is brought together, the context information could be modeled. The term "social pooling" became popular [66,92], describing the interdependencies of all agents in the scene, which was originally developed for human trajectory prediction [93–95]. Furthermore, multi-modality came into focus, in which possible outcomes of the same initial context were modeled. Either of these aspects became a standard ingredient for most of the following models [62,64,96,97]:

The DESIRE framework [72] models the trajectories of each agent in RNN encoders, which are then concatenated and combined with the scene context from a standard convolutional layer. They found that a 2 s history is enough to predict 4 s ahead. It is one of the first papers that propose an RNN-based encoder–decoder structure. The RNN encoder captures the motion pattern of the observation period and determines a latent representation of the temporal data. This latent representation can now easily be concatenated with other data, e.g., feature maps resulting from a CNN that processes an image representation of the static scene context. Finally, either the original latent representation of the RNN encoder or a concatenated tensor is used as input for the RNN-based decoder, which predicts the future waypoints of vehicles in a recurrent and auto-regressive fashion. If the latent representation from the encoder is extended by additional information, the decoder is able to condition its prediction not only on the previous motion of a certain vehicle but also takes into account further information which enables much more accurate predictions.

Another feature that has become very popular in trajectory prediction during the last few years is the Conditional Variational Autoencoder framework [33,72]. The basic idea for using this kind of generative model in trajectory prediction, is to capture uncertainty, ambiguity, and multi-modality. Due to the lacking of knowledge about the intention and inner state of drivers controlling the surrounding vehicles, the future trajectory of those vehicles can only be estimated but not precalculated exactly. In many situations given a specific motion of a vehicle during the observation period, different future evolutions of that trajectory are possible and in accordance to the behavior of other traffic participants and the traffic rules in general. Such situations require more than a single prediction to better capture the entire probability distribution over the future trajectory. Therefore, generative models are able to predict multiple possible trajectory predictions.

Such an approach was followed by Xu et al. [98], who worked on large scale crowd-sourced video data, using an LSTM temporal encoder fused with a fully convolutional visual encoder and a semantic segmentation as side task. Further works combining an LSTM encoder–decoder structure with semantic information followed, the semantic information in form of a maneuver classification as input for trajectory modeling [99] or combination of LSTM and CNN layers [67,68]. Park et al. combined their encoder–decoder model with a beam search algorithm to keep the k best models as trajectory candidates [63].

Time series can also be forecasted well with the CNNs. The input to such a network is similar to LSTMs, a sequence of state representations. The historical context information is filtered with the 1-dimensional temporal CNN kernel, with its size determining the length of the time window. Some authors favor this approach [100], since CNNs are easier to train than LSTMs. A model using a pure CNN model was realized by Luo et al. [101], who used a 4D tensor as input to their model where only one dimension is reserved for the temporal information, and the three remaining dimensions contain spatial information. 3D point clouds measured by the sensors were processed in a tracking model and assigned to an occupancy grid. The output of their model is a detection of bounding boxes for n timesteps. The authors claim that their model is more robust to occlusion.

To summarize, at this stage models were aiming at better imitating the human behavior.

4.2.3. GNNs, Attention and New Use Cases

A new trend in representation of the acting agents for neural network models, were graph models [102–109]. Describing the connections between objects in a graph structure in opposition to an occupancy grid, is a huge advantage if there are only sparse connections between the objects. The input to such a neural network is a graph as visualized in Figure 9. The nodes represent the objects, with feature vectors holding, e.g., positions, velocities etc., and the edges are the connections between the objects. This reflects interactions between vehicles and the interdependencies in their motion naturally and allows for efficient implementation as well as calculation. The information aggregation step (similar to message passing in graphical models) in such GNN models can also be considered as the approximation of negotiation processes that result from manual driving of humans sharing the available space or even the same lane on the street. The input is usually turned to an embedding which then serves for further evaluation with RNN- or CNN-based models. An unusual approch was taken by Yu et al. [110], who processed time information in a spectral graph convolution. Later models usually work on the time data itself, e.g., [111]. Since 2019 the models were getting very complex with a lot of combinations of different building blocks.

In 2019, attention models were introduced in the field of trajectory prediction [112–117]. In [113,118,119], the attention module is used to prevent to pre-define an exact graph structure of a given traffic situation. Instead all the traffic participants are considered and the attention modules determines the attention weights that correspond to the degree one vehicle determines the motion of another one while predicting the vehicle's trajectories at the same time. Since usually the attention weights are calculated based on the hidden states of an RNN-based encoder, attention is not able to focus on the spatial dimension by means of the position of those vehicle but also on the temporal dimension because the attention weights are calculated for every single time step. Therefore, it learns to give more importance to the relevant latent features and thus, pushed the performance of the interaction-aware models.

Later, graph convolutions were used as input to the attention model, where the attention was given to both, spatial and temporal data [69,111]. Tang et al. [69] stress the importance of multi-modality in closed-form in opposition to Monte Carlo models which are not capable of this. They claim the strong achievement of their contribution being a model that is scalable for any number of agents in a scene.

Additionally, a few first transformer models have entered the field of scene prediction, currently with the focus more on human traffic prediction [120,121] or long-term traffic flow (time interval on several minutes) [122]. As in Natural Language Processing (NLP) transformers partially begin to challenge RNN-based sequence-to-sequence models in (human) trajectory too, demonstrating the importance of the attention module in those tasks. Different from NLP, where the order of words in a sentence may change without modifying its meaning, the order of trajectory points cannot be changed without entirely confusing its plausibility. However, the successive processing of the time series input data gets lost when transformers are used for trajectory prediction in contrast to RNN-based models. This could be one reason why Transformers are still relatively rare in the landscape of trajectory prediction models while RNNs continue to be used as an integral part of most approaches. At the same time, this lack of sequence information provides the benefit of reduced computation time due to the possibility of parallelization.

Multi-modality was also brought more into focus by Chai et al. [74]. The authors used a 3D fully convolutional network in space and time, combined with static information for all agents, which generates a feature map. In a second state, this information is evaluated in an agent-centric network, outputting anchors for trajectory per agent. The output is in form of a bi-variate Gaussian model, which gives different weights to the samples to better approximate the distribution. A similar approach for the urban use case was followed by Phan-Minh et al., where map information was used as additional input [80].

A shift towards urban use cases was observable [76,78,82,123,124]: the difficulties are the variety of traffic participants and shorter time scales. A work that especially focuses on heterogeneous traffic is the 4D LSTM approach by Ma et al. [125]. Casas et al. [79] use 17 binary channels as input to a CNN for encoding different semantics, combining this information with a graphical representation of the agents' states and map information. Further semantics were included in the paper by Chandra et al. [71], by predicting if surrounding vehicles are overspeeding or underspeeding. This is used as additional input to the network, as well as regularization methods based on spectral clustering. Li et al. [83]

use a graph double attention network with a kinematic contstraint layer on top for assessing the physical feasibility of predicted maneuvers. Model robustness was tested for missing sensor input and found performing well by Mohta et al. [84].

Recent advances try to integrate the planning step into the traffic prediction. A standard pipeline takes the motion forecasts from scene prediction as output and ego planning is done separately, decoupled from the forecasting step. The new approach takes a hypothetical ego trajectory and integrates this information in the multi-agent forecasting. Such a coupled approach was used in [70], taking as input the future planning of the ego vehicle and past trajectories of the agents in the proximity of the ego vehicle. The output is a distribution of likely trajectories. Using a LSTM encoder to capture the temporal structure with social pooling, abstraction of interactions is done in a fusion module based on CNN architecture. The same group [77] suggested a different model with a top layer with explicit kinematic constraints by working in a curvilinear frame. Using a 1D CNN for temporal encoding and several LSTMs as well as attention for interaction modeling, their model is robust to missing information due to occluded objects. The problem with such approaches is still the real-time performance since the integrated task requires a lot of computation.

A completely different approach, for which we do not want to go into the details but nevertheless mention, is imitation learning. The idea is that a (driving) policy shall be learned that imitates the actions of an expert. The interested reader can refer to, e.g., [126,127].

5. Public Datasets

First, a quick overview of publicly available datasets which are most frequently used for training and evaluation of scene prediction models is given. The purpose of the training data in the context of learning-based models is obvious. After training (and validation) a separate test dataset is needed to evaluate the trained models. This enables advanced analysis, continuous development, and a comparison to competitive and baseline approaches. Therefore, datasets are an essential piece in the entire development pipeline for designing and engineering scene prediction models.

The Next Generation SIMulation (NGSIM) dataset [128] was recorded in California on the highways US 101 and I-80. Data collection was done from the top of nearby buildings through a set of synchronized cameras with a sampling rate of 10 Hz. Thus, the exact location of each object is known.

The Karlsruhe Institute of Technology and Toyota Technological Institute (KITTI) dataset [129] was recorded in Germany on rural roads, highways, and inner-cities. It contains several hours of traffic scenarios, taken with stereo cameras and 3D laser scanner mounted on a measurement vehicle.

The Argoverse dataset [130] was recorded in the United States, in Pittsburgh and Miami. It contains more than 30,000 scenarios sampled at a rate of 10 Hz with 360° camera and Lidar, which were aligned with map information. The data is already split into training, validation and test sets, each with a prediction horizon of 3 s.

The nuScenes dataset [131] was recorded in Boston and Singapore (left- and right-handed traffic) with a sensor system of 6 cameras, 5 radars, 1 Lidar, IMU and GPS. Furthermore, it provides detailed map information. Each of the 1000 recorded scenes has a length of 20 s. It contains 3D manually labeled bounding boxes for several object classes, with additional annotations regarding the scene conditions and human activity and pose information. This dataset is thus useful for urban use cases.

6. Discussion

The dominance of deep learning models for fully self-consistent scene prediction is obvious when comparing the methods in Table 1. Only few approaches based on models other than Neural Networks can be found among the papers with highest impact in the field. To note is that it is not possible to quantitatively compare the model results which have been evaluted only on private datasets [59–64]. Especially the very low root mean squared error of 0.45 m in the paper of Houenou et al. [60] is difficult to evaluate. The model performance depends heavily on the scenarios that are evaluated. Straight highway driving with little traffic is a simple task and should be tackled well by any model, whereas on the other hand curvy roads with lots of traffic participants doing multiple maneuvers are very challenging.

For the standardized datasets in Table 1, the models all use either LSTM-, CNN- or Attention-based models. Additionally, here, due to the different time horizons, only a qualitative performance estimate is given. There exist predefined test datasets only for Argoverse and nuScenes, which are thus the only datasets allowing a quantitative model performance comparison. An intra-dataset comparison shows that the Attention Network gives especially good performance [69,75,76].

Please note that at the time of submission, the Argoverse leaderboard [132] has a minimum average displacement error of 0.7897 m. Therefore, the result of minADE = 0.73 m in [76] is questionable.

It is also interesting to see that the same model sometimes performs well on one dataset but rather poorly on another, e.g., for Tang et al. [69] with a RMSE of 3.80 m on the NGSIM dataset being among the top results and the minADE = 1.40 m on the Argoverse dataset is just an average result.

One observes a competition of model improvement which is taking place on the cm scale by now. However, it is questionable if time and resources are invested at the right place since such performance improvements might not make a difference when deploying a scene prediction model in the final AD architecture. It is rather important to understand the differences in the model performance on different datasets. The distribution of cases could be different, one dataset containing many situations of straight highway driving with little traffic, while the other dataset contains dense traffic with many corner cases and unpredictable situations. However, it could be an issue in the model itself that is trimmed to a latent feature in a specific dataset. Therefore more effort should be spent on this question.

7. Conclusions and Outlook

It is difficult to estimate the performance of some earlier model results since they have often been evaluted on private datasets. The model performance depends heavily on the constitution of scenarios. Maneuvers and scenarios are not standardised, thus datasets with long intervals of straight highway driving will give better performance than highly crowded or curvy scenarios. Future research will benefit if methods are evaluated on public datasets, whose number will hopefully grow, since they facilitate inter-model comparisons.

Often a significant difference in the displacement error can be observed when comparing the model performance on two different datasets. This poses questions regarding the generalisation of models: if a model was developed on a public dataset, how will it perform in a different environment or with a new sensor model? This should be tackled by increasing the amount of data on which a model is tested. Realistic simulations can support testing but here one is facing the challenging task of modeling a realistic environment.

A further issue, which is addressed seldomly in the referenced publications, is online learning. Updating a model on the fly can be dangerous since the model cannot be tested rigorously. Nevertheless, an algorithm should be running in parallel which can detect anomalies for further analysis and collects data for updating the model in a postprocessing step. This way, corner cases can be identified. Anomaly detection is furthermore necessary in order to protect against adversarial attacks [60]. Guaranteeing a safe system is one of the most central research questions these days. Everyday huge amount of data is collected which can be used for model updates. Even large cloud storages will reach their limitation. Thus, it is important to wisely select data for storing, most efficiently in a preprocessing step already in the vehicle before sending the data to the cloud.

In the context of efficient data usage, research opportunities are to be found in strategies such as transfer learning and active learning. Transfer learning makes use of already well performing models from a different domain which can be fine-tuned for the specific task. It was shown initially for computer vision applications that this approach leads to amazing performances [133]. Active learning is an incremental learning process, which is especially useful if data is sparse. The data is rated on a trained model concerning its information content, which is then used for model updates [134].

Scene prediction models based on neural networks often use a combination of many different architectures, i.e., a graphical representation, recurrent models and attention as well as convolutions for semantic information. For urban use cases map information is often integrated. The observed trend towards a complete scene prediction is not far from the disruptive approach. The argument against the disruptive approach was that such an algorithm cannot be tested for functional safety. Thus, it is necessary to think about the degree of complexity that is still manageable for testing.

When scene prediction models are robust enough, the next step will be the industrialisation of the product for a broad target market. Currently required sensor equipment is cost-intensive, using HD (high density) maps, camera, radar, Lidar, GPS and many more, and often can fill up the entire trunk of a car. Strategies for a minimalization of the costs and material require new approaches such as discretization of neural networks for deployment in embedded systems [135,136]. Alternative strategies follow using only a subset of the sensors, e.g., neglecting the costly Lidar technology, while still aiming at a comparable perception performance. The product costs eventually determine the target group of the self-driving vehicle. If it is not affordable for persons with a regular income, the business model will address high-value customers and mobility as a service.

A further streamline in the deep learning community is the development of a The so-called weak artificial intelligence is limited to specific tasks. A strong or general AI shall be able to handle multiple tasks, similar to a human [137]. Fitting into this picture, the latest approaches focus on a coupling of prediction and planning, as presented in the previous section. For such models, the functional safety aspect can only be satisfied if this is done in a multi-stage approach. From the complex model intermediate results need to be extracted which can be interpreted and evaluated, such as optimization of implicit layers [138].

Author Contributions: Conceptualization, A.S.N. and M.K.; writing—original draft preparation, A.S.N., M.K., M.S. and T.B.; writing—review and editing, A.S.N., M.K. and M.S.; visualization, A.S.N.; supervision, A.S.N. and T.B. All authors have read and agreed to the published version of the manuscript.

Funding: This research received no external funding.

Data Availability Statement: Not applicable.

Acknowledgments: Personal acknowledgements from A.S.N. to Reinhard Schlickeiser: This paper is written for the occasion of Reinhard's 70th birthday who has been a great supervisor during my, Anne Stockem Novo, early career. As a PhD student in Reinhard's group he taught me the joy of analytical calculations, wrangling equations and function approximations. Because of their excellent scientific network I came into contact with highly recognized researchers from the computer simulation domain with whom I made first contact with programming. The analytical mindset and critical thinking that I developed under Reinhard's supervision, were the basis for a new challenge in industry at the ZF group in algorithm development for automated driving with machine learning techniques, a field not directly connected with my previous research. This paved the way for becoming a professor for applied artificial intelligence.

Conflicts of Interest: The authors declare no conflict of interest.

Abbreviations

The following abbreviations are used in this manuscript:

AD	Automated Driving
ADAS	Advanced Driver Assistance System
ADE	Average Displacement Error
API	Application Programming Interface
CNN	Convolutional Neural Network
CVAE	Conditional Variational Auto Encoder
CYRA	Constant Yaw Rate and Acceleration
ELBO	Evidence Lower Bound
FDE	Final Displacement Error
FFNN	Feed-Forward Neural Network
GNN	Graph Neural Network
GPS	Global Positioning System
GRU	Gated Recurrent Unit
HD	High Density
HMM	Hidden Markov Model
IMU	Integrated Motion Unit
LSTM	Long Short-Term Memory Network
MAE	Mean Absolute Error
NLP	Natural Language Processing
OEM	Original Equipment Manufacture
RNN	Recurrent Neural Network
RMSE	Root Mean Squared Error
SAE	Society of Automotive Engineers

Appendix A. Long Short-Term Memory Networks

There are several gates in which the information is processed. The *forget gate* determines which information shall be kept in memory by adapting the weight matrices W_f, U_f and bias B_f,

$$f_t = \sigma \left[W_f X_t + U_f h_{t-1} + B_f \right]. \tag{A1}$$

The non-linear activation function is usually a sigmoid function, $\sigma(x) = 1/(1 + \exp(-x))$, limiting the output f_t to the range $[0, 1]$.

Similarly, the input gate determines which information shall be added to the long-term memory,

$$i_t = \sigma[W_i X_t + U_i h_{t-1} + B_i] \tag{A2}$$

with corresponding weight matrices W_i and U_i, as well as bias B_i.

The output gate receives similar input,

$$o_t = \sigma[W_o X_t + U_o h_{t-1} + B_o]. \tag{A3}$$

One can also identify a simple RNN cell, which usually applies tanh as activation function,

$$\tilde{c}_t = \tanh[W_c X_t + U_c h_{t-1} + B_c]. \tag{A4}$$

The long-term memory is then calculated as

$$c_t = f_t \cdot c_{t-1} + i_t \cdot \tilde{c}_t, \tag{A5}$$

A combination of the forget gate, f_t, and input gate, i_t, as well as the long-term state vector of the previous time step, c_{t-1}, and the output of the simple RNN cell at the current time step, $\tilde{c}_t$.

The short-term memory and cell output are updated with the output gate, o_t, and long-term state vector, c_t,

$$h_t = o_t \cdot \tanh(c_t). \tag{A6}$$

Additionally, in this case, the output to the following layer, y_t, is identical to the state vector, h_t.

Appendix B. Gated Recurrent Unit

The GRU is a simpler form with only two gates. The original GRU architecture [31] consists of a reset gate,

$$r_t = \sigma[W_r X_t + U_r h_{t-1} + B_r]$$ (A7)

with weight matrices W_r and U_r, as well as an update gate,

$$z_t = \sigma[W_z X_t + U_z h_{t-1} + B_z]$$ (A8)

with weight matrices W_z and U_z. The hidden state vector is calculated as

$$h_t = z_t \cdot h_{t-1} + (1 - z_t) \cdot \tilde{h}_t$$ (A9)

with

$$\tilde{h}_t = \tanh[W_h X_t + U_h(r_t \cdot h_{t-1}) + B_h].$$ (A10)

References

1. Autopilot and Full Self-Driving Capability. Available online: https://www.tesla.com/support/autopilot/ (accessed on 31 October 2021).
2. The Society of Automotive Engineers (AES). Available online: https://www.sae.org (accessed on 31 October 2021).
3. Available online: https://twitter.com/elonmusk/status/1148070210412265473 (accessed on 31 October 2021).
4. Scanlon, J.M.; Kusano, K.D.; Daniel, T.; Alderson, C.; Ogle, A.; Victor, T. Waymo Simulated Driving Behavior in Reconstructed Fatal Crashes within an Autonomous Vehicle Operating Domain. 2021. Available online: http://bit.ly/3bu5VcM (accessed on 31 October 2021).
5. Jiang, R.; Hu, M.B.; Jia, B.; Wang, R.L.; Wu, Q.S. The effects of reaction delay in the Nagel-Schreckenberg traffic flow model. *Eur. Phys. J. B* **2006**, *54*, 267–273. [CrossRef]
6. Preliminary Regularoty Impact Analysis. Available online: https://www.nhtsa.gov/sites/nhtsa.gov/files/documents/v2v_pria_12-12-16_clean.pdf (accessed on 31 October 2021).
7. Huang, X.; Kroening, D.; Ruan, W.; Sharp, J.; Sun, Y.; Thamo, E.; Wu, M.; Yi, X. A survey of safety and trustworthiness of deep neural networks: Verification, testing, adversarial attack and defence, and interpretability. *Comput. Sci. Rev.* **2020**, *37*, 100270. [CrossRef]
8. Szegedy, C.; Zaremba, W.; Sutskever, I.; Bruna, J.; Erhan, D.; Goodfellow, I.; Fergus, R. Intriguing properties of neural networks. *arXiv* **2013**, arXiv:1312.6199.
9. Marti, E.; de Miguel, M.A.; Garcia, F.; Perez, J. A Review of sensor technologies for perception in automated driving. *IEEE Intell. Transp. Syst. Mag.* **2019**, *11*, 94–108. [CrossRef]
10. Wang, Z.; Wu, Y.; Niu, Q. Multi-sensor fusion in automated driving: A survey. *IEEE Access* **2020**, *8*, 2847–2868. [CrossRef]
11. Yurtsever, E.; Lambert, J.; Carballo, A.; Takeda, K. A Survey of autonomous driving: Common practices and emerging technologies. *IEEE Access* **2020**, *8*, 58443–58469. [CrossRef]
12. Lu, Z.; Happee, R.; Cabrall, C.D.D.; Kyriakidis, M.; de Winter, J.C.F. Human factors of transitions in automated driving: A general framework and literature survey. *Transp. Res. Part F Traffic Psychol. Behav.* **2016**, *43*, 183–198. [CrossRef]
13. Petermeijer, S.M.; de Winter, J.C.F.; Bengler, K.J. Vibrotactile displays: A survey with a view on highly automated driving. *IEEE Trans. Intell. Transp. Syst.* **2016**, *17*, 897–907. [CrossRef]
14. Pfleging, B.; Rang, M. Investigating user needs for non-driving-related activities during automated driving. In Proceedings of the 15th International Conference on Mobile and Ubiquitous Multimedia MUM '16, Rovaniemi, Finland, 12–15 December 2016; Association for Computing Machinery: New York, NY, USA, 2016; pp. 91–99. [CrossRef]
15. Lefèvre, S.; Vasquez, D.; Laugier, C. A survey on motion prediction and risk assessment for intelligent vehicle. *Robomech J.* **2014**, *1*, 1. [CrossRef]
16. Toledo, T. Driving behaviour: Models and challenges. *Transp. Rev.* **2007**, *27*, 65–84. [CrossRef]
17. Xue, J.R.; Fang, J.W.; Zhang, P.A. Survey of scene understanding by event reasoning in autonomous driving. *Int. J. Autom. Comput.* **2018**, *15*, 249–266. [CrossRef]
18. The Web of Science. Available online: https://www.webofscience.com/ (accessed on 5 May 2021).
19. Hirakawa, T.; Yamashita, T.; Tamaki, T.; Fujiyoshi, H. Survey on vision-based path prediction. In Proceedings of the 6th International Conference on Distributed, Ambient, and Pervasive Interactions: Technologies and Contexts (DAPI 2018), Las Vegas, NV, USA, 15–20 July 2018; Springer: Cham, Switzerland, 2018; pp. 48–64. [CrossRef]

20. Yuan, J.; Abdul-Rashid, H.; Li, B. A survey of recent 3D scene analysis and processing methods. *Multimed. Tools Appl.* **2021**, *80*, 19491–19511. [CrossRef]

21. Yin, X.; Wu, G.; Wei, J.; Shen, Y.; Qi, H.; Yin, B. Deep learning on traffic prediction: Methods, analysis and future directions. *IEEE Trans. Intell. Transp. Syst.* **2021**, in press. [CrossRef]

22. Tedjopurnomo, D.A.; Bao, Z.; Zheng, B.; Choudhury, F.; Qin, A.K. A Survey on Modern Deep Neural Network for Traffic Prediction: Trends, Methods and Challenges. *IEEE Trans. Knowl. Data Eng.* **2020**, *1*, 1. [CrossRef]

23. Rasouli, A. Deep learning for Vision-based prediction: A survey. *arXiv* **2020**, arXiv:2007.00095.

24. Kalman, R.E. A new approach to linear filtering and prediction problems. *Trans. ASME-J. Basic Eng.* **1960**, *82*, 35–45. [CrossRef]

25. Dellaert, F.; Fox, D.; Burgard, W.; Thrun, S. Monte Carlo localization for mobile robots. In Proceedings of the 1999 IEEE International Conference on Robotics and Automation, Detroit, MI, USA, 10–15 May 1999; pp. 1322–1328. [CrossRef]

26. Rabiner, L.R. A tutorial on hidden Markov models and selected applications in speech recognition. *Proc. IEEE* **1989**, *77*, 257–286. [CrossRef]

27. Rumelhart, D.; Hinton, G.E.; Williams, R.J. Learning representations by back-propagating errors. *Nature* **1986**, *323*, 533–536. [CrossRef]

28. Hochreiter, J. Untersuchungen zu Dynamischen Neuronalen Netzen. Master's Thesis, Universität München, München, Germany, 1991. (In Germany)

29. Pascanu, R.; Mikolov, T.; Bengio, Y. On the difficulty of training recurrent neural networks. In Proceedings of the 30th International Conference on Machine Learning (ICML'13), Atlanta, GA, USA, 16–21 June 2013; Volume 28, pp. 1310–1318. Available online: http://proceedings.mlr.press/v28/pascanu13.pdf (accessed on 31 October 2021).

30. Hochreiter, S.; Schmidhuber, J. Long short-term memory. *Neural Comput.* **1997**, *9*, 1735–1780. [CrossRef]

31. Cho, K.; van Merrienboer, B.; Gulcehre, C.; Bahdanau, D.; Bougares, F.; Schwenk, H.; Bengio, Y. Learning phrase representations using RNN encoder–decoder for statistical machine translation. In Proceedings of the 2014 Conference on Empirical Methods in Natural Language Processing (EMNLP), Doha, Qatar, 25–29 October 2014; pp. 1724–1734. [CrossRef]

32. Bahdanau, D.; Cho, K.H.; Bengio, Y. Neural machine translation by jointly learning to align and translate. *arXiv* **2014**, arXiv:1409.0473.

33. Kingma, D.P.; Welling, M. Auto-encoding variational Bayes. *arXiv* **2014**, arXiv:1312.6114.

34. Karras, T.; Laine, S.; Aittala, M.; Hellsten, J.; Lehtinen, J.; Aila, T. Analyzing and improving the image quality of StyleGAN. In Proceedings of the IEEE/CVF Conference on Computer Vision and Pattern Recognition (CVPR), Salt Lake City, UT, USA, 18–23 June 2018. [CrossRef]

35. Goodfellow, I.; Bengio, Y.; Courville, A. *Deep Learning*; MIT Press: Cambridge, MA, USA, 2016. [CrossRef]

36. Vaswani, A.; Shazeer, N.; Parmar, N.; Uszkoreit, J.; Jones, L.; Gomez, A.N.; Kaiser, L.; Polosukhin, I. Attention is all you need. *Adv. Neural Inf. Process. Syst.* **2017**, *30*, 5998–6008.

37. Zhou, J.; Cui, G.; Hu, S.; Zhang, Z.; Yang, C.; Liu, Z.; Wang, L.; Li, C.; Sun, M. Graph neural networks: A review of methods and applications. *AI Open* **2020**, *1*, 57–81. [CrossRef]

38. Hamilton, W.L.; Ying, R.; Leskovec, J. Inductive representation learning on large graphs. In Proceedings of the 31st International Conference on Neural Information Processing Systems, Long Beach, CA, USA, 4–9 December 2017; Volume 30.

39. Gipps, P.G. A model for the structure of lane-changing decisions. *Transp. Res. Part B Methodol.* **1986**, *20B*, 403–414. [CrossRef]

40. Toledo, T.; Koutsopoulos, H.N.; Ben-Akiva, M.E. Modeling integrated Lane-changing behavior. *Transp. Res. Rec. J. Transp. Res. Board* **2003**, *1857*, 30–38. [CrossRef]

41. Hunt, J.G.; Lyons, G.D. Modelling dual carriageway lane changing using neural networks. *Transp. Res. Part C Emerg. Technol.* **1994**, *2*, 231–245. [CrossRef]

42. Salvucci, D.D.; Liu, A. The time course of a lane change: Driver control and eye-movement behavior. *Transp. Res. Part F Traffic Psychol. Behav.* **2002**, *5*, 123–132. [CrossRef]

43. Oliver, N.; Pentland, A.P. Graphical models for driver behavior recognition in a SmartCar. In Proceedings of the IEEE Intelligent Vehicles Symposium 2000, Dearborn, MI, USA, 3–5 October 2000; pp. 7–12. [CrossRef]

44. Toledo-Moreo, R.; Zamora-Izquierdo, M.A. IMM-based lane-change prediction in highways with low-cost GPS/INS. *IEEE Trans. Intell. Transp. Syst.* **2009**, *10*, 180–185. [CrossRef]

45. Kasper, D.; Weidl, G.; Dang, T.; Breuel, G.; Tamke, A.; Wedel, A.; Rosenstiel, W. Object-oriented Bayesian networks for detection of lane change maneuvers. *IEEE Intell. Transp. Syst. Mag.* **2012**, *4*, 19–31. [CrossRef]

46. Lefèvre, S.; Laugier, C.; Ibañez-Guzmán, J. Exploiting map information for driver intention estimation at road intersections. In Proceedings of the 2011 IEEE Intelligent Vehicles Symposium (IV), Baden-Baden, Germany, 5–9 June 2011; pp. 583–588. [CrossRef]

47. Liebner, M.; Baumann, M.; Klanner, F.; Stiller, C. Driver intent inference at urban intersections using the intelligent driver model. In Proceedings of the 2012 IEEE Intelligent Vehicles Symposium, Madrid, Spain, 3–7 June 2012; pp. 1162–1167. [CrossRef]

48. Schlechtriemen, J.; Wedel, A.; Hillenbrand, J.; Breuel, G.; Kuhnert, K. A lane change detection approach using feature ranking with maximized predictive power. In Proceedings of the 2014 IEEE Intelligent Vehicles Symposium Proceedings, Dearborn, MI, USA, 8–11 June 2014; pp. 108–114. [CrossRef]

49. Laval, J.A.; Daganzo, C.F. Lane-changing in traffic streams. *Transp. Res. Part B Methodol.* **2006**, *40*, 251–264. [CrossRef]

50. Aoude, G.S.; Desaraju, V.R.; Stephens, L.H.; How, J.P. Driver behavior classification at intersections and validation on large naturalistic data set. *IEEE Trans. Intell. Transp. Syst.* **2012**, *13*, 724–736. [CrossRef]

51. Kumar, P.; Perrollaz, M.; Lefèvre, S.; Laugier, C. Learning-based approach for online lane change intention prediction. In Proceedings of the 2013 IEEE Intelligent Vehicles Symposium (IV), Gold Coast City, Australia, 23 June 2013; pp. 797–802. [CrossRef]

52. Mandalia, H.M.; Salvucci, M.D.D. Using support vector machines for lane change detection. In *Proceedings of the Human Factors and Ergonomics Society Annual Meeting*; SAGE Publications: Los Angeles, CA, USA, 2005; Volume 49, pp. 1965–1969. [CrossRef]

53. Bahram, M.; Hubmann, C.; Lawitzky, A.; Aeberhard, M.; Wollherr, D. A Combined Model- and Learning-Based Framework for Interaction-Aware Maneuver Prediction. *IEEE Trans. Intell. Transp. Syst.* **2016**, *17*, 1538–1550. [CrossRef]

54. Khosroshahi, A.; Ohn-Bar, E.; Trivedi, M.M. Surround vehicles trajectory analysis with recurrent neural networks. In Proceedings of the 2016 IEEE 19th International Conference on Intelligent Transportation Systems (ITSC), Rio de Janeiro, Brazil, 1–4 November 2016; pp. 2267–2272. [CrossRef]

55. Mammar, S.; Glaser, S.; Netto, M. Time to line crossing for lane departure avoidance: A theoretical study and an experimental setting. *IEEE Trans. Intell. Transp. Syst.* **2006**, *7*, 226–241. [CrossRef]

56. Ahmed, K.I. Modeling Drivers' Acceleration and Lane Changing Behavior. Sc.D. Thesis, Massachusetts Institute of Technology, Department of Civil and Environmental Engineering, Cambridge, MA, USA, 1999. Available online: https://dspace.mit.edu/handle/1721.1/9662 (accessed on 31 October 2021).

57. Helbing, D.; Tilch, B. Generalized force model of traffic dynamics. *Phys. Rev. E* **1998**, *58*, 133. [CrossRef]

58. Available online: https://paperswithcode.com/ (accessed on 31 October 2021).

59. Hermes, C.; Wohler, C.; Schenk, K.; Kummert, F. Long-term vehicle motion prediction. In Proceedings of the 2009 IEEE Intelligent Vehicles Symposium, Xi'an, China, 3–5 June 2009; pp. 652–657. [CrossRef]

60. Houenou, A.; Bonnifait, P.; Cherfaoui, V.; Yao, W. Vehicle trajectory prediction based on motion model and maneuver recognition. In Proceedings of the 2013 IEEE/RSJ International Conference on Intelligent Robots and Systems, Tokyo, Japan, 3–7 November 2013; pp. 4363–4369. [CrossRef]

61. Deo, N.; Rangesh, A.; Trivedi, M.M. How would surround vehicles move? A unified framework for maneuver classification and motion prediction. *IEEE Trans. Intell. Veh.* **2018**, *3*, 129–140. [CrossRef]

62. Casas, S.; Luo, W.; Urtasun, R. IntentNet: Learning to predict intention from raw sensor data. In Proceedings of the 2nd Conference on Robot Learning (CoRL 2018), Zürich, Switzerland, 29–31 October 2018; pp. 947–956. Available online: http://proceedings.mlr.press/v87/casas18a/casas18a.pdf (accessed on 31 October 2021).

63. Park, S.H.; Kim, B.; Kang, C.M.; Chung, C.C.; Choi, J.W. Sequence-to-sequence prediction of vehicle trajectory via LSTM encoder-decoder architecture. In Proceedings of the 2018 IEEE Intelligent Vehicles Symposium (IV), Suzhou, China, 26–30 June 2018; pp. 1672–1678. [CrossRef]

64. Cui, H.; Radosavljevic, F.; Chou, F.-C.; Lin, T.-H.; Nguyen, T.; Huang, T.-K.; Schneider, J.; Djuric, N. Multimodal trajectory predictions for autonomous driving using deep convolutional networks. In Proceedings of the 2019 International Conference on Robotics and Automation (ICRA), Montreal, QC, Canada, 20–24 May 2019; pp. 2090–2096. [CrossRef]

65. Altché, F.; de la Fortelle, A. An LSTM network for highway trajectory prediction. In Proceedings of the 2017 IEEE 20th International Conference on Intelligent Transportation Systems (ITSC), Las Vegas, NV, USA, 27–30 June 2016; pp. 353–359. [CrossRef]

66. Deo, N.; Trivedi, M.M. Convolutional social pooling for vehicle trajectory prediction. In Proceedings of the 2018 IEEE/CVF Conference on Computer Vision and Pattern Recognition Workshops (CVPRW), Salt Lake City, UT, USA, 18–22 June 2018; pp. 1549–15498. [CrossRef]

67. Chandra, R.; Bhattacharya, U.; Bera, A.; Manocha, D. TraPHic: Trajectory prediction in dense and heterogeneous traffic using weighted interactions. In Proceedings of the 2019 IEEE/CVF Conference on Computer Vision and Pattern Recognition (CVPR), Long Beach, CA, USA, 15–20 June 2019; pp. 8475–8484. [CrossRef]

68. Zhao, T.; Xu, Y.; Monfort, M.; Choi, W.; Baker, C.; Zhao, Y.; Wang, Y.; Wu, Y.N. Multi-agent tensor fusion for contextual trajectorypPrediction. In Proceedings of the 2019 IEEE/CVF Conference on Computer Vision and Pattern Recognition (CVPR), Long Beach, CA, USA, 15–20 June 2019; pp. 12118–12126. [CrossRef]

69. Tang, Y.C.; Salakhutdinov, R. Multiple futures prediction. *arXiv* **2019**, arXiv:1911.00997.

70. Song, H.; Ding, W.; Chen, Y.; Shen, S.; Wang, M.Y.; Chen, Q. PiP Planning-informed trajectory prediction for autonomous driving. In *European Conference on Computer Vision—ECCV 2020*; Vedaldi, A., Bischof, H., Brox, T., Frahm, J.M., Eds.; Springer: Cham, Switzerland, 2020; pp. 598–614. [CrossRef]

71. Chandra, R.; Guan, T.; Panuganti, S. Forecasting trajectory and behavior of road-agents using spectral clustering in graph-LSTMs. *IEEE Robot. Autom. Lett.* **2020**, *5*, 4882–4890. [CrossRef]

72. Lee, N.; Choi, W.; Vernaza, P.; Choy, C.B.; Torr, P.H.S.; Chandraker, M. DESIRE: Distant future prediction in dynamic scenes with interacting agents. In Proceedings of the 2017 IEEE Conference on Computer Vision and Pattern Recognition (CVPR), Honolulu, HI, USA, 21–26 July 2017; pp. 2165–2174. [CrossRef]

73. Choi, C.; Choi, J.H.; Li, J.; Malla, S. Shared cross-modal trajectory prediction for autonomous driving. *arXiv* **2020**, arXiv:2004.00202.

74. Chai, Y.; Sapp, B.; Bansal, M.; Anguelov, D. MultiPath: Multiple probabilistic anchor trajectory hypotheses for behavior prediction. In *Conference on Robot Learning*; PMLR: New York, NY, USA, 2020; Volume 100, pp. 86–99.

75. Mangalam, K.; Girase, H.; Agarwal, S.; Lee, K.H.; Adeli, E.; Malik, J.; Gaidon, A. It is not the journey but the destination: Endpoint conditioned trajectory prediction. In Proceedings of the European Conference on Computer Vision—ECCV 2020, Glasgow, UK, 23–28 August 2020; Vedaldi, A., Bischof, H., Brox, T., Frahm, J.M., Eds.; Springer: Cham, Switzerland, 2020. [CrossRef]

76. Park, S.H.; Lee, G.; Seo, J.; Bhat, M.; Kang, M.; Francis, J.; Jadhav, A.; Liang, P.P.; Morency, L.P. Diverse and admissible trajectory forecasting through multimodal context understanding. In Proceedings of the European Conference on Computer Vision—ECCV 2020; Vedaldi A., Bischof, H., Brox, T., Frahm, J.M., Eds.; Springer: Cham, Switzerland, 2020; pp. 282–298. [CrossRef]

77. Song, H.; Luan, D.; Ding, W.; Wang, M.Y.; Chen, Q. Learning to predict vehicle trajectories with model-based planning. *arXiv* **2021**, arXiv:2103.04027.

78. Zeng, W.; Liang, M.; Liao, R.; Urtasun, R. LaneRCNN: Distributed representations for graph-centric motion forecasting. *arXiv* **2021**, arXiv:2101.06653.

79. Casas, S.; Gulino, C.; Liao, R.; Urtasun, R. SpAGNN: Spatially-Aware Graph Neural Networks for relational behavior forecasting from sensor data. In Proceedings of the 2020 IEEE International Conference on Robotics and Automation (ICRA), Paris, France, 31 May–31 August 2020; pp. 9491–9497. [CrossRef]

80. Phan-Minh, T.; Grigore, E.C.; Boulton, F.A.; Beijbom, O.; Wolff, E.M. CoverNet: Multimodal behavior prediction using trajectory sets. In Proceedings of the 2020 IEEE/CVF Conference on Computer Vision and Pattern Recognition (CVPR), Seattle, WA, USA, 14–19 June 2020; pp. 14062–14071. [CrossRef]

81. Liang, M.; Yang, B.; Zeng, W.; Chen, Y.; Hu, R.; Casas, S.; Urtasun, R. PnPNet: End-to-end perception and prediction with tracking in theLoop. In Proceedings of the 2020 IEEE/CVF Conference on Computer Vision and Pattern Recognition (CVPR), Seattle, WA, USA, 14–19 June 2020; pp. 11550–11559. [CrossRef]

82. Narayanan, S.; Moslemi, R.; Pittaluga, F.; Liu, B.; Chandraker, M. Divide-and-conquer for lane-aware diverse trajectory prediction In Proceedings of the IEEE/CVF Conference on Computer Vision and Pattern Recognition (CVPR), Nashville, TN, USA, 20–25 June 2021. [CrossRef]

83. Li, J.; Ma, H.; Zhang, Z.; Tomizuka, M. Social-WaGDAT interaction-aware trajectory prediction via Wasserstein graph double-attention network. *arXiv* **2020**, arXiv:2002.06241.

84. Mohta, A.; Chou, F.C.; Becker, B.C.; Vallespi-Gonzalez, C.; Djuric, N. Investigating the effect of sensor modalities in multi-sensor detection-prediction models. *arXiv* **2021**, arXiv:2101.03279.

85. Pomerleau, D.A. ALVINN: An autonomous land vehicle in a neural network. *Adv. Neural Inf. Process. Syst.* **1989**, *1*, 305–313.

86. Fix, E.; Armstrong, H.G. Modeling human performance with neural networks. *Int. Jt. Conf. Neural Netw.* **1990**, *1*, 247–252.

87. Vasquez, D.; Fraichard, T. Motion prediction for moving objects: A statistical approach. *Proc. IEEE Int. Conf. Robot. Autom.* **2004**, *4*, 3931–3936. [CrossRef]

88. Li, X.; Hu, W.; Hu, W. A coarse-to-fine strategy for vehicle motion trajectory clustering. In Proceedings of the 18th International Conference on Pattern Recognition, Hong Kong, China, 20–24 August 2006, Volume 1, pp. 591–594. [CrossRef]

89. Vasquez, D.; Fraichard, T.; Laugier, C. Incremental learning of statistical motion patterns with growing hidden Markov models. *IEEE Trans. Intell. Transp. Syst.* **2009**, *10*, 403–416. [CrossRef]

90. Morton, J.; Wheeler, T.A.; Kochenderfer, M.J. Analysis of recurrent neural networks for probabilistic modeling of driver behavior. *IEEE Trans. Intell. Transp. Syst.* **2017**, *18*, 1289–1298. [CrossRef]

91. Kim, B.; Kang, C.M.; Kim, J.; Lee, S.H.; Chung, C.C.; Choi, J.W. Probabilistic vehicle trajectory prediction over occupancy grid map via recurrent neural network. In Proceedings of the 2017 IEEE 20th International Conference on Intelligent Transportation Systems (ITSC), Yokohama, Japan, 16–19 October 2017; pp. 399–404. [CrossRef]

92. Krüger, M.; Stockem Novo, A.; Nattermann, T.; Bertram, T. Interaction-aware trajectory prediction based on a 3D spatio-temporal tensor representation using convolutional–recurrent neural networks. In Proceedings of the 2020 IEEE Intelligent Vehicles Symposium (IV), Las Vegas, NV, USA, 19 October–13 November 2020; pp. 1122–1127. [CrossRef]

93. Alahi, A.; Goel, K.; Ramanathan, V.; Robicquet, A.; Fei-Fei, L.; Savarese, S. Social LSTM: Human trajectory prediction in crowded spaces. In Proceedings of the 29th IEEE Conference on Computer Vision and Pattern Recognition (CVPR), Las Vegas, NV, USA, 27–30 June 2016; pp. 961–971. [CrossRef]

94. Robicquet, A.; Sadeghian, A.; Alahi, A.; Savarese, S. Learning social etiquette: Human trajectory understanding in crowded scenes. In Proceedings of the Computer Vision—ECCV 2016, , Glasgow, UK, 23–28 August 2020; Leibe, B., Matas, J., Sebe, N., Welling, M., Eds.; Springer: Cham, Switzerland, 2016; pp. 549–565. [CrossRef]

95. Rudenko, A.; Palmieri, L.; Herman, M.; Kitani, K.M.; Gavrila, D.; Arras, K. Human motion trajectory prediction: A survey. *Int. J. Robot. Res.* **2020**, *39*, 895–935. [CrossRef]

96. Li, X.; Ying, X.; Chuah, M.C. GRIP: Graph-based interaction-aware trajectory prediction. In Proceedings of the 2019 IEEE Intelligent Transportation Systems Conference (ITSC), Auckland, New Zealand, 27–30 October 2019; pp. 3960–3966. [CrossRef]

97. Zhang, L.; Su, P.-H.; Hoang, J.; Haynes, G.C.; Marchetti-Bowick, M. Map-adaptive goal-based trajectory prediction. *arXiv* **2020**, arXiv:2009.04450.

98. Xu, H.; Gao, Y.; Yu, F.; Darrell, T. End-to-end learning of driving models from large-scale video datasets. In Proceedings of the 2017 IEEE Conference on Computer Vision and Pattern Recognition (CVPR), Honolulu, HI, USA, 21–26 July 2017; pp. 3530–3538. [CrossRef]

99. Deo, N.; Trivedi, M.M. Multi-modal trajectory prediction of surrounding vehicles with maneuver based LSTMs. In Proceedings of the 2018 IEEE Intelligent Vehicles Symposium (IV), Suzhou, China, 26–30 June 2018; pp. 1179–1184. [CrossRef]

100. Hoermann, S.; Bach, M.; Dietmayer, K. Dynamic occupancy grid prediction for urban autonomous driving: A deep learning approach with fully automatic labeling. In Proceedings of the 2018 IEEE International Conference on Robotics and Automation (ICRA), Brisbane, Australia, 21–25 May 2018; pp. 2056–2063. [CrossRef]

101. Luo, W.; Yang, B.; Urtasun, R. Fast and furious: Real time end-to-end 3D detection, tracking and motion forecasting with a single convolutional net. In Proceedings of the 2018 IEEE/CVF Conference on Computer Vision and Pattern Recognition Workshops (CVPRW), Salt Lake City, UT, USA, 18–22 June 2018; pp. 3569–3577. [CrossRef]
102. Choi, C.; Malla, S.; Patil, A.; Choi, J.H. DROGON: A trajectory prediction model based on intention-conditioned behavior reasoning. *arXiv* **2020**, arXiv:1908.00024.
103. Gao, J.; Sun, C.; Zhao, H. VectorNet: Encoding HD maps and agent dynamics from vectorized representation. *arXiv* **2020**, arXiv:2005.04259.
104. Gomez-Gonzalez, S.; Prokudin, S.; Schölkopf, B.; Peters, J. Real time trajectory prediction using deep conditional generative models. *IEEE Robot. Autom. Lett.* **2020**, *5*, 970–976. [CrossRef]
105. Hu, Y.; Chen, S.; Zhang, Y.; Gu, X. Collaborative motion prediction via neural motion message passing. In Proceedings of the 2020 IEEE/CVF Conference on Computer Vision and Pattern Recognition (CVPR), Seattle, WA, USA, 14–19 June 2020; pp. 6318–6327. [CrossRef]
106. Mohamed, A.; Qian, K.; Elhoseiny, M.; Claudel, C. Social-STGCNN: A social spatio-temporal graph convolutional neural network for human trajectory prediction. In Proceedings of the 2020 IEEE/CVF Conference on Computer Vision and Pattern Recognition (CVPR), Seattle, WA, USA, 14–19 June 2020; pp. 14412–14420. [CrossRef]
107. Pareja, A.; Domeniconi, G.; Chen, J.; Ma, T.; Suzumura, T.; Kanezashi, H.; Kaler, T.; Schardl, T.B.; Leiserson, C.E. EvolveGCN: Evolving Graph Convolutional Networks for dynamic graphs. *arXiv* **2019**, arXiv:1902.10191.
108. Roh, J.; Mavrogiannis, C.; Madan, R.; Fox, D.; Srinivasa, S.S. Multimodal trajectory prediction via yopological nvariance for navigation at uncontrolled intersections. *arXiv* **2011**, arXiv:2011.03894.
109. Zhang, Z.; Gao, J.; Mao, J.; Liu, Y.; Anguelov, D.; Li, C. STINet: Spatio-Temporal-Interactive Network for Pedestrian Detection and Trajectory Prediction. *arXiv* **2020**, arXiv:2005.04255.
110. Yu, B.; Yin, H.; Zhu, Z. Spatio-temporal graph convolutional networks: A deep learning framework for traffic forecasting. In Proceedings of the Twenty-Seventh International Joint Conference on Artificial Intelligence, Stockholm, Sweden, 13–19 July 2018; pp. 3634–3640.
111. Guo, S.; Lin, Y.; Feng, N.; Song, C.; Wan, H. Attention based spatial-temporal graph convolutional networks for traffic flow forecasting. *IJCAI-19* **2019**, *33*, 922–929. [CrossRef]
112. Diehl, F.; Brunner, T.; Le, M.T.; Knoll, A. Graph neural networks for modelling traffic participant interaction. In Proceedings of the 2019 IEEE Intelligent Vehicles Symposium (IV), Paris, France, 9–12 June 2019; pp. 695–701.
113. Ivanovic, B.; Pavone, M. The Trajectron: Probabilistic multi-agent trajectory modeling with dynamic spatiotemporal graphs. In Proceedings of the 2019 IEEE/CVF International Conference on Computer Vision (ICCV), Seoul, Korea, 27–28 October 2019; pp. 2375–2384. [CrossRef]
114. Huang, Y.; Bi, H.; Li, Z.; Mao, T.; Wang, Z. STGAT: Modeling spatial-temporal interactions for human trajectory prediction. In Proceedings of the 2019 IEEE/CVF International Conference on Computer Vision (ICCV), Seoul, Korea, 27–28 October 2019; pp. 6271–6280. [CrossRef]
115. Khandelwal, S.; Qi, W.; Sing, J.; Hartnett, A.; Ramanan, D. What-if motion prediction for autonomous driving. *arXiv* **2020**, arXiv:2008.10587.
116. Kosaraju, V.; Sadeghian, A.; Martin-Martin, R. Social-BiGAT: Multimodal trajectory forecasting using bicycle-GAN and graph attention networks. *arXiv* **2019**, arXiv:1907.03395.
117. Liang, M.; Yang, B.; Hu, R. Learning lane graph representations for motion forecasting. *European Conference on Computer Vision*; Springer: Cham, Switzerland, 2020; Volume 12347, pp. 541–556.
118. Messaoud, K.; Yahiaoui, I.; Verroust-Blondet, A.; Nashashibi, F. Attention based vehicle trajectory prediction. *IEEE Trans. Intell. Veh.* **2021**, *6*, 175–185. [CrossRef]
119. Salzmann, T.; Ivanovic, B.; Chakravarty, P.; Pavone, M. Trajectron++: Multi-agent generative trajectory forecasting with heterogeneous data for control. *arXiv* **2020**, arXiv:2001.03093.
120. Giuliari, F.; Hasan, I.; Cristani, M.; Galasso, F. Transformer networks for trajectory forecasting. In Proceedings of the 2020 25th International Conference on Pattern Recognition (ICPR), Milan, Italy, 10–15 January 2021; pp. 10335–10342. [CrossRef]
121. Xue, H.; Salim, F.D. TERMCast: Temporal relation modeling for effective urban flow forecasting. In *Advances in Knowledge Discovery and Data Mining, Proceedings of the Pacific-Asia Conference on Knowledge Discovery and Data Mining 2021 (PAKDD 2021), Virtual Event, 11–14 May 2021*; Karlapalem, K., Cheng, H., Ramakrishnan, N., Agrawal, R.K., Krishna Reddy, P., Srivastava, J., Chakraborty, T., Eds.; Springer: Cham, Switzerland, 2021; pp. 741–753. [CrossRef]
122. Cai, L.; Janowicz, K.; Mai, G.; Yan, B.; Zhu, R. Traffic transformer: Capturing the continuity and periodicity of time series for traffic forecasting. *Trans. GIS* **2020**, *24*, 736–755. [CrossRef]
123. Bogaerts, T.; Masegosa, A.; Angarita-Zapata, J.S.; Onieva, E.; Hellinckx, P. A graph CNN-LSTM neural network for short and long-term traffic forecasting based on trajectory data. *Transp. Res. Part C Emerg. Technol.* **2020**, *112*, 62–77. [CrossRef]
124. Zhao, H.; Gao, J.; Lan, T.; Sun, C.; Sapp, B.; Varadarajan, B.; Shen, Y.; Chai, Y.; Schmid, C.; Congcong, L.; et al. TNT: Target-driveN Trajectory prediction. *arXiv* **2020**, arXiv:2008.08294.
125. Ma, Y.; Zhu, X.; Zhang, S.; Yang, R.; Wang, W.; Manocha, D. TrafficPredict: Trajectory prediction for heterogeneous traffic-agents. *arXiv* **2019**, arXiv:1811.02146.

126. Kuefler, A.; Morton, J.; Wheeler, T.; Kochenderfer, M. Imitating driver behavior with generative adversarial networks. In Proceedings of the 2017 IEEE Intelligent Vehicles Symposium (IV), Los Angeles, CA, USA, 11–14 June 2017; pp. 204–211. [CrossRef]

127. Rhinehart, N.; Kitani, K.M.; Vernaza, P. r2p2: A Reparameterized pushforward policy for diverse, precise generative path forecasting. In Proceedings of the European Conference on Computer Vision—ECCV 2018, Munich, Germany, 8–14 September 2018; Ferrari, V., Hebert, M., Sminchisescu, C., Weiss, Y., Eds.; Springer: Cham, Switzerland, 2018.

128. FHWA, U.S. Department of Transportation. NGSIMNext Generation SIMulation. Available online: http://ops.fhwa.dot.gov/trafficanalysistools/ngsim.htm (accessed on 31 October 2021).

129. Geiger, A.; Lenz, P.; Urtasun, R. Are we ready for autonomous driving? The KITTI vision benchmark suite. In Proceedings of the 2012 IEEE Conference on Computer Vision and Pattern Recognition (CVPR), Providence, RI, USA, 16–21 June 2012; pp. 3354–3361. [CrossRef]

130. Chang, M.-F.; Lambert, J.; Sangkloy, P.; Singh, J.; Bak, S.; Hartnett, A.; Wang, D.; Carr, P.; Lucey, S.; Ramanan, D.; et al. Argoverse: 3D tracking and forecasting with rich maps. In Proceedings of the 2019 IEEE/CVF Conference on Computer Vision and Pattern Recognition (CVPR), Long Beach, CA, USA, 15–20 June 2019; pp. 8748–8757. [CrossRef]

131. Caesar, H.; Bankiti, V.; Lang, A.H.; Vora, S.; Lion, V.E.;Xu, Q.; Krishnan, A.; Pan, Y.; Baldan, G.; Beijbom, O. nuScenes: A multimodal dataset for autonomous driving. *arXiv* **2020**, arXiv:1903.11027.

132. Argoverse Motion Forecasting Competition Leaderboard. Available online: https://eval.ai/web/challenges/challenge-page/454/leaderboard/1279 (accessed on 31 October 2021).

133. Zhuang, F.; Qi, Z.; Duan, K. A Comprehensive survey on transfer learning. *Proc. IEEE* **2021**, *109*, 43–76. [CrossRef]

134. Mahapatra, D.; bozorgtabar, B.; Thiran, J.P.; Reyes, M. Efficient active learning for image classification and segmentation using a sample selection and conditional generative adversarial network. In *Medical Image Computing and Computer-Assisted Intervention—MICCAI 2018*; Frangi, A., Schnabel, J., Davatzikos, C., Alberola-López, C., Fichtinger, G., Eds.; Springer: Cham, Switzerland, 2018; pp. 580–588.

135. Louizos, C.; Reisser, M.; Blankevoort, T.; Gavves, E.; Welling, M. Relaxed Quantization for discretized neural networks. In Proceedings of the International Conference on Learning Representations, Vancouver, BC, Canada, 30 April–3 May 2018.

136. Wu, S.; Li, G.; Chen, F.; Shi, L. Training and inference with integers in deep neural networks. In Proceedings of the International Conference on Learning Representations, Vancouver, BC, Canada, 30 April–3 May 2018. Available online: https://openreview.net/forum?id=HJGXzmspb (accessed on 31 October 2021).

137. Baker, C.L.; Jara-Ettinger, J.; Saxe, R.; Tenenbaum, J.B. Rational quantitative attribution of beliefs, desires and percepts in human mentalizing. *Nat. Hum. Behav.* **2017**, *1*, 0064. [CrossRef]

138. Look, A.; Doneva, S.; Kandemir, M.; Gemulla, R.; Peters, J. Differentiable implicit layers. *arXiv* **2020**, arXiv:2010.07078.

Article

Simulating Three-Wave Interactions and the Resulting Particle Transport Coefficients in a Magnetic Loop

Seve Nyberg * and Rami Vainio *

Department of Physics and Astronomy, University of Turku, FI-20014 Turku, Finland
* Correspondence: shonyb@utu.fi (S.N.); rami.vainio@utu.fi (R.V.)

Abstract: In this paper, the effects of wave–wave interactions of the lowest order, i.e., three-wave interactions, on parallel-propagating Alfvén wave spectra on a closed magnetic field line are considered. The spectra are then used to evaluate the transport parameters of energetic particles in a coronal loop. The wave spectral density is the main variable investigated, and it is modelled using a diffusionless numerical scheme. A model, where high-frequency Alfvén waves are emitted from the two footpoints of the loop and interact with each other as they pass by, is considered. The wave spectrum evolution shows the erosion of wave energy starting from higher frequencies so that the wave mode emitted from the closer footpoint of the loop dominates the wave energy density. Consistent with the cross-helicity state of the waves, the bulk velocity of energetic protons is from the loop footpoints towards the loop apex. Protons can be turbulently trapped in the loop, and Fermi acceleration is possible near the loop apex, as long as the partial pressure of the particles does not exceed that of the resonant waves. The erosion of the Alfvén wave energy density should also lead to the heating of the loop.

Keywords: Alfvén waves; wave–wave interactions; magnetic loops; solar corona; particle transport; numerical methods; modelling

Citation: Nyberg, S.; Vainio, R. Simulating Three-Wave Interactions and the Resulting Particle Transport Coefficients in a Magnetic Loop . *Physics* **2022**, *4*, 394–409. https://doi.org/10.3390/physics4020026

Received: 10 January 2022
Accepted: 18 March 2022
Published: 31 March 2022

Publisher's Note: MDPI stays neutral with regard to jurisdictional claims in published maps and institutional affiliations.

1. Introduction

Coronal loops are closed magnetic field structures in the solar atmosphere, filled with dense plasma [1]. They vary greatly in length, ranging from tens to tens of thousands of kilometers, or even further. Loops are important in many regions of the corona, including the active regions where the solar active phenomena, such as flares, are hosted [2].

Non-thermal spectral line broadening, most likely caused by high-frequency Alfvén waves, has been observed in coronal loops; see, e.g., [3–6], so the transport of these waves in loops is of interest. Under the Wentzel-Kramers-Brillouin (WKB) approximation, a magnetohydrodynamic (MHD) wave propagates in a static background plasma conserving its frequency and refracting under the laws of geometric optics. Alfvén waves have the property that their group velocity is directed along the magnetic field and the refraction of the wave vector is towards the lower values of the Alfvén speed. If the field line is anchored to the solar surface only at one end, all waves injected at the Sun propagate in the same direction along the field, i.e., outward from the Sun. However, in the case of a closed field line, i.e., a loop, both ends of the field line can inject waves into the loop. This leads us to a situation where counter-propagating waves are present. Thus, wave–wave interactions between counter-propagating waves become a topic of investigation to find out how the spectral energy density of the waves is evolving beyond the WKB approximation. Three-wave interactions have been studied theoretically (see, e.g., [7–9]), numerically (see, e.g., [10–12]), and experimentally (see, e.g., [13,14]). Their descriptions in coronal (see, e.g., [15]), and solar-wind (see, e.g., [16,17]) plasmas, have been a topic of recent research.

MHD wave spectra in magnetized plasmas determine the transport parameters of energetic charged particles [18], which are an important element of solar eruptions such as solar flares and coronal mass ejections; see, e.g., [19–21]. Under the weak turbulence approximation, wave–wave interactions can, to the lowest order, be treated as three-wave interactions [22], where two waves either coalesce to produce a single wave or a single wave decays to produce two waves.

In this paper, the effect of three-wave interactions on Alfvén waves propagating in coronal loops is considered. The influence of these on the wave spectra is studied, and the spectra to compute the transport parameters of energetic particles in these loops is used. The modification of the spectra by the three-wave interactions are discussed, and the consequences of the resulting spectra on particle transport inside the loops are considered. It is shown that under plausible assumptions of injected wave spectral energy densities, three-wave interactions lead to a significant modification of wave spectra and the mode of particle transport inside the loop. The study concludes that wave–wave interactions in coronal plasmas should be considered when the coupled evolution of waves and particles is modelled.

2. Theory

To investigate the three-wave interactions of two Alfvén waves and a magnetosonic wave, the model of Chin and Wentzel [22,23] is used which has been extended, e.g., by Zhao et al. [24] and Modi and Sharma [25], and applied to the case of two Alfvén waves and a sound wave by Vainio and Spanier [26] to calculate the cross-helicity of Alfvén waves downstream of a fast-mode shock. Only the low-plasma-β regime, where the Alfvén speed is larger than the sound speed, a condition applicable to most of the corona and inner solar wind [27], is studied. For simplicity, the study is limited to the discussion of waves propagating parallel and anti-parallel to the magnetic field. Eigenmodes propagating along the mean magnetic field are circularly polarized and the conservation of angular momentum is equivalent to conservation of polarization in three-wave interactions [28]. In the non-dispersive limit, the same evolution equation applies to both left-handed and right-handed waves. When computing the particle transport coefficients, it is assumed that both circular polarization states have the same intensity, which would be the case for linearly polarized waves. The three-wave interactions applicable to such conditions are an Alfvén wave decaying into a sound wave and a counter-propagating Alfvén wave, or the coalescence of an Alfvén wave and a sound wave into an Alfvén wave, i.e.,

$$A_\pm \longleftrightarrow A'_\mp + S_\pm, \tag{1}$$

where A and A' denote Alfvén waves, S denotes sound waves, and the $+ (-)$ sign denotes a wave propagating parallel (anti-parallel) to the magnetic field. The three-wave interaction requires the resonance conditions,

$$\begin{aligned}
\omega_A^\pm &= \omega_{A'}^\mp + \omega_S^\pm \\
k_A^\pm &= k_{A'}^\mp + k_S^\pm,
\end{aligned} \tag{2}$$

to be met, where ω is the wave angular frequency, k is the wave number, $\omega_A^\pm = \pm k_A^\pm v_A$ and $\omega_S^\pm = \pm k_S^\pm c_S$ are the dispersion relations of the waves (applicable to the primed ones, as well), $v_A = B/\sqrt{\mu_0 \rho}$ is the Alfvén speed, $c_S = \sqrt{\gamma P/\rho}$ is the sound speed, B is the magnetic field magnitude, ρ is the mass density, P is the thermal pressure, and γ is the adiabatic index of the plasma. In the convention used, $\omega > 0$ and the sign of k denotes the propagation direction of the wave. Using the resonance conditions and dispersion relations, one can solve the wave numbers and frequencies of the decay products. In particular, for the Alfvén waves

$$\frac{\omega_A}{\omega_{A'}} = -\frac{k_A}{k_{A'}} = \frac{v_A + c_S}{v_A - c_S} \equiv R_A, \tag{3}$$

where primed (unprimed) quantities are for waves on the right-hand (left-hand) side of Equation (1).

The model considered uses the simplified coronal loop presented in Figure 1 to eliminate the effects of field-line geometry on the wave distribution. In particular, it is assumed that the wave speeds remain constant over the whole length of the loop and that the loop has an Alfvén speed lower than its surroundings to allow waves to be ducted and maintain their wavevector as parallel/anti-parallel to the field. Additionally, gravitational stratification is neglected to have a constant density and temperature along the loop length.

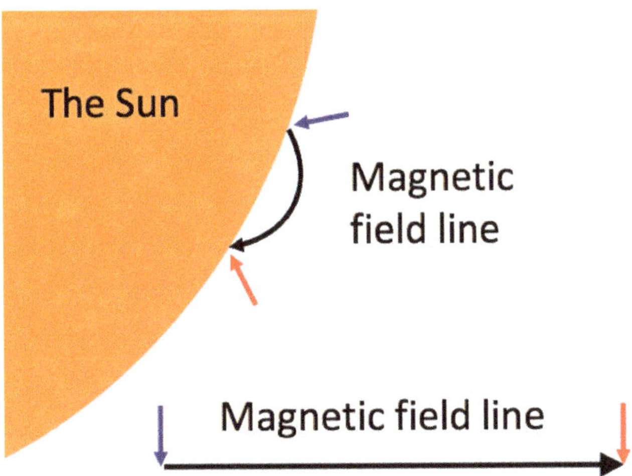

Figure 1. A depiction of the simplification of the closed field lines of the magnetic field of the Sun. The curved magnetic field line is straightened so that geometrical factors can be ignored in the simulation. The colored arrows indicate the injection points of wave energy to the magnetic field line, blue being the parallel wave mode injection point and red being the anti-parallel wave mode injection point.

Finally, sound waves in a coronal plasma are assumed to quickly damp by thermal particles, leading to only the reactions proceeding from left to right in Equation (1) being relevant. This simplifies the three-wave interaction term in the wave evolution equation so that all terms proportional to sound-wave intensity vanish. Thus, the equations governing the evolution of the Alfvén waves are [26]

$$\frac{\partial \mathfrak{E}^{\pm}(x,\omega,t)}{\partial t} \pm v_{\mathrm{A}} \frac{\partial \mathfrak{E}^{\pm}(x,\omega,t)}{\partial x} = \Gamma^{\pm}(x,\omega,t)\mathfrak{E}^{\pm}(x,\omega,t), \tag{4}$$

where

$$\mathfrak{E}^{\pm} = |k|\omega^{\pm}N_A^{\pm}/U_B \tag{5}$$

are the normalized spectral energy densities of the Alfvén waves propagating parallel ($+$) and anti-parallel ($-$) to the field, $\omega^{\pm}$ are the wave frequencies in the respective directions, $N_A^{\pm}$ are the wave action densities in the respective directions, and U_B is the background magnetic field energy density. Here, the left-hand side implements the spatial transport of wave-energy density along the magnetic field and the right-hand side the evolution of wave-energy densities due to wave–wave interactions. The net growth rates due to three-wave interactions are given as [26]

$$\Gamma^{\pm}(x,\omega,t) = \frac{\pi\omega v_{\mathrm{A}}^3}{c_S(v_{\mathrm{A}}^2 - c_S^2)}[\mathfrak{E}^{\mp}(x,\omega R_A,t) - \mathfrak{E}^{\mp}(x,\omega/R_A,t)].$$

The rate is dependent only on the energy density of the counter-propagating wave mode. In particular, the term describes counter-propagating waves from angular frequency ωR_A decaying into waves at angular frequency ω and those, in turn, decaying into counter-propagating waves at ω/R_A. The frequency domain under consideration extends to proton cyclotron frequency only, since dispersive effects [7,8,28] are not considered in the model.

Wave–particle interactions between energetic charged particles and MHD waves can be described as a resonant elastic scattering of particles in the frame co-moving with the waves [29–31]. In a homogeneous mean magnetic field, when scattering is strong enough to keep the particle distribution close to isotropic, the spatial transport of particles is a diffusion–advection process, where the advection velocity is determined by the mean velocity of the waves and the diffusion coefficient is determined by the rate of scattering. In addition, adiabatic particle energy changes result from gradients of the advection velocity and stochastic acceleration from counter-propagating waves scattering particles in the same volume [18,29]. To analyze the effect of the Alfvén waves on charged particles, here, the transport equations are used, as derived by Schlickeiser [18]:

$$\frac{\partial F}{\partial t} = \frac{\partial}{\partial x}\left(\kappa\frac{\partial F}{\partial x}\right) - \frac{1}{4p^2}\frac{\partial}{\partial p}(p^2 v a_1)\frac{\partial F}{\partial x} + \frac{1}{4}v\frac{\partial a_1}{\partial x}\frac{\partial F}{\partial p} + \frac{1}{p^2}\frac{\partial}{\partial p}\left(p^2 a_2\frac{\partial F}{\partial p}\right) + S_0, \qquad (6)$$

$$\kappa = \frac{v^2}{8}\int_{-1}^{+1}\frac{\mathrm{d}\mu(1-\mu^2)^2}{D_{\mu\mu}}, \qquad (7)$$

$$a_1 = \int_{-1}^{+1}\mathrm{d}\mu\frac{(1-\mu^2)D_{\mu p}}{D_{\mu\mu}}, \qquad (8)$$

$$a_2 = \frac{1}{2}\int_{-1}^{+1}\mathrm{d}\mu\left(D_{pp} - \frac{D_{\mu p}^2}{D_{\mu\mu}}\right), \qquad (9)$$

where $F = F(x,p,t)$ is the pitch-angle averaged distribution function of the energetic particles, p and v are the particle momentum and speed, κ is the spatial diffusion coefficient, a_1 is the coefficient of advection and adiabatic cooling, a_2 is the coefficient of momentum diffusion, and S_0 is the particle source. In the integrals, μ is the cosine of pitch angle and $D_{\mu\mu}$, $D_{\mu p}$, and D_{pp} are the Fokker–Planck coefficients related to momentum-space diffusion produced by wave–particle interactions [18]. For the normalized wave energy density, $\mathcal{E}$, they have the following forms:

$$D_{\mu\mu} = \frac{\pi}{4}\Omega(1-\mu^2)\left[\left(1-\frac{\mu v_A}{v}\right)^2\mathcal{E}_+(\omega_+) + \left(1+\frac{\mu v_A}{v}\right)^2\mathcal{E}_-(\omega_-)\right], \qquad (10)$$

$$D_{\mu p} = \frac{\pi}{4}\Omega(1-\mu^2)\frac{v_A p}{v}\left[\left(1-\frac{\mu v_A}{v}\right)\mathcal{E}_+(\omega_+) - \left(1+\frac{\mu v_A}{v}\right)\mathcal{E}_-(\omega_-)\right], \qquad (11)$$

$$D_{pp} = \frac{\pi}{4}\Omega(1-\mu^2)\frac{v_A^2 p^2}{v^2}\left[\mathcal{E}_+(\omega_+) + \mathcal{E}_-(\omega_-)\right], \qquad (12)$$

where

$$\omega_\pm = \Omega\frac{v_A}{|v\mu \mp v_A|} \qquad (13)$$

is the wave frequency of parallel/anti-parallel $(+/-)$ propagating waves resonant with particles of speed v and pitch-angle cosine μ and Ω is the (relativistic) particle cyclotron frequency.

Because of the use of MHD dispersion relations, only ion transport can be considered. Additionally, since the discussion is limited to the frequency range below the ion-cyclotron frequency, one needs to deal with the resonance gap resulting from the cut-off in the spectrum. This is achieved by extrapolating the spectra $\mathcal{E}^\pm(\omega)$ with a spectral index of δ' from the last frequency considered below the cyclotron frequency, where δ' is defined as

$$\delta' = \frac{\delta+2}{3}, $$

and δ is the spectral index ($\mathcal{E} \propto \omega^{\delta}$) solved from the last two cells below the cyclotron frequency. This produces Fokker–Planck coefficients that are approximately consistent with ion-cyclotron waves, with the same spectral index in wavenumber as the highest-frequency Alfvén waves, being responsible for scattering across the gap [32].

3. Numerical Methods

The wave evolution equations are solved on a two-dimensional (x, ω) grid as a function of time. The waves are transported in the spatial dimension using a diffusionless scheme, where wave spectral energy densities are propagated by one grid-cell of size Δx per each time step, Δt. Thus, as the Alfvén speed is constant in the presented model, the time step is defined as

$$\Delta t = \frac{\Delta x}{v_{\mathrm{A}}}. \tag{14}$$

In the angular frequency direction, a logarithmic grid spacing with the last cell ω_N being equal to the proton cyclotron frequency Ω_p is employed. Solving the rest of the cells from there onwards:

$$\omega_N = \Omega_p, \\ \omega_{j-1} = \omega_j R_A^{-1}, \tag{15}$$

where N denotes the last cell of the grid and $j = 1, \ldots, N$. The lowest frequency on the grid is set to a sufficiently low value to prevent boundary effects from altering the results of the simulation.

Operator splitting is used so that the spatial transport half-cycle is followed by a half-cycle implementing the wave–wave interactions in a time-explicit scheme. Thus, the approximate solution to the evolution Equation (4) can be written as a single step as

$$\mathcal{E}^{\pm}(x_i, \omega_j, t + \Delta t) = \mathcal{E}^{\pm}(x_{i\mp1}, \omega_j, t) \\ \times \exp\left\{ \frac{\Delta t \pi \omega_j v_{\mathrm{A}}^3}{c_S(v_{\mathrm{A}}^2 - c_S^2)} [\mathcal{E}^{\mp}(x_i, \omega_{j+1}, t) - \mathcal{E}^{\mp}(x_i, \omega_{j-1}, t)] \right\}. \tag{16}$$

Waves are injected from both ends of the spatial domain and those incident on the spatial boundaries from inside the domain are absorbed. The boundary values for the frequency domain are determined as follows: The lower boundary condition is set as

$$\mathcal{E}^{\pm}(x, \omega_0, t) = \mathcal{E}^{\pm}(x, \omega_1, t). \tag{17}$$

The upper boundary is set by assuming that cyclotron waves are damped by thermal plasma ions at the lowest wavelengths. That is,

$$\mathcal{E}^{\pm}(x, \omega_N, t) = 0 \tag{18}$$

is taken.

The parameters used in the simulations are chosen from ranges provided by [33] for the magnetic field strength and [1] for the rest:

$$v_{\mathrm{A}} = 2000 \text{ km/s}, \\ c_S = 200 \text{ km/s}, \\ B = 0.03 \text{ T}, \\ x \in [0, L], \quad L = 10{,}000 \text{ km}, \ 30{,}000 \text{ km} \\ \Delta x = 20 \text{ km}, 5 \text{ km}, \\ \omega \in [\omega_0, \Omega_p],$$

from which one can solve for the cyclotron frequency used as the upper boundary for the angular frequency axis,

$$\Omega_p = \frac{eB}{m_p} \approx 2.87365 \times 10^6 \text{ s}^{-1}, \tag{19}$$

where e and m_p are the charge and mass of a proton.

The wave spectra injected at the spatial boundaries are formulated so that the spectrum is flat at low frequencies, below ω_{in}, and obeys a power-law with a spectral index of $-(q-1)$ thereafter. The spectrum is also set to equal zero at the cyclotron frequency to synergize with the frequency boundary condition. The resulting form is

$$\mathfrak{E}^+(0,\omega) = \alpha \sqrt{\left(\frac{\omega_{\text{in}}^{q-1}}{(\omega_{\text{in}}^2 + \omega^2)^{(q-1)/2}} \right)^2 - \left(\frac{\omega_{\text{in}}^{q-1}}{(\omega_{\text{in}}^2 + \omega_N^2)^{(q-1)/2}} \right)^2} \tag{20}$$

where α is a normalization value. Here, $q = 5/3$ is used throughout the study.

Additionally, the injection spectrum contains a temporal ramp to provide a gradual increase in the wave spectral energy densities. The simulation reaches the steady-state injection after the time t_{ramp}. Thereafter, once a full wave-transport time has passed ($t > t_{\text{ramp}} + L/v_A$), a steady state emerges.

Thus, the injection spectrum for the wave mode propagating in the positive x direction is of the form

$$\mathfrak{E}^+(0,\omega,t) = \mathfrak{E}^+(0,\omega) \times \left[1 - \frac{t_{\text{ramp}} - t}{t_{\text{ramp}}} \mathcal{H}(t_{\text{ramp}} - t) \right], \tag{21}$$

where t_{ramp} is a parameter that controls the duration of the ramping up of the injection and $\mathcal{H}$ the Heaviside step function. The injection of the wave mode propagating in the negative x direction is identical, except that it is set at the other boundary, $x = L$.

The assumption of broadband wave emission is not based on a particular physical model of Alfvén wave generation in the footpoints of the coronal loop. To note, however, is that similar modelling has been applied, e.g., in connection with cyclotron heating on open field lines; see, e.g., [34,35]. Additionally, the observed heliospheric turbulence evolution suggests a broadband power spectrum of the form $P(f) \sim 1/f$, identical to those considered here at low frequencies, being eroded by turbulent processes (see, e.g., [36]). The heliospheric fluctuations are, of course, observed at lower frequencies but the spectrum below the correlation scale still spans several orders of magnitude. Thus, we regard the assumption of broadband wave emission from the footpoints of the loop as a reasonable, albeit not the only possible, assumption.

4. Results

The results on wave spectra are presented as a set of contour plots depicting the wave spectral energy density $\mathfrak{E}$ in x and ω, and plots of $\mathfrak{E}$ as a function of the wave frequency ω from three locations in the simulation box. The angular frequency range to be investigated was chosen to correspond to wavelengths above 10 km, and up to the proton cyclotron frequency. Figure 2 presents the spectra with a normalization value of $\alpha = 3 \times 10^{-5}$. One can see the three-wave interactions having the largest effect at high frequencies. The wave spectra evolve so that the wave modes propagating away from the closer end of the loop (the surface of the Sun) dominate. In the centre of the loop, both wave spectra are equal, as required by symmetry.

To compute the Fokker–Planck coefficients, the value of $\mathfrak{E}_\pm(\omega_\pm)$ is logarithmically interpolated from the simulation data at the resonant frequency $\omega_\pm(\mu)$. At values of μ close to $\pm v_A/v$, the resonant frequency falls outside the grid, producing the well-known resonance gap. In this study, it is assumed that processes such as wave dispersion [32] and/or resonance broadening [37] fill the gap and prevent the integrals of the Fokker–Planck coefficients from diverging. In particular, the gap is filled by extrapolating the spectral energy density values from the last simulated frequency value on the grid (ω_{N-1}) to the region above that resonant frequency, as if scattering in the resonance gap would be caused by ion-cyclotron waves with a wavenumber spectrum extrapolated from the Alfvénic range to the ion-cyclotron range (see Section 2 above).

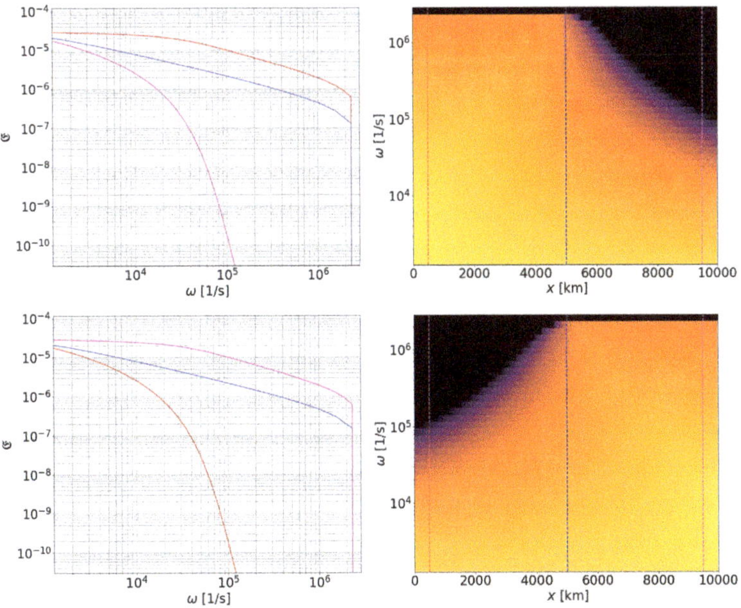

Figure 2. Steady state of the simulation of a coronal loop with equal wave spectral energy density injections at both footpoints. Upper plots contain the parallel-propagating wave spectrum and bottom plots contain the anti-parallel-propagating wave spectrum. The plots on the left contain slices of the wave spectral energy density data at three locations, which are color coded to match the dashed lines on the contour plots. The normalization value for this simulation is $\alpha = 3 \times 10^{-5}$, the loop size $L = 10,000$ km, and the spatial cell size $\Delta x = 20$ km.

Let us consider protons with speeds $v = 10\,v_{\rm A}$. The calculations are non-relativistic, in line with $v \ll c$ (with c being the speed of light) applicable for the parameters studied in this paper.

The bulk velocity of the energetic particles is derived using Equations (6) and (8). If only the leading order dependence of a_1 on momentum ($a_1 \sim p/v$) is accounted for, the bulk velocity of energetic protons becomes

$$u = \frac{3v}{4p} a_1 = \frac{3v}{4p} \int_{-1}^{+1} \mathrm{d}\mu \frac{(1 - \mu^2) D_{\mu p}}{D_{\mu\mu}}. \tag{22}$$

The integral is evaluated from the Fokker–Planck coefficients using the trapezoidal rule.

The results for simulations with different normalization values α can be seen in Figure 3. In higher injection spectral energy density simulations, particles move away from the surface of the Sun at the Alfvén velocity in almost all of the spatial domain. With lower energy densities, the region in the middle of the loop with a lower bulk speed broadens. The bulk speed approaches zero as the injection spectral energy density vanishes, as expected for non-interacting waves injected at equal intensities from both ends of the loop.

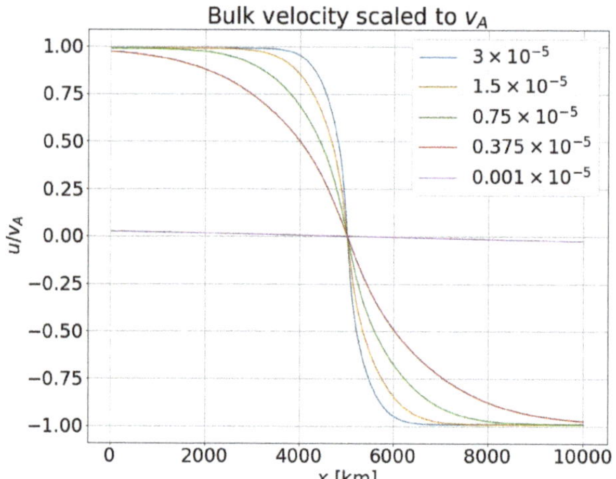

Figure 3. The calculated bulk velocity of energetic particles normalized to the Alfvén velocity in simulations of a coronal loop with the size L = 10,000 km and the spatial cell size Δx = 20 km. Different normalization values α are denoted in the legend.

Next, spatial diffusion is considered. The scattering mean free path for the particles is derived using Equation (7) as

$$\lambda \equiv \frac{3}{v}\kappa = \frac{3v}{8}\int_{-1}^{+1}\frac{\mathrm{d}\mu(1-\mu^2)^2}{D_{\mu\mu}}. \tag{23}$$

The mean free paths of simulations with different spectral energy densities are presented in Figure 4. In the two lowest presented wave spectral energy densities, mean free paths exceed the loop size, but in the two higher spectral energy densities, the mean free paths are smaller than the loop length. This allows the particles to be efficiently isotropized in the loop, a requirement for the diffusion approximation employed in the transport equation to be valid.

The actual mode of transport of particles inside the loop is determined by the diffusion length, and the length scale particles can diffuse upstream in a flow. For a constant spatial diffusion coefficient and bulk speed (and neglecting momentum diffusion), the transport equation reads

$$u\frac{\partial F}{\partial x} = \frac{\partial}{\partial x}\left(\kappa\frac{\partial F}{\partial x}\right)$$

and can be solved as

$$F = A + B\,\exp\left\{\frac{ux}{\kappa}\right\},$$

where A and B are constants to be determined by boundary conditions. One can see that the diffusion length, $l = \kappa/u = (v/3u)\lambda$, is the scale height of the distribution. Thus, a loop with a size certainly larger than the diffusion length can turbulently trap particles with a converging flow of the scattering centres towards the loop apex.

The diffusion lengths of different wave spectral energy density simulations are presented in Figure 5. As the diffusion length diverges in the middle of the loop (since $u = 0$ there), the diverging part is cut off in the figure. The diffusion length exceeds the loop length in all injection cases considered, leading to the conclusion that with these wave spectral energy densities the particles would not be confined in the loop but would be able to diffuse out of it quite efficiently.

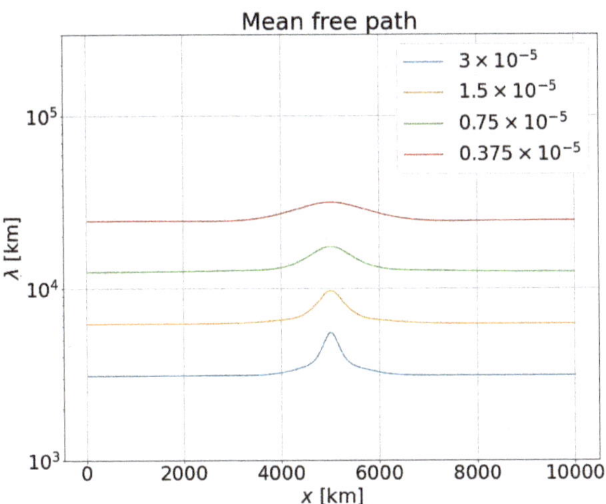

Figure 4. The calculated mean free path of particles in simulations of a coronal loop with the size $L = 10{,}000$ km and the spatial cell size $\Delta x = 20$ km. Different normalization values α are denoted in the legend.

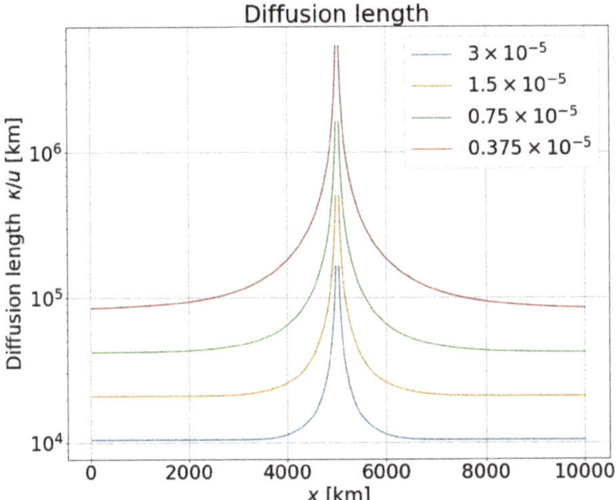

Figure 5. The calculated diffusion lengths of energetic particles for simulations of a coronal loop with the size $L = 10{,}000$ km and the spatial cell size $\Delta x = 20$ km. The curves are cut off to avoid presenting the diverging values of the diffusion length in the middle. Different normalization values α are denoted in the legend.

In order to study this further, longer loops were then investigated. In Figure 6, a 30,000 km loop with a lower injection spectral energy density is presented and compared with the 10,000 km loop with a higher injection spectral energy density presented in Figure 2. An interesting result is the invariant nature of the product between the wave spectral energy density and the loop length: scaling the injected wave spectral energy inversely with the loop length produces identical-shaped wave spectral energy densities in a scaled spatial coordinate and frequency, i.e., in steady state,

$$\mathcal{E}^{\pm}_{L_1}(\zeta L_1, \omega) L_1 = \mathcal{E}^{\pm}_{L_2}(\zeta L_2, \omega) L_2,\tag{24}$$

where $\xi = x/L \in [0, 1]$ is the dimensionless spatial coordinate scaled to the length of the loop. This can be seen to hold also by applying these scaling laws to the wave transport Equations (4) and assuming a steady state.

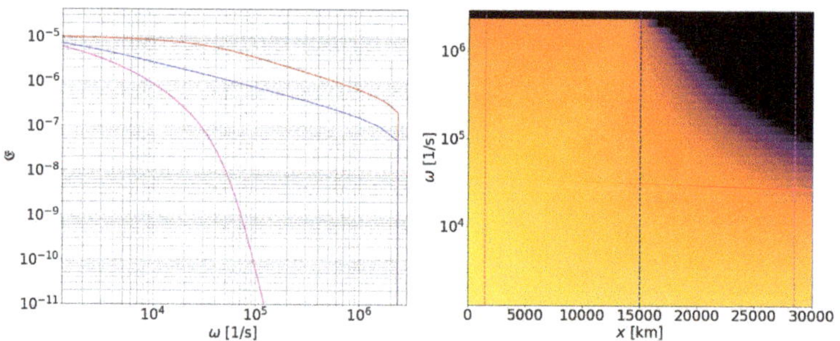

Figure 6. Steady state of the parallel-propagating wave mode of a simulation of a $L = 30{,}000$ km coronal loop with equal wave spectral energy density injections at both edges. The anti-parallel-propagating wave information is omitted due to symmetry with the parallel-propagating wave information. The plot on the left contains slices of the wave spectral energy density data at three locations, which are color coded to match to the dashed lines on the contour plots. The normalization value is $\alpha = 1 \times 10^{-5}$.

To investigate the evolution of the particle transport properties, analyses of the bulk velocity, mean free path, and diffusion length were conducted on the 30,000 km loop with injection normalization parameters $\alpha = 1 \times 10^{-5}$, 3×10^{-5} and to the 10,000 km loop with identical injections. The results are presented in Figures 7–10, where the particle transport coefficients are plotted along the dimensionless spatial coordinate ξ. The curves, representing the spectra, given in Figures 2 and 6, lie on top of each other, as expected. The other cases are ordered so that the higher injection spectral energy density corresponds to the steeper gradient of the bulk speed. The case with a longer loop and higher spectral energy density of the waves now demonstrates a regime where the loop length exceeds the diffusion length and the turbulent trapping of protons in the loop apex seems plausible.

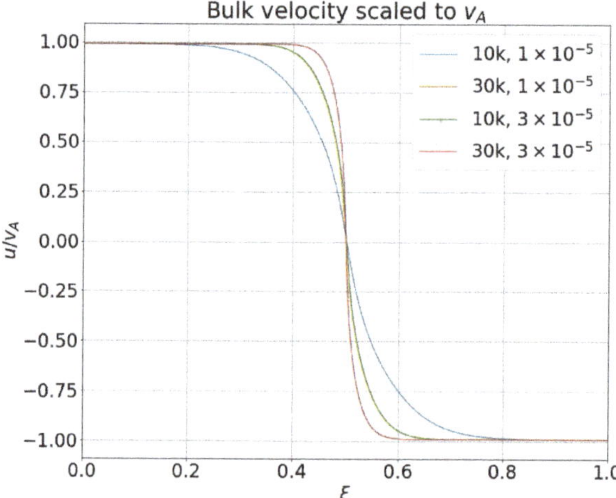

Figure 7. The calculated energetic particle bulk velocities for simulations of coronal loops with the sizes $L = 10{,}000$ km and $L = 30{,}000$ km. The 30,000 km loop, $\alpha = 3 \times 10^{-5}$ simulation has a spatial cell size $\Delta x = 5$ km and the rest have a spatial cell size $\Delta x = 20$ km. The loop lengths in kilometers and normalization values for the curves are displayed in the legend in the mentioned order.

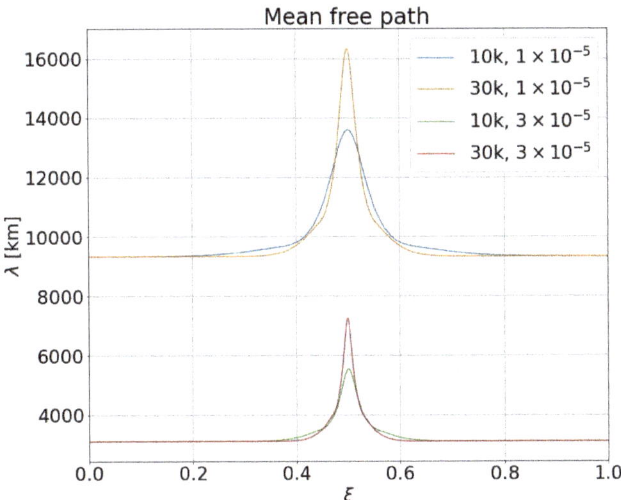

Figure 8. The calculated energetic particle mean free paths for simulations of coronal loops with the sizes L = 10,000 km and L = 30,000 km. The 30,000 km loop, $\alpha = 3 \times 10^{-5}$ simulation has a spatial cell size $\Delta x = 5$ km and the rest have a spatial cell size $\Delta x = 20$ km. The loop lengths in kilometers and normalization values for the curves are displayed in the legend in the mentioned order.

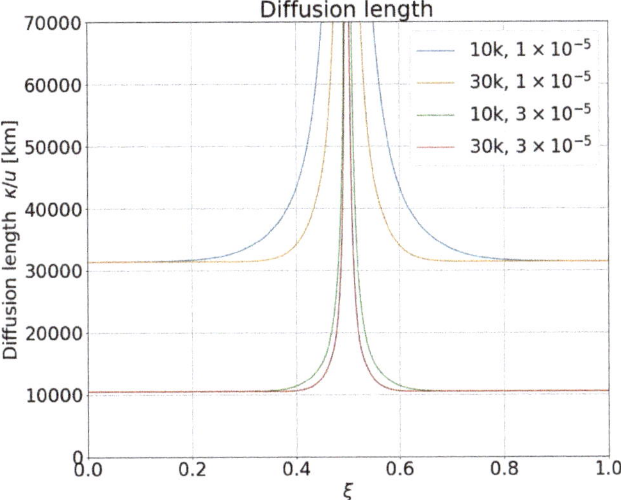

Figure 9. The calculated energetic particle diffusion lengths for simulations of coronal loops with the sizes L = 10,000 km and L = 30,000 km. The 30,000 km loop $\alpha = 3 \times 10^{-5}$ simulation has a spatial cell size $\Delta x = 5$ km and the rest have a spatial cell size $\Delta x = 20$ km. The loop lengths in kilometers and normalization values for the curves are displayed in the legend in the mentioned order.

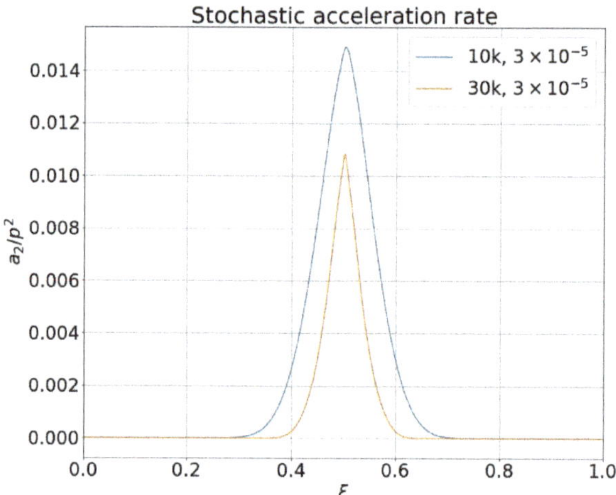

Figure 10. The calculated stochastic acceleration rate of particles for simulations of coronal loops with the sizes L = 10,000 km and L = 30,000 km. The normalization value used for both simulations is $\alpha = 3 \times 10^{-5}$. The longer loop has a spatial cell size $\Delta x = 5$ km and the smaller loop $\Delta x = 20$ km.

For these cases, where the bulk velocity is directed towards the loop apex and particles can be confined there diffusively, the model predicts an interesting situation, where first-order Fermi acceleration operates in the loop apex. The requirement of the diffusion length being longer than the gradient scale of the velocity while being shorter than the two "upstream regions" is a shock-like situation where particles can scatter back and forth from converging scattering centres and gain energy while being confined. The particle acceleration time scale in diffusive shock acceleration is [38]

$$\tau_{\mathrm{DSA}} = \frac{3}{\Delta u}\left(\frac{\kappa_1}{u_1} + \frac{\kappa_2}{u_2} \right),\tag{25}$$

where $\kappa_{1(2)}$ and $u_{1(2)}$ are the upstream (downstream) diffusion coefficient and bulk speed and $\Delta u = u_1 - u_2$. Applying the analogy of the setup to diffusive shock acceleration, i.e., replacing $\kappa_1 = \kappa_2 = \kappa$, $u_1 = u_2 = v_A$ and $\Delta u = 2v_A$, one can express the first-order Fermi acceleration time scale as

$$\tau_{\mathrm{I}} = \frac{3\kappa}{v_A^2} = \frac{\lambda v}{v_A^2},\tag{26}$$

which for the considered case of $v = 10\,v_A$, $v_A = 2000$ km/s and $\lambda \approx 3000$ km gives about $\tau_{\mathrm{I}} = 15$ s.

The loop apex is also a potential location for second-order Fermi acceleration, as waves propagating in both directions along the field are present there. Thus, the rate of stochastic acceleration of particles (Figure 10) is also investigated. Let us define it using the coefficient of momentum diffusion a_2 (Equation (9)), as $\tau_{\mathrm{II}}^{-1} = a_2/p^2$, where p is the particle momentum i.e., τ_{II} is the time scale needed for the particles to diffuse a "distance" p in momentum space. The values for the second-order Fermi acceleration rate τ_{II}^{-1} displayed in Figure 10 are visibly lower than the rate of the first-order Fermi acceleration, $\tau_{\mathrm{I}}^{-1} \approx 0.07\ \mathrm{s}^{-1}$.

5. Discussion and Conclusions

Above, wave transport in a coronal magnetic loop was considered including the effect of three-wave interactions between the counter-propagating Alfvén waves. For wave energy densities still in the weak-turbulence regime (i.e., with values of $\mathcal{E}^{\pm}$ far lower than unity), it was demonstrated that the interaction of counter-propagating waves erodes the spectrum of the wave mode emitted from the larger distance and leaves the wave field dominated by waves emitted from the closer footpoint. The spectrum erodes in particular at high frequencies. Additionally, the compatibility of the wave amplitudes of the simulation to spectral observations of nonthermal velocities in active regions is investigated. The Alfvén wave amplitude fulfills [39]

$$\delta V = v_{\mathrm{A}} \frac{\delta B}{B}, \tag{27}$$

$$\Rightarrow \frac{\langle (\delta B)^2 \rangle}{B^2} = \frac{\langle (\delta V)^2 \rangle}{v_{\mathrm{A}}^2}, \tag{28}$$

where δV is the amplitude of the velocity fluctuations and δB is the amplitude of the magnetic field fluctuations. On the other hand, one can write:

$$\frac{\langle (\delta B)^2 \rangle}{B^2} = \frac{1}{2} \int \frac{\mathcal{E}(\omega)}{\omega} d\omega \tag{29}$$

due to the definition of $\mathcal{E}(\omega)$ as the energy density of the waves (per $\ln k$ or $\ln \omega$, equivalently) divided by the magnetic field energy density. Using an Alfvén wave amplitude of 30 km/s from nonthermal velocity measurements [40] and the Alfvén speed of 2000 km/s from the simulation parameters used, one obtains $\langle (\delta B)^2 \rangle / B^2 = 2.25 \times 10^{-4}$ as an upper limit consistent with observations. Estimating the integral (29) throughout the simulation box for the highest injection value case of $\alpha = 3 \times 10^{-5}$ and $L = 30{,}000$ km gives $\langle (\delta B)^2 \rangle / B^2 < 6.2 \times 10^{-5}$, showing that the highest values of α used in the presented simulations are still comfortably below the limit set by the observations.

One of the major simplifications in the modelling here was the restriction to parallel-propagating waves. If this simplifying assumption is dropped, wave–wave interactions will lead to the development of the spectrum in the perpendicular direction, which may take over the parallel evolution modelled in our study [41]. Wave transport processes such as ducting [42] could allow for the parallel nature of the high frequency Alfvén waves to be maintained, but further studies should be performed to identify the parameters of the coronal loops that could sustain the parallel evolution of the three-wave interactions. Wave–particle interaction processes that produce the highest growth rates for the field-aligned wave modes could also lead to parallel waves dominating but such systems would naturally have more complicated wave transport than this simplified model, where the waves are generated in the loop legs.

In the present study, loops of intermediate lengths were considered corresponding to scales applicable to active regions [1]. It was shown that wave–wave interactions can lead to wave distributions that yield very interesting parameters for charged-particle transport: (1) The bulk velocity of ions interacting with the Alfvén waves is generally directed from the loop legs to the loop apex and is close to the Alfvén speed for most part of the loop if the wave spectral energy density injected into the loop exceeds a threshold value. (2) The diffusion length of energetic ($\sim$MeV) protons can be shorter than the loop length, allowing particles to be turbulently trapped in the loop. (3) The converging velocity of the scattering centres can make the loop apex act as a Fermi acceleration site; the first-order acceleration process seems to dominate over the second-order process, at least with the parameters studied in the modelling here.

However, to point out is that no back-reaction effects of the accelerated particles on the waves were considered in our brief analysis [43,44]. If the accelerated particle pressure greatly exceeds the resonant wave pressure (half of the wave spectral energy density for Alfvén waves), particles streaming away from the acceleration site at the loop apex would actually dampen the waves propagating towards the apex and amplify the waves that propagate towards the footpoints of the loop [43]. Thus, the theory presented here only applies in situations where the externally driven waves are strong enough that their pressure exceeds the partial pressure of the accelerated particles resonant with them. This implies that a full model with all weak-turbulence effects (wave–wave and wave–particle interactions) included would have a natural way of quenching the Fermi acceleration mechanism at the apex.

Let us also point out that as the model predicts the erosion of Alfvén wave spectra by the emission and subsequent absorption of sound waves by the thermal plasma, it leads to the heating of the loop, which might seed the acceleration process with particles from the thermal tail. This is, however, to be a subject of further investigations.

Author Contributions: Conceptualization, R.V.; methodology, S.N. and R.V.; software, S.N.; validation: S.N.; formal analysis, S.N. and R.V.; investigation, S.N. and R.V.; resources, R.V.; data curation, S.N.; writing—original draft preparation, S.N. and R.V.; writing—review and editing, S.N.; visualization, S.N.; supervision, R.V.; project administration, R.V.; funding acquisition, R.V. All authors have read and agreed to the published version of the manuscript.

Funding: This research was funded by the European Union's Horizon 2020 Programme grant number 870405 (EUHFORIA 2.0).

Data Availability Statement: Not applicable.

Acknowledgments: The authors acknowledge fruitful discussions with A. Afanasiev. This study was performed in the framework of Finnish Centre of Excellence in Research of Sustainable Space (FORESAIL, 2018–2025).

Conflicts of Interest: The authors declare no conflict of interest.

References

1. Reale, F. Coronal loops: Observations and modeling of confined plasma. *Living Rev. Sol. Phys.* **2010**, *7*, 5. lrsp-2010-5. [CrossRef]
2. Aschwanden, M.J. *Physics of the Solar Corona. An Introduction with Problems and Solutions*; Springer: Berlin/Heidelberg, Germany, 2005. [CrossRef]
3. Tian, H.; McIntosh, S.W.; De Pontieu, B.; Martinez-Sykora, J.; Sechler, M.; Wang, X. Two components of the coronal emission revealed by EUV spectroscopic observations. *Astrophys. J.* **2011**, *738*, 18. [CrossRef]
4. Tian, H.; McIntosh, S.W.; Wang, T.; Ofman, L.; De Pontieu, B.; Innes, D.E.; Peter, H. Persistent doppler shift oscillations observed with hinode/eis in the solar corona: Spectroscopic signatures of alfvénic waves and recurring upflows. *Astrophys. J.* **2012**, *759*, 144. [CrossRef]
5. Tian, H.; McIntosh, S.W.; Xia, L.; He, J.; Wang, X. What can we learn about solar coronal mass ejections, coronal dimmings, and Extreme-Ultraviolet jets through spectroscopic observations? *Astrophys. J.* **2012**, *748*, 106. [CrossRef]
6. McIntosh, S.W. The inconvenient truth about coronal dimmings. *Astrophys. J.* **2009**, *693*, 1306–1309. [CrossRef]
7. Shi, Y.; Qin, H.; Fisch, N.J. Three-wave scattering in magnetized plasmas: From cold fluid to quantized Lagrangian. *Phys. Rev. E* **2017**, *96*, 023204. [CrossRef]
8. Shi, Y. Three-wave interactions in magnetized warm-fluid plasmas: General theory with evaluable coupling coefficient. *Phys. Rev. E* **2019**, *99*, 063212. [CrossRef]
9. TenBarge, J.M.; Ripperda, B.; Chernoglazov, A.; Bhattacharjee, A.; Mahlmann, J.F.; Most, E.R.; Juno, J.; Yuan, Y.; Philippov, A.A. Weak Alfvénic turbulence in relativistic plasmas. Part 1. Dynamical equations and basic dynamics of interacting resonant triads. *J. Plasma Phys.* **2021**, *87*, 905870614. [CrossRef]
10. Matteini, L.; Landi, S.; Velli, M.; Hellinger, P. Kinetics of parametric instabilities of Alfvén waves: Evolution of ion distribution functions. *J. Geophys. Res. Space Phys.* **2010**, *115*, A09106. [CrossRef]
11. Matteini, L.; Landi, S.; Zanna, L.D.; Velli, M.; Hellinger, P. Parametric decay of linearly polarized shear Alfvén waves in oblique propagation: One and two-dimensional hybrid simulations. *Geophys. Res. Lett.* **2010**, *37*, L20101.
12. Chang, O.; Peter Gary, S.; Wang, J. Energy dissipation by whistler turbulence: Three-dimensional particle-in-cell simulations. *J. Plasma Phys.* **2014**, *21*, 052305. [CrossRef]
13. Dorfman, S.; Carter, T.A. Nonlinear excitation of acoustic modes by large-amplitude Alfvén waves in a laboratory plasma. *Phys. Rev. Lett.* **2013**, *110*, 195001. [CrossRef]

14. Shi, P.; Yang, Z.; Luo, M.; Wang, R.; Lu, Q.; Sun, X. Observation of spontaneous decay of Alfvénic fluctuations into co- and counter-propagating magnetosonic waves in a laboratory plasma. *J. Plasma Phys.* **2019**, *26*, 032105. [CrossRef]

15. Sharma, P.; Yadav, N.; Sharma, R.P. Nonlinear interaction of kinetic Alfvén waves and ion acoustic waves in coronal loops. *J. Plasma Phys.* **2016**, *23*, 052304. [CrossRef]

16. Zanna, L.D.; Matteini, L.; Landi, S.; Verdini, A.; Velli, M. Parametric decay of parallel and oblique Alfvén waves in the expanding solar wind. *J. Plasma Phys.* **2015**, *81*, 325810102. [CrossRef]

17. Shoda, M.; Yokoyama, T. Nonlinear reflection process of linearly polarized, broadband Alfvén waves in the fast solar wind. *Astrophys. J.* **2016**, *820*, 123. [CrossRef]

18. Schlickeiser, R. Cosmic-ray transport and acceleration. I. Derivation of the kinetic equation and application to cosmic rays in static cold media. *Astrophys. J.* **1989**, *336*, 243. [CrossRef]

19. Le Roux, J.A.; Arthur, A.D. Acceleration of solar energetic particles at a fast traveling shock in non-uniform coronal conditions. *J. Phys. Conf. Ser.* **2017**, *900*, 012013. [CrossRef]

20. Dröge, W.; Kartavykh, Y.Y.; Klecker, B.; Kovaltsov, G.A. Anisotropic three-dimensional focused transport of solar energetic particles in the inner heliosphere. *Astrophys. J.* **2010**, *709*, 912–919. [CrossRef]

21. Effenberger, F.; Petrosian, V. The Relation between escape and scattering times of energetic particles in a turbulent magnetized plasma: Application to solar flares. *Astrophys. J. Lett.* **2018**, *868*, L28. [CrossRef]

22. Chin, Y.C.; Wentzel, D.G. Nonlinear dissipation of Alfvén waves. *Astrophys. Space Sci.* **1972**, *16*, 465–477. doi: 10.1007/BF00642346. [CrossRef]

23. Wentzel, D.G. Coronal heating by Alfvén waves. *Sol. Phys.* **1974**, *39*, 129–140. [CrossRef]

24. Zhao, J.; Wu, D.; Lu, J. A nonlocal wave-wave interaction among Alfven waves in an intermediate-beta plasma. *J. Plasma Phys.* **2011**, *18*, 032903. [CrossRef]

25. Modi, K.V.; Sharma, R.P. Nonlinear interaction of inertial Alfvén wave with magnetosonic wave and cavitation phenomena. *Astrophys. Space Sci.* **2014**, *350*, 223–229. [CrossRef]

26. Vainio, R.; Spanier, F. Evolution of Alfvén waves by three-wave interactions in super-Alfvénic shocks. *Astron. Astrophys.* **2005**, *437*, 1–8. [CrossRef]

27. Gary, G.A. Plasma beta above a solar active region: Rethinking the paradigm. *Sol. Phys.* **2001**, *203*, 71–86. doi: 10.1023/A:1012722021820. [CrossRef]

28. Spanier, F.; Vainio, R. Three-wave interactions of dispersive plasma waves propagating parallel to the magnetic field. *Adv. Sci. Lett.* **2008**, *2*, 337–346. [CrossRef]

29. Skilling, J. Cosmic ray streaming—I. Effect of Alfvén waves on particles. *Mon. Not. R. Astron. Soc.* **1975**, *172*, 557–566. [CrossRef]

30. Ruffolo, D. Effect of Adiabatic deceleration on the focused transport of solar cosmic rays. *Astrophys. J.* **1995**, *442*, 861. [CrossRef]

31. Kocharov, L.; Vainio, R.; Kovaltsov, G.A.; Torsti, J. Adiabatic deceleration of solar energetic particles as deduced from Monte Carlo simulations of interplanetary transport. *Sol. Phys.* **1998**, *182*, 195–215. [CrossRef]

32. Vainio, R. Charged-particle resonance conditions and transport coefficients in slab-mode waves. *Astrophys. J. Suppl. Ser.* **2000**, *131*, 519–529. [CrossRef]

33. Kuridze, D.; Mathioudakis, M.; Morgan, H.; Oliver, R.; Kleint, L.; Zaqarashvili, T.V.; Reid, A.; Koza, J.; Löfdahl, M.G.; Hillberg, T. Mapping the magnetic field of flare coronal loops. *Astrophys. J.* **2019**, *874*, 126. [CrossRef]

34. Tu, C.-Y.; Marsch, E. Two-fluid model for heating of the solar corona and acceleration of the solar wind by high-frequency Alfvén waves. *Sol. Phys.* **1997**, *171*, 363–391. [CrossRef]

35. Vainio, R.; Laitinen, T.; Fichtner, H. A simple analytical expression for the power spectrum of cascading Alfvén waves in the solar wind. *Astron. Astrophys.* **2003**, *407*, 713–723. [CrossRef]

36. Bruno, R.; Carbone, V. The solar wind as a turbulence laboratory. *Living Rev. Sol. Phys.* **2013**, *10*, 2. [CrossRef]

37. Schlickeiser, R.; Achatz, U. Cosmic-ray particle transport in weakly turbulent plasmas. Part 1. Theory. *J. Plasma Phys.* **1993**, *49*, 63–77. [CrossRef]

38. Drury, L.O. Review Article: An introduction to the theory of diffusive shock acceleration of energetic particles in tenuous plasmas. *Rep. Prog. Phys.* **1983**, *46*, 973–1027. [CrossRef]

39. Zhao, J.S.; Voitenko, Y.M.; Wu, D.J.; Yu, M.Y. Kinetic Alfvén turbulence below and above ion cyclotron frequency. *J. Geophys. Res. Space Phys.* **2016**, *121*, 5–18. [CrossRef]

40. Doschek, G.A.; Warren, H.P.; Mariska, J.T.; Muglach, K.; Culhane, J.L.; Hara, H.; Watanabe, T. Flows and nonthermal velocities in solar active regions observed with the EUV Imaging Spectrometer on Hinode: A tracer of active region sources of heliospheric magnetic fields? *Astrophys. J.* **2008**, *686*, 1362–1371. [CrossRef]

41. Zhao, J.S.; Voitenko, Y.; De Keyser, J.; Wu, D.J. Scalar and vector nonlinear decays of low-frequency Alfvén waves. *Astrophys. J.* **2015**, *799*, 222. [CrossRef]

42. Mazur, V.A.; Stepanov, A.V. The role of Alfven wave dispersion in the dynamics of energetic protons in the solar corona. *Astron. Astrophys.* **1984**, *139*, 467–473.

43. Vainio, R.; Kocharov, L. Proton transport through self-generated waves in impulsive flares. *Astron. Astrophys.* **2001**, *375*, 251–259. [CrossRef]

44. Afanasiev, A.; Vainio, R. Monte Carlo simulation model of energetic proton transport through self-generated Alfvén waves. *Astrophys. J. Suppl. Ser.* **2013**, *207*, 29. [CrossRef]

*physics*

MDPI

Article

Propagation of Cosmic Rays in Plasmoids of AGN Jets-Implications for Multimessenger Predictions

Julia Becker Tjus [1,2,*], Mario Hörbe [1,2], Ilja Jaroschewski [1,2], Patrick Reichherzer [1,2,3], Wolfgang Rhode [4], Marcel Schroller [1,2] and Fabian Schüssler [3]

1   Theoretische Physik IV: Plasma-Astroteilchenphysik, Ruhr-Universität Bochum, Universitätsstrasse 150, 44801 Bochum, Germany; mario.hoerbe@ruhr-uni-bochum.de (M.H.); ilja@tp4.ruhr-uni-bochum.de (I.J.); patrick.reichherzer@rub.de (P.R.); marcel.schroller@rub.de (M.S.)
2   Ruhr Astroparticle and Plasma Physics Center (RAPP Center), Ruhr-Universität Bochum, 44780 Bochum, Germany
3   IRFU, CEA, Université Paris-Saclay, F-91191 Gif-sur-Yvette, France; fabian.schussler@cea.fr
4   Department of Physics, TU Dortmund University, 44221 Dortmund, Germany; wolfgang.rhode@tu-dortmund.de
*   Correspondence: julia.tjus@ruhr-uni-bochum.de

Abstract: After the successful detection of cosmic high-energy neutrinos, the field of multiwavelength photon studies of active galactic nuclei (AGN) is entering an exciting new phase. The first hint of a possible neutrino signal from the blazar TXS 0506+056 leads to the anticipation that AGN could soon be identified as point sources of high-energy neutrino radiation, representing another messenger signature besides the established photon signature. To understand the complex flaring behavior at multiwavelengths, a genuine theoretical understanding needs to be developed. These observations of the electromagnetic spectrum and neutrinos can only be interpreted fully when the charged, relativistic particles responsible for the different emissions are modeled properly. The description of the propagation of cosmic rays in a magnetized plasma is a complex question that can only be answered when analyzing the transport regimes of cosmic rays in a quantitative way. In this paper, therefore, a quantitative analysis of the propagation regimes of cosmic rays is presented in the approach that is most commonly used to model non-thermal emission signatures from blazars, i.e., the existence of a high-energy cosmic-ray population in a relativistic plasmoid traveling along the jet axis. It is shown that in the considered energy range of high-energy photon and neutrino emission, the transition between diffusive and ballistic propagation takes place, significantly influencing not only the spectral energy distribution, but also the lightcurve of blazar flares.

Keywords: cosmic rays; cosmic-ray diffusion; active galactic nuclei (AGN)

Citation: Becker Tjus, J.; Hörbe, M.; Jaroschewski, I.; Reichherzer, P.; Rhode, W.; Schroller, M.; Schüssler, F. Propagation of Cosmic Rays in Plasmoids of AGN Jets-Implications for Multimessenger Predictions. *Physics* **2022**, *4*, 473–490. https://doi.org/10.3390/physics4020032

Received: 31 January 2022
Accepted: 1 April 2022
Published: 28 April 2022

Publisher's Note: MDPI stays neutral with regard to jurisdictional claims in published maps and institutional affiliations.

1. Introduction

Active galactic nuclei (AGN) are among the most enigmatic objects in the universe. With luminosities in excess of 10^{37} W (10^{44} erg s^{-1}), they represent the most luminous, continuous sources of radiation. With their central supermassive black holes (SMBHs), they provide an environment that can help us to understand black holes at work. The question of how energy is transferred from the black hole and/or the accretion disk to launch gigantic radio jets is still largely unsolved and subject to ongoing research. AGN also constitute one of the few sites in the universe that provide enough energy on total to serve as a candidate for the observed flux of ultra-high energy cosmic rays (UHECRs) and might be a key to understand particle acceleration up to the highest observed particle energies; see, e.g., [1,2] for summaries. As one of the very few source classes, active galaxies provide an astrophysical, extreme environment which might be suited to accelerate particles to an incredible amount of 10^{20} eV (see, e.g., [3]). These sources are therefore also considered to contribute to the diffuse astrophysical high-energy neutrino flux as measured by IceCube at Earth [4]. In particular, the sub-class of blazars is known for their strong short- and long-term variability, especially (but not exclusively) at gamma-ray energies. In the literature, blazar jets are often discussed to be dominated by leptonic particle processes, which fit observational quiescent data of blazar spectral energy densities (SEDs; see [5–7]). Yet, the understanding of the complex, time-variable structure of the SEDs is still far from being understood in full detail. Additionally, the picture of an electron-positron plasma in the jet is not the only possibility, and an electron–proton (hadron) plasma is certainly a realistic option. Thus, in the past decades, AGN jets and particularly blazar emissions have also been argued to naturally contain hadronic components, which would not only contribute to the emission of electromagnetic radiation in blazars [8], but also lead to the production of secondary high-energy neutrinos [9–11]. The general detection of up to PeV high-energy neutrinos of astrophysical origin observed by IceCube [4] has started to shed more light on the non-thermal high-energy universe. From the distribution of events in the projected sky, it is clear that the flux is not focused in the galactic plane, and that it is therefore likely to contain a significant extragalactic component, see, e.g., Becker Tjus and Merten [12] for a summary. In the past few years, several possible associations of neutrinos with astrophysical objects have been identified. Each of these is at a $\sim 3\sigma$ confidence level so far and serves as first evidence. A first hint for an association of a high-energy neutrino with a blazar comes from the source TXS 0506+056. In September 2017, a neutrino of ~ 300 TeV could be associated with an exceptional gamma-ray flare at GeV energies from this source. This gamma-neutrino correlation can be estimated to have a significance of 3σ by combining data of the IceCube neutrino observatory and Fermi-LAT [13]. This detection initiated a dedicated search for neutrino clustering in the ~ 10 years of IceCube data around the position of TXS 0506+056, with the result that there was an enhanced flux of neutrinos in a half-year period from September 2014 to March 2015, also with a statistical significance of $\sim 3\sigma$ of being incompatible with the background hypothesis [14].

In order to explain the neutrino signatures detected from the direction of TXS 0506+065, hadronic jet models have been applied (e.g., [15–19]). The two potential flares appear quite different in their evolution: the neutrino signal above the atmospheric background detected in 2014/2015 lasted about 100 days and consisted of about 8–18 neutrinos with energies of 10–100 TeV, and the gamma-ray light curve at GeV is in a minimum state [14]. The 2017 detection is based on one single high-energy neutrino with extreme energy (~ 300 TeV). A gamma-ray flare was observed in coincidence with the arrival of the neutrino. It has been noted by Kun et al. [20], however, that at the exact time of the neutrino detection, even here the GeV gamma-ray flux was at a local minimum and only rose to a high emission state shortly after the neutrino detection. It was argued in [20] that this observation is consistent with a model for which the neutrinos are produced in a high-density medium in which gamma-rays are absorbed, which either becomes less dense with time or for which the gamma-rays take some time to cascade down to GeV energies before escaping.

By now, there are tens of possible associations of neutrinos and AGN jets [21–23]. With IceCube in continued operation, this number will further increase in the upcoming years, with the expectation to finally confirm at least some of these sources at the >5 sigma level to be neutrino emitters. One common conundrum in all of the different high-energy neutrino detections (diffuse and potential sources) is the lack of TeV, but even GeV gamma-ray emission. Hadronic interactions that are responsible for neutrino production inevitably lead to the co-production of high-energy gamma rays. The reason is that neutrinos are produced from the subsequent decay of charged pions and kaons, which in turn are produced in a fixed ratio of neutral pions and kaons, leading to the production of gamma rays with an energy threshold at the mass of the pion, $E_{\gamma,\,\min} = m_{\pi_0} c^2/2 = 70\,\text{MeV}$, with c being the speed of light. Comparing the detected diffuse neutrino flux with the measured extragalactic, diffuse component of gamma rays leads to the conjecture of a source environment that must absorb the gamma rays at energies >GeV [24,25]. The absorption of photons can be due to a strong accretion disk, as it has been discussed in, e.g., Brodatzki et al. [26] for the case of TeV emission. The potential neutrino source fluxes from the 2014/2015 TXS signature also indicates that, in order to match the observed gamma-ray flux, it must be diminished significantly at >GeV energies to make the neutrino production model work. Such a model of gamma-ray absorption can even be used as a tracer in the searches for associations of high-energy neutrinos with blazars. That is, rather than searching for an enhanced gamma-ray flux, the neutrinos can actually arrive at times of reduced gamma-ray activity [20]. Such a scenario can be produced for regions of extreme gas or photon densities. In the first case, the photons will interact with the dense gas via Compton scattering. In the second case, gamma–gamma interactions will lead to electromagnetic cascades. It is clear that in both scenarios, the energy of the high-energy gamma rays will be deposited at much lower energies. If these environments only become transparent at MeV energies as suggested by, e.g., Halzen et al. [15], this is not observable now, as a dedicated mission for MeV detection of the gamma-ray sky is currently missing. Future missions like e-Astrogam, MeVCube, or AMEGO will shed more light on these questions. At this point, the theoretical models are being challenged by GeV-TeV measurements, which indicate that there is no significant increase in the energy output connected to the neutrinos.

In general, the modeling of the steady-state emission, but even more the modeling of the flares is complex and requires a complete consideration of the jet physics, including different scenarios for the acceleration region, gas and photon targets, as well as the magnetic field structure. The latter is highly important for the proper modeling of the diffusive cosmic-ray transport, which is relevant for the evolution of the flares, both for the leptonic and hadronic signatures [19]. Quantitative theoretical modeling is necessary to establish a physical connection of the neutrinos to the blazars. Mere directional coincidence is not enough, because the angular uncertainty of the neutrino events is larger than $1°$, making source identification difficult without theoretical input.

Models of high-energy neutrino and electromagnetic up to gamma-ray emission in the jets of AGN cover a variety of scenarios and parameter spaces. Blazars are known to be highly variable across the electromagnetic spectrum, with a variety of models put forward to explain these flares. Such models include, among others, particle acceleration via internal shocks in the jet (see, e.g., [3,9,10,27]) and reconnection driven plasmoids (see, e.g., [19,28,29]). These latter relativistic and compact structures have been discussed in the literature since the 1960s [30,31]. What here is referred to as *plasmoid* is often called *blob* in the literature, meaning compact, dense structures traveling with relativistic speeds along the jet axis. The term "plasmoid" is preferred here, as it is typically used in the context of the plasmoid creation via magnetic reconnection events. In this scenario, the injection of a relativistic plasma into the system (here the AGN jet) can lead to reconnection events that, under certain circumstances, lead to the plasmoid instability that breaks down the streaming plasma into small blobs, i.e., the plasmoids. In this scenario, charged particles can be pre-accelerated in the reconnection events. While non-relativistic reconnection is limited to below-knee energies [32], relativistic reconnection can be much more efficient [33], also by further acceleration via a Fermi second-order process when the particles scatter in between the plasmoids. Such an acceleration scenario can solve the long-standing injection-problem. They also justify the assumption that the cosmic-ray population is distributed homogeneously in the plasmoid, as the turbulent field in the plasmoid is used to isotropize the direction of the incoming particles. The assumption of a homogeneously distributed population is implicit in those models that do not resolve the blob, but work with timescales. In test-particle simulations, it is a reasonable approach to start with a homogeneous distribution, as done in this paper.

The modeled hadronic component of proton-proton interaction was discussed by Eichmann et al. [27]; the modeling of leptonic and (lepto-)hadronic emission by Christie et al. [34], Keivani et al. [35] for the case of TXS 0506+056. Other flare models based on external factors include gas clouds entering the base of jets [36–38] and jets of former binary AGN drilling through their own dust tori and/or accretion disks after being redirected by the merger of their host black holes [39]. Such scenarios of high density all tend to be more hadron-dominated due to the nature of their occurrence.

Figure 1 shows a sketch of an AGN with a jet. Also shown are the photon and/or gas targets (yellow/red). The structure of the large-scale magnetic field is shown in blue. A turbulent component of the magnetic field exists as well and is not drawn in the figure, but only indicated by purple text. The structure of the gas/photon fields is highly relevant for the particle interactions and therefore needs to be included in the models in three dimensions. The same is true for the magnetic field structure as the synchrotron radiation is sensitive to the direction of the field. For a diffusive transport description, it is also highly relevant as it is often assumed that the propagation is dominantly along the field lines with a smaller component perpendicular to the field. In fact, the perpendicular diffusion coefficient can even dominate the picture if the magnetic field turbulence level is $\delta B/B > 1$, something that is certainly possible in these extreme environments.

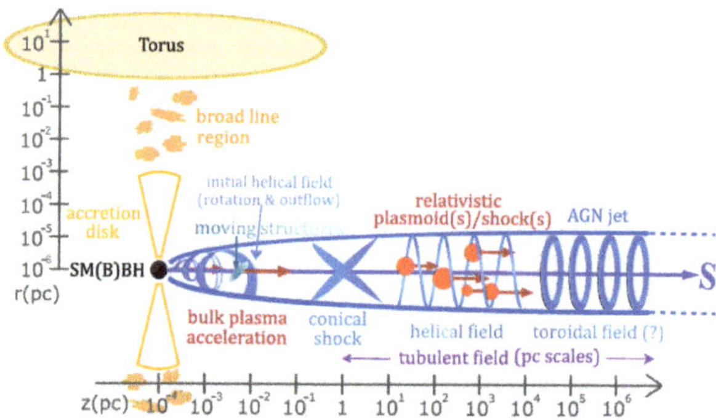

Figure 1. Structure of an active galactic nucleus (AGN) with a jet. Yellow/red components are targets for cosmic-ray interactions with gamma rays or gas. Blue colors indicate the likely structure of the magnetic field along the jet. Particle acceleration happens either at the shock fronts or in context with the relativistic plasmoids.

Current state-of-the-art of numerical codes include many of the necessary features. The codes and their most important properties are summarized in Table 1. Cerruti et al. [40] compared the codes for blazar hadronic models (with the exception of the CRPropa code). The models are typically designed to numerically solve the transport equation including loss processes from which the secondary particle radiation from electrons and protons can be calculated. A loss term for the particle transport is usually included via a timescale, i.e., a term $-n/\tau$, where n is the particle density and τ is the characteristic timescale. In [8,41,42], the timescale is chosen to be the ballistic one, $\tau = c/R$ (where R is the radius of plasmoid), thus being energy-independent. Gao et al. [43] has argued that the propagation is of diffusive nature so that the escape time of particles is assumed to be a factor of 10 times longer than the ballistic one, but still assumed to be energy-independent. However, there are models that implement the escape of particles with a diffusive, energy dependent approach Rodrigues et al. [44] modeled the particle escape by distinguishing between the two most extreme cases, i.e., escape via diffusion and escape via advection and argue that the physical escape spectrum will emerge in between these two scenarios. All of these models are designed to model propagation and particle interaction in blazars. Due to the strong variability of the sources, it can be deduced that the signal must come from the very compact region on the order of 10^{14} m. This makes the plasmoid model favorable over an approach of shock acceleration and explains why all codes focus on such an approach.

Table 1. Basic properties of state-of-the-art blazar propagation codes. A detailed comparison of the codes in this table for blazar hadronic models was presented by Cerruti et al. [40], with the exception of CRPropa code.

Code	AM^3 Gao et al. [43]	PARIS Cerruti et al. [41]	ATHEνA Dimitrakoudis et al. [42]	Böttcher Böttcher et al. [8]	CRPropa Hörbe et al. [19]
Transport equation	yes	yes	yes	yes	yes
Ballistic	no	no	no	no	yes
Steady state	yes	yes	yes	yes	yes
Time dependent	yes	no	yes	no	yes
magnetic-field	turbulent (isotropic)	turbulent (isotropic)	turbulent (isotropic)	turbulent (isotropic)	turbulent (isotropic), regular (helical)
Diffusion	1-dim	1-dim	1-dim	1-dim	3-dim
Photohadron	yes	yes	yes	yes	yes
Hadron-hadron	no	no	no	no	yes

A new code for propagation of particles in relativistic plasmoids has been developed recently [19]. This new framework has been derived from the public transport code CR-Propa 3.1 [45]. The advantage with this numerical approach is that ballistic propagation can be performed in a test particle approach, thus not relying on the simplifying assumption of time-scales. Further, a second transport framework is integrated in CRPropa 3.1, which is the solution of the transport equation via the stochastic differential equations (SDEs) approach. This approach enables us to solve the transport equation via pseudo-particle propagation, which makes it compatible to be used in the ballistic test-particle propagation of CRPropa. The SDE approach is designed to include a full diffusion tensor, which is also an improvement when compared to other codes. A CRPropa modification presented in Hörbe et al. [19] makes use of the modular structure of CRPropa to create a propagation environment of a plasmoid traveling along the jet axis. The photon field of a thin accretion disk is implemented at the foot of the jet for gamma–gamma and proton–gamma interactions. Technically, the propagation is done in the reference frame of the plasmoid and then transferred into the observer's frame. The plasmoid itself contains a plasma with a constant density n_{plasma}, which is considered as a target for cosmic-ray interactions as well. The magnetic field, in which the particles propagate, was assumed to be of purely turbulent nature of Kolmogorov type in [19]. Due to the modular structure of the code, it can easily be changed to include a regular component as well, to change the nature of the turbulence, etc.

In this paper, the spotlight is put on the propagation regime in the plasmoids of blazars. In Section 2, the energy, at which a transition between diffusive and ballistic propagation happens and the consequences such a transition has for the description of SEDs and lightcurves of blazars are quantified. In Section 3, the influence of a first phase of ballistic propagation before the limit of diffusion is reached in time is investigated and the necessity to go from a diffusive approach to the description via the telegraph equation is discussed. In Section 4, first test simulations to investigate a possible difference in the flaring behavior in the diffusive vs. ballistic description is performed. Conclusions and outlook are given in Section 5.

2. The Space-Domain: Diffusive vs. Ballistic Propagation

As discussed above, propagation of charged particles in a turbulent (plus regular) magnetic field can be of fundamentally different nature depending on the astrophysical setting, in particular concerning the parameters of the particle energy, E, the ratio, $\delta B/B$, of the turbulent to regular magnetic field, and the correlation length, l_c, of the field as the lower boundary for the deterministic description of the magnetic field lines. Reichherzer et al. [46] quantified five propagation regimes with respect to the particle's reduced rigidity $\rho = r_g/l_c$, with $r_g = E/(c q B)$ as the relativistic gyro radius and $l_c \approx l_{\text{max}}/5$ as the correlation length. Here, $l_{\text{max}} = 2\pi/k_{\text{min}}$ is the maximum scale of the magnetic turbulence spectrum, connected to the lowest wave number k_{min}, defined by the turbulence injection scale. Diffusive propagation corresponds to the resonant-scattering regime

(RSR). This regime is only valid for particles that can scatter with the entire spectrum of wavelengths $k_{min} < k < k_{max}$, where $k_{max} = 2\pi/l_{min}$ is the dissipation scale. The general scheme of resonant scattering then breaks down toward the lowest and highest reduced rigidities. At the lower boundary, when scattering does not happen with the entire angular spectrum anymore, mirroring occurs more and more often, altering the diffusion coefficient. It is expected that this regime is not relevant for particles in AGN plasmoids, as the gyro radius of $\sim$TeV-PeV particles that are considered here in magnetic fields of $\sim$G strength fulfills the boundary condition $\rho > l_{min}/(\pi l_c \, \delta B/B)$ [46]. The situation is different toward large reduced rigidities, for which particles propagate in a *quasi-ballistic* way: for gyro radii that start to reach the correlation length of the system, meaning for plasmoids also coming closer to the actual size of the system, only a few gyrations are performed by the particles before leaving the source. This is happening close to the Hillas limit of the source. It is clear that the number of gyrations then does not suffice for a diffusive description.

The quasi-ballistic regime becomes relevant at a reduced rigidity of $\rho = r_g/l_c \gtrsim 5/(2\pi)$ [46]. Inserting the relativistic gyro radius, the energy at which the ballistic regime becomes dominant is given as

$$E \gtrsim \frac{5}{2\pi} \cdot l_c \cdot c \cdot q \cdot B. \tag{1}$$

For a given parameter set of magnetic field strengths, coherence lengths of the turbulence, and particle energies, Equation (1) can be applied to determine in what regimes the particles are propagating in typical astrophysical sources of cosmic rays [47]. Normalized to a standard set of parameters, the equation becomes

$$E \gtrsim Z \cdot \left(\frac{l_c}{10^{11}\,\mathrm{m}}\right) \cdot \left(\frac{B}{0.42\,\mathrm{G}}\right) \cdot 10^{15}\,\mathrm{eV}. \tag{2}$$

This implies that for protons (the atomic number $Z = 1$) in a source region with a parameter set $l_c = 10^{11}$ m and $B = 0.42$ G, diffusive propagation is happening below energies of 10^{15} eV, ballistic propagation needs to be applied above 10^{15} eV. For other parameter combinations, this transition energy can be calculated accordingly, always with ballistic propagation above the energy, diffusive transport below.

Figure 2 shows the energy limits for protons as a function of the product $B \cdot l_c$. The grey shaded area illustrates diffusive and the blue area ballistic propagation, with the transition between the resonant scattering regime and the quasi-ballistic regime indicated as the solid line in between, following Equation (1). The area of ballistic propagation in blue is bounded by the maximum possible proton energies according to the Hillas-Limit in each parameter space. The horizontal lines indicate the energies for the knee (10^{15} eV), ankle ($10^{18.5}$ eV), and maximum observed energy (10^{20} eV).

The parameter space covered by the plasmoids is approximated to be in the range 10^{10} m $< l_c < 10^{14}$ m. This range is based on the assumption that the plasmoids have a radius on the order of $R \sim 10^{12}$–10^{16} m, using a correlation length of $l_c = 0.01 \cdot R$ as described above. As the plasmoids are launched at the foot of the jet, magnetic fields are large, on the order of 10^{-3} G $< B < 10$ G. What we want to understand is how the propagation of particles needs to be performed to describe the multimessenger emission from AGN in the plasmoid-model. The energy range of interest for high-energy photons reaches from GeV energies up to approximately 10^{16} eV, neutrino detection happens in an energy range corresponding to proton energies of approximately $2 \cdot 10^{13}$ eV to 10^{17} eV. Figure 2 shows the relevant parameter space, displayed as $B \cdot l_c$ on the x-axis, a fraction will be diffusive (grey area) at lower energies. The high-energy part before reaching the Hillas limit (colored, thick line) needs to be performed in the ballistic limit (blue area). A first extreme example is a combination $B \cdot l_c = 10^8$ G m, where diffusive propagation happens up to $\sim 10^{12.5}$ eV, ballistic propagation up to the Hillas limit at around $10^{14.5}$ eV. These would be sources with a relatively low acceleration limit, as the combination of R and B only allows for maximum energies below the knee. More realistic parameter combinations that would allow the sources to reach the maximum observed energy would be a combination of $B \cdot l_c = 10^{14}$ G m. In this case, diffusive propagation needs to be performed up to $10^{18.5}$ eV, ballistic propagation needs to be performed up to the Hillas limit at $10^{20.5}$ eV.

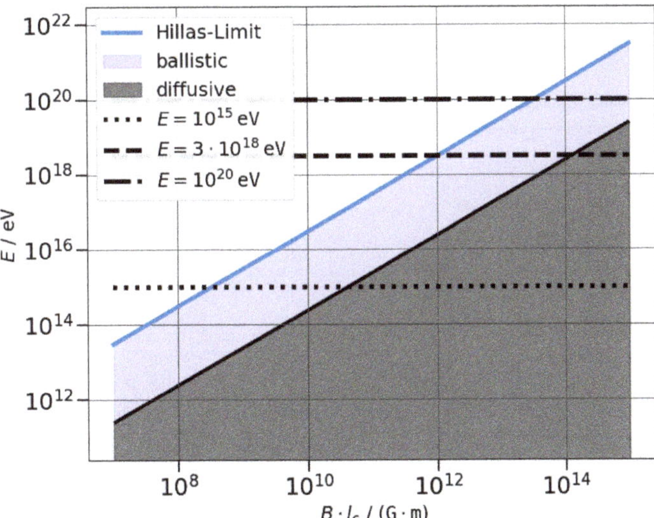

Figure 2. Illustration of the transition between the quasi-ballistic regime (in blue) and the resonant scattering regime (in grey) in the dependence of the magnetic field strength, B, correlation length of the magnetic field, l_c, and particle energy, E. In a simplified approach, it is assumed here that particles in the quasi-ballistic regime propagate ballistically and in the resonant scattering regime diffusively. The position of the diagonal line represents the transition energy from diffusive (below) to ballistic (above) for protons (the atomic number $Z = 1$), determined by Equation (1). Diffusive propagation is expected for the result of a parameter combination below the line, while ballistic propagation lies above the line.

This result has immediate consequences on the observed energy spectra: a break in the energy behavior of the timescales applied in simplified transport equation approaches, where the term $-n/\tau_{\mathrm{esc}}$ describes the escape, is expected to be observed: For ballistic transport, the timescale needs to be chosen as a constant value, $\tau_{\mathrm{esc}}^{\mathrm{ballistic}} = R/c$, while it becomes energy dependent in the case of diffusive propagation, $\tau_{\mathrm{esc}}^{\mathrm{diffusive}} = R^2/\kappa \propto \sqrt{R}\,E^{-\delta}$, with $\kappa = \kappa_0 \cdot E^\delta$ as the diffusion coefficient, for which the energy dependence can be approximated as a power-law behavior with an index δ that depends on the underlying magnetic turbulence, in this description being in the limit of quasi-linear theory, i.e., $\delta B/B \ll 1$. Such a change in the timescale directly induces a change in the shape of the spectral energy distribution: applying the leaky box model, the emitted proton spectrum follows approximately $n(E) \sim Q(E) \cdot \tau_{\mathrm{esc}}(E)$. Those secondary photons and neutrinos that are induced by (photo-)hadronic interactions basically mirror that behavior in the energy range above the threshold for the process, so that even these are in first approximation proportional to the escape time and the primary injection spectrum $Q(E)$, $n_{\gamma,\nu} \propto Q(E) \cdot \tau_{\mathrm{esc}}$. Assuming an injection spectrum $Q(E) \propto E^{-2.3}$ and Kolmogorov-type turbulence, $\tau_{\mathrm{esc}}^{\mathrm{diffusive}} \propto E^{-0.3}$, the SED is expected to behave as

$$n_{\gamma,\nu}(E) \propto \begin{cases} E_{\gamma,\nu}^{-2.6} & E < E_{\mathrm{transition}}^{\gamma,\nu} \\ E_{\gamma,\nu}^{-2.3} & E > E_{\mathrm{transition}}^{\gamma,\nu} \end{cases}. \tag{3}$$

That means for a typical parameter combination of $l_c = 10^{11}$ m and $B = 0.42$ G, the transition energy for protons is $E_{\mathrm{transition}}^{\mathrm{CR}} = 10^{15}$ eV. This translates (see, e.g., [1]) into a transition energy for photons of $E^\gamma_\text{transition} \approx 1/10 \cdot E^\text{CR}_\text{transition} \approx 10^{14}$ eV. These 100 TeV photons are currently barely accessible for gamma-ray telescopes, and so the propagation in a purely diffusive regime is reasonable. Purely ballistic transport, however, leads to a spectrum that is too flat, as the steepening by the diffusive escape timescale due to the diffusion tensor is neglected.

For neutrinos, the transition energy is $E^\nu_\text{transition} \approx 1/20 \cdot E^\text{CR}_\text{transition} \approx 5 \cdot 10^{13}$ eV. As the neutrinos detected by IceCube are in the energy range 10 TeV to a few PeV, this transition region can fall right into the relevant parameter range, so that a combination of ballistic and diffusive propagation needs to be considered. A break in the observed neutrino energy spectrum from a steeper to a flatter behavior is therefore expected in such a scenario. If such a break is observed, it can be used to estimate the parameter combination $B \cdot l_c$. To summarize the above result in the context of the propagation of particles in the plasmoids of blazars, it is shown that it is of high importance to evaluate the transport regime and adjust it accordingly to the problem under consideration in order to receive reliable results.

3. The Time-Domain

The result from the previous section applies to steady-state sources with $\delta n/\delta t \approx 0$, where it is implicitly assumed that all particles have already had the time to reach a steady-state limit in their propagation; here, t denotes the time. However, blazars are highly variable objects and individual flares are often modeled by injecting a high-energy particle population on short timescales. Particle acceleration time-scales in reconnection events responsible for the blob creation in relativistic sources are on order of $\tau_\text{acc} \sim E/(qBc^2)$, which is typically much shorter than the escape timescales (see, e.g., [48]), motivating a two-zone scenario where acceleration is typically performed in a first step, propagation afterwards.

While there is scientific consensus that the description of the transport process of particles in turbulent fields is ensured by the general concept, discussed in Section 2 in the limit of infinitely large times, the question arises under which conditions and on which timescale such a limit consideration is appropriate. In this section, criteria are derived for which, in a given plasmoid setup, the diffusive approximation still holds.

The general problem of assuming diffusive propagation during the initial propagation process, for which the diffusive limit is not yet granted, is expressed in the following points:

- The solution of the diffusion equation results in a Gaussian spatial distribution of the particles in the plasmoid. However, especially in the initial transport phase, this has the consequence that the particles are granted a non-vanishing probability of reaching positions in the plasmoid, which they cannot reach at the initial time due to their finite speed.
- Numerical simulations show a linear increase of the running diffusion coefficient, caused by ballistic particle trajectories until a constant value is reached that is known as the final diffusion coefficient κ and used within the numerical and theoretical computations of diffusive transport. The discrepancy between the final diffusion coefficient and the actual running diffusion coefficient approaches zero as the propagation length increases, but is significant at the beginning.

3.1. Timescale for Transition to Diffusive Propagation

Whereas the consideration of particle transport via the diffusion equation and its solution of a Gaussian particle distribution cannot distinguish between the initial, ballistic propagation and the subsequent diffusive propagation, the telegraph equation has recently been attributed this ability [49–52]:

$$\frac{\partial f}{\partial t} + \tau \frac{\partial^2 f}{\partial t^2} = \kappa \left(\frac{\partial^2 f}{\partial x^2} + \frac{\partial^2 f}{\partial y^2} + \frac{\partial^2 f}{\partial z^2} \right), \tag{4}$$

where x, y and z are space coordinates.

The telegraph timescale τ describes the transition between these two propagation phases. This timescale enables us to make a statement about when the diffusive phase is established and when the description of the particle transport via the diffusion equation is sufficiently accurate. If the initial ballistic phase is neglected, τ disappears and the telegraph equation turns into the known diffusion equation. By neglecting adiabatic focusing, the telegraph timescale yields [49]:

$$\tau = \frac{3v}{8\lambda_{\parallel}} \int_{-1}^{1} d\mu \left(\int_{0}^{\mu} d\mu' \frac{1 - \mu'^2}{D_{\mu\mu}(\mu')} \right)^2 , \tag{5}$$

with μ being the cosine of the pitch angle and $D_{\mu\mu}$ being the pitch-angle Fokker-Planck coefficient that, in a negligible background field ($\delta B \gg B_0$), yields [53]

$$D_{\mu\mu} = (1 - \mu^2)D. \tag{6}$$

Here, D is the pitch-angle Fokker–Planck coefficient at $90°$. The mean-free path $\lambda_{\parallel}$ is defined as

$$\lambda_{\parallel} = \frac{3v}{8} \int_{-1}^{1} d\mu \frac{(1 - \mu^2)^2}{D_{\mu\mu}(\mu)} , \tag{7}$$

where v denotes the particle velocity.

3.2. Quantifying the Time Needed to Achieve Certain Levels of Diffusivity

Since the diffusion equation assumes diffusive transport at all times and, furthermore, the solution uses the final diffusion coefficient, κ, over all timescales, the norm N of the solution remains constant in time:

$$N = 4\pi \int_{0}^{\infty} dr\, r^2 f_{\mathrm{diff}}(r) = 4\pi \int_{0}^{\infty} dr\, \frac{r^2}{(4\pi t \kappa)^{3/2}} \exp\left(-\frac{r^2}{4\kappa t} \right) = 1. \tag{8}$$

The norm may be interpreted as the fraction of particles participating in diffusion [52]. On the other hand, the solution of the isotropic telegraph equation yields:

$$f_{\mathrm{telegraph}}(r, t) = \frac{e^{-t/2\tau}}{4\pi\kappa^{3/2}} \left[\frac{\delta\left(t - r\sqrt{\tau/\kappa}\right)}{r/\sqrt{\kappa}} I_0\left(\frac{1}{2}\sqrt{\frac{t^2}{\tau^2} - \frac{r^2}{\kappa\tau}} \right) \right.$$
$$\left. + \frac{\Theta\left(t/\sqrt{\tau} - r/\sqrt{\kappa}\right)}{2\tau^{3/2}\sqrt{\frac{t^2}{\tau^2} - \frac{r^2}{\kappa\tau}}} I_1\left(\frac{1}{2}\sqrt{\frac{t^2}{\tau^2} - \frac{r^2}{\kappa\tau}} \right) \right]. \tag{9}$$

Here, $\delta(\ldots)$ is the Dirac's delta function, $\Theta(\ldots)$ is the Heaviside step function, and $I_\nu(\ldots)$ is the modified Bessel function. The norm can be computed individually on each of the two terms of the function:

$$N = \int f_{\mathrm{telegraph}}(r, t)\, dr = \int \frac{e^{-t/2\tau}}{4\pi\kappa^{3/2}} \left[\frac{\delta\left(t - r\sqrt{\tau/\kappa}\right)}{r/\sqrt{\kappa}} I_0\left(\frac{1}{2}\sqrt{\frac{t^2}{\tau^2} - \frac{r^2}{\kappa\tau}} \right) \right] dr$$
$$+ \int \frac{e^{-t/2\tau}}{4\pi\kappa^{3/2}} \left[\frac{\Theta\left(t/\sqrt{\tau} - r/\sqrt{\kappa}\right)}{2\tau^{3/2}\sqrt{\frac{t^2}{\tau^2} - \frac{r^2}{\kappa\tau}}} I_1\left(\frac{1}{2}\sqrt{\frac{t^2}{\tau^2} - \frac{r^2}{\kappa\tau}} \right) \right] dr. \tag{10}$$

The delta function simplifies the first part to t/τ and the second part can be solved by employing Bessel function integration rules. The details of the calculation are omitted. The norm yields

$$N = 1 - \exp\left(-\frac{t}{\tau}\right). \tag{11}$$

In this scenario, no particles are diffusive at the beginning. With time, the number of diffusive particles increases exponentially until the value for large propagation times approaches the maximum value with all particles in a diffusive state.

Rearranging the equation leads to a calculation rule for the propagation time $t_{\mathrm{diff},N}$ required to establish a certain diffusion level:

$$t_{\mathrm{diff},N} = -\ln\left(1 - N\right)\tau. \tag{12}$$

The relation (11) is shown in Figure 3 in comparison with the constant number of diffusing particles in the case of modeling the transport with the diffusion equation. The initial phase of a flare of cosmic rays is therefore in a non-diffusive state until the steady-state limit is reached as shown in Figure 3. Note that the derivations made here for a purely turbulent field also apply to the generalized case of an additional directional magnetic field component, since the timescales for reaching the diffusive phase during transport parallel and perpendicular to the directional background field are identical for a large parameter space [54].

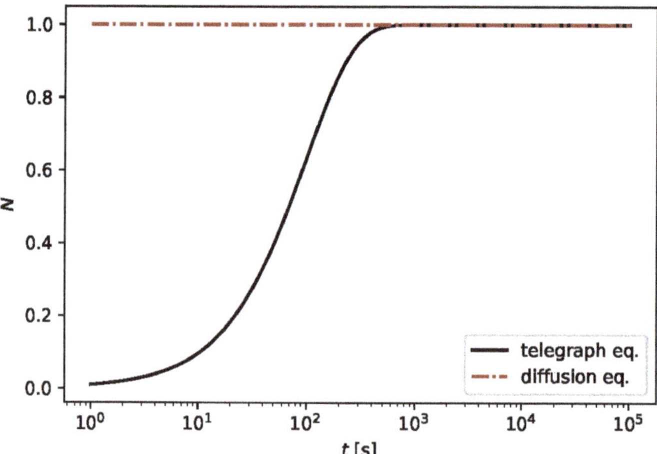

Figure 3. Comparison of the time evolution of the ratio of particles that are already diffusively propagating on average. The solution of the diffusion equation leads to the fact that particles always propagate diffusively. The solution of the telegraph equation shows an increase in the diffusively propagating particles with time and an approach to the maximum value.

In Section 4, this scenario is evaluated for the conditions in a plasmoid.

3.3. Conditions for Plasmoid Settings

In what follows, the critical time to reach diffusive propagation is expressed as a function of the blob properties and the particle energy. The resulting estimates give an overview of the expected type of propagation for special parameter combinations. For this purpose, let us start with the definition of the timescale connected to the mean free path, λ, of the particle,

$$\tau = \frac{\lambda}{v} = \frac{3\kappa}{v^2},$$ (13)

where $\lambda = 3\kappa/v$ is expressed as a function of the diffusion coefficient. Here, the case of isotropic turbulence without background field is considered in which case Bohm diffusion to be applied with a linear energy dependence in the resonant-scattering regime, $\kappa \propto E$. In the quasi-ballistic regime, the diffusion coefficient becomes $\kappa \propto E^2$. The transition between the two regimes is given at a reduced rigidity of $\rho \approx 5/(2\pi)$, so that the diffusion coefficient can be written as

$$\kappa = \kappa_0 \left(\frac{\rho}{\rho_0}\right)^{\epsilon} \text{ with } \begin{cases} \epsilon = 1 & \text{for } \rho \lesssim 5/(2\pi), \\ \epsilon = 2 & \text{for } \rho \gg 5/(2\pi), \end{cases}$$ (14)

where κ_0 and ρ_0 are normalisation constants and $\rho_0 = 5/(2\pi)$ is chosen, so that κ_0 becomes the value of the diffusion coefficient at the chosen value of ρ_0.

It follows for the timescale:

$$\tau = \frac{3\kappa_0}{v^2} \left(\frac{\rho}{\rho_0}\right)^{\epsilon} \text{ with } \begin{cases} \epsilon = 1 & \text{for } \rho \lesssim 5/(2\pi), \\ \epsilon = 2 & \text{for } \rho \gg 5/(2\pi), \end{cases}$$ (15)

finally leading to an expression for the propagation time required to reach a certain diffusion level N (from Equation (12))

$$t_{\text{diff},N} = -\ln(1-N)\frac{3\kappa_0}{v^2}\left(\frac{2\pi\rho}{5}\right)^{\epsilon} \text{ with } \begin{cases} \epsilon = 1 & \text{for } \rho \lesssim 5/(2\pi), \\ \epsilon = 2 & \text{for } \rho \gg 5/(2\pi). \end{cases}$$ (16)

By inserting the definition of the reduced rigidity, this relation can be expressed as functions of E, B and l_c:

$$t_{\text{diff},N} = -\ln(1-N)\frac{3\kappa_0}{v^2}\left(\frac{2\pi E}{5qcBl_c}\right)^{\epsilon} \text{ with } \begin{cases} \epsilon = 1 & \text{for } \rho \lesssim 5/(2\pi), \\ \epsilon = 2 & \text{for } \rho \gg 5/(2\pi). \end{cases}$$ (17)

The time $t_{\text{diff},N}$ needed for the fraction N of the particles to be diffusive depends on the parameters E, B and l_c. Figure 4 shows this condition for different plasmoid parameters. The figure shows the influence of the particle energy, the magnetic field properties, and the trajectory on the fraction of already diffusively propagating particles. The vertical lines indicate the timescales required for particles on ballistic trajectories to travel one plasmoid radius. This timescale has the same order as typical escape times of charged particles during initial ballistic propagation. If there is not yet a significant fraction of diffusive particles at the vertical lines for the respective plasmoid radii, the particles must be considered completely ballistic. For example, charged particles with $E = 1\,\text{TeV}$, $B = 1\,\text{G}$, and $l_c = 10^{10}\,\text{m}$ can be treated diffusively in plasmoids with $R = 10^{16}\,\text{m}$, but must be treated via equation-of-motion approaches at smaller radii such as $R = 10^{14}\,\text{m}$, $R = 10^{12}\,\text{m}$, and $R = 10^{10}\,\text{m}$.

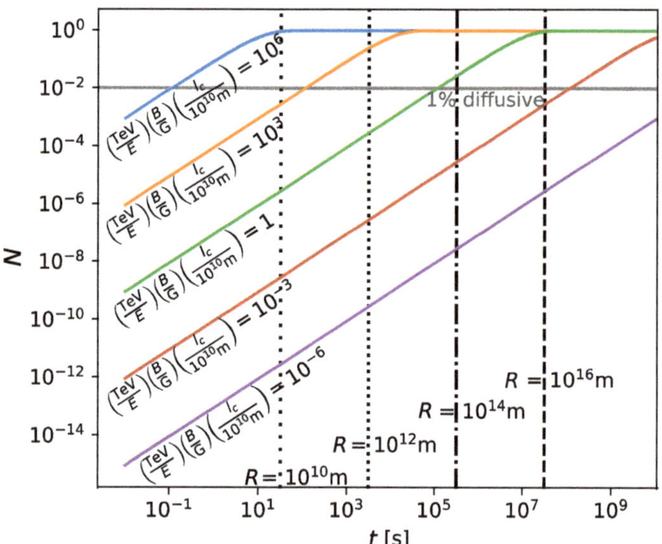

Figure 4. Fraction of particles, N, that are diffusive as a function of propagation time, t, for different blob parameters and particle energies. The vertical lines illustrate the time required for ballistic particle propagation to traverse the respective blob radii.

4. Simulations: Ballistic vs. Diffusive Simulation Results

In this Section, the effect of ballistic vs. diffusive propagation is investigated by performing simulations of cosmic-ray transport in the plasmoid of is applied an AGN traveling along the jet axis. Here, interactions with photon and gas targets are considered to be switched off in order to focus on the effects coming from cosmic-ray propagation, but otherwise follow the procedure described in [19]. This is done as a first step to understand the influence of the change in the propagation regime, so that in future studies, one can calculate the neutrino and gamma-ray spectra, adding the effect of hadronic cascades. This allows us to disentangle the two effects of the change in the flare due to propagation vs. interaction. For now, focus is on the effect of the change in propagation regime, future work will then also quantify changes through interactions. The parameter set used in the simulation is summarized in Table 2. In addition, not only simulations with the equation-of-motion [55] are performed, but diffusive propagation applied by using the module DiffusionSDE in CRPropa 3.1 [45], which solves the transport equation with a diffusion term $\kappa \Delta n$. In order to have a quantitative comparison, first, one needs to calculate the diffusion coefficient κ as an input to the diffusive simulation from the ballistic part. This is done here for energies from 10^5 GeV up to 10^8 GeV. Here, the Taylor Green Kubo (TGK) formalism is applied as described in [46] (see also references therein). The simulation parameters from Table 2 are chosen to minimize numerical errors such as the interpolation effect of turbulence [56,57].

Table 2. Simulation parameters, given in the rest frame of the plasmoid. All simulations are performed in a modified version of CRPropa 3.1 as presented in Hörbe et al. in [19] with further additions made for this paper as described above.

Parameter	Value
Proton energy, E_p	10^5 GeV–10^8 GeV
Plasmoid radius, R	10^{13} m
Plasmoid Lorentz factor, Γ	10
Magnetic field: Initial root-mean-square (RMS) value B_0	1 G
Magnetic field: Turbulence & spectral index α	Kolmogorov-type, $\alpha = -5/3$
Magnetic field: Correlation length, l_c	$(2/3) \cdot 10^{11}$ m
Magnetic field: Grid points	$(512)^3$
Magnetic field: Spacing, Δs_G	$R/(256 \cdot 32 \cdot 2)$
Magnetic field: l_{min}	$2\Delta s_G$
Propagation module (CRPropa intern): Ballistic	PropagationBP
Propagation module (CRPropa intern): Diffusive	DiffusionSDE
Propagation step size	$10^{-3}R$

Figure 5 shows the result of the running diffusion coefficient $\kappa(t)$. For low energies, i.e., 10^5 GeV (brown), $10^{5.5}$ GeV (purple), 10^6 GeV (red), and $10^{6.5}$ GeV (green), particles reach a plateau after an initial ballistic phase as described in Section 3, which can be used as the steady-state diffusion coefficient. For larger energies (10^7 GeV, orange, and 10^8 GeV, blue), such a convergence is not observed. The reason is that the particles leave the plasmoid before being able to reach a steady-state diffusion limit. Therefore, the first three values of the steady-state diffusion coefficient are used to determine the energy dependence $\kappa(E)$.

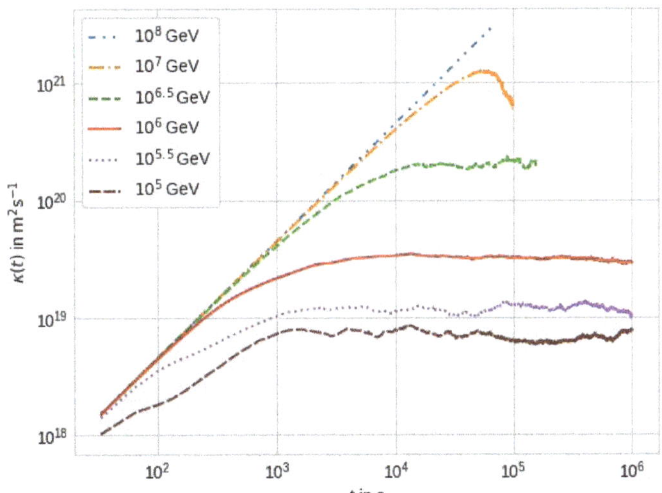

Figure 5. Running diffusion coefficient, κ, for energies 10^5 GeV to 10^8 GeV. A plateau is built up for energies between 10^5 GeV $< E < 10^6$ GeV. Toward higher energies, the coefficient breaks off as the particles leave the plasmoid before they can reach the steady-state diffusion coefficient. This is consistent with the calculated energy for a transition between a ballistic and diffusive behavior at 10^6 GeV.

In Figure 6, the values averaged from the plateau in Figure 5 are shown with the corresponding error bars. In the simulation setup, the magnetic field is a purely turbulent one. That means Bohm diffusion is at work, and the energy dependence is expected to be $\kappa = \kappa_0 \cdot (E/\text{GeV})$; see, e.g., [12]. We therefore perform a linear regression and find $\kappa_0 = 10^{13.54 \pm 0.06}$. For the simulations considered, the calculated values are used for the three energies, where this was possible in a reliable way (10^5 GeV to 10^6 GeV). For larger values, the result from the linear regression is used to determine the diffusion coefficient. From the findings of Section 2, one expects the ballistic and diffusive flares to provide approximately the same results until the transition energy is reached according to Equation (1), for the set of parameters used (see Table 2); this happens at around 10^6 GeV. From the results of Section 3, a deviation between the diffusive and ballistic approach is expected to be largest at small times and the two approaches should converge toward large times, when the steady-state diffusion coefficient is reached. As can be seen from Figure 5, this effect is energy dependent and for low energies (10^5 GeV), the diffusive steady-state is reached at around 10^3 s, while it takes $\gg 10^4$ s for the highest energies ($E > 10^7$ GeV).

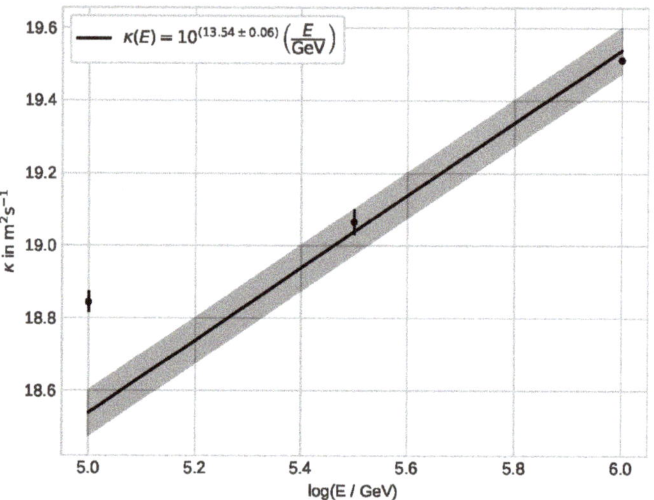

Figure 6. Steady-state diffusion coefficient as a function of energy for particles between 10^5 GeV and 10^6 GeV. A linear regression for the function $\kappa(E) = \kappa_0 \cdot (E/\mathrm{GeV})$ is performed, with the uncertainty band depicted as gray shaded area. Here, κ_0 is the normalisation constant. The linear behavior with energy is based on the assumption that Bohm diffusion is dominant in the purely turbulent field.

Figure 7 shows the flaring behavior for a monochromatic energy flare at $E = 10^5$ GeV. The diffusive description shows an especially large enhancement with respect to the equation-of-motion approach at early times below 10^3 s, in accordance with the findings of Section 3. The behavior of $\mathrm{d}N/\mathrm{d}t$ is similar for both approaches, with a small shift that can be explained by the uncertainties in the numerical determination of the diffusion coefficient used here for the diffusive approach. In the diffusive transport regime for a constant diffusion coefficient, one expects $\langle \Delta x \rangle \propto t^{1/2}$. This results in the differential particle number $\mathrm{d}N/\mathrm{d}t \propto t^{-1/2}$ of escaping particles. Note that, due to the steady escape, the decrease in the number of remaining particles in the plasmoid leads to a strong cut-off at large times. Since, in the diffusive approach, more particles initially leave the plasmoid, the particle density in the plasmoid is lower than in the equation-of-motion approach, so that an earlier cut-off is visible.

Figure 8 shows the flare for diffusive and equation-of-motion behavior at 10^8 GeV. Here, one obtains a visible difference between the two flares, with the diffusive approach yielding a dominant contribution at early times. In contrast to the diffusive regime with $\mathrm{d}N/\mathrm{d}t \propto t^{-1/2}$, one expects a constant differential particle number for the ballistic transport regime with $\langle \Delta x \rangle \propto t$. The initial slight drop for the equation-of-motion may be explained by statistical deviations from the initial homogeneous particle distribution in the plasmoid, especially when slightly more particles are in the outer spheres of the plasmoid at $t = 0$. Since, in the diffusive case, significantly more particles initially leave the plasmoid, the particle density in the plasmoid is much lower than in the equation-of-motion approach, so that a cut-off is visible much earlier in the diffusive approach. Thus, these test simulations emphasize the importance of propagating the particles in the proper transport regime. Only a thorough analysis of the transport properties will lead to a prediction that can be compared to the observation of non-thermal emission from blazars.

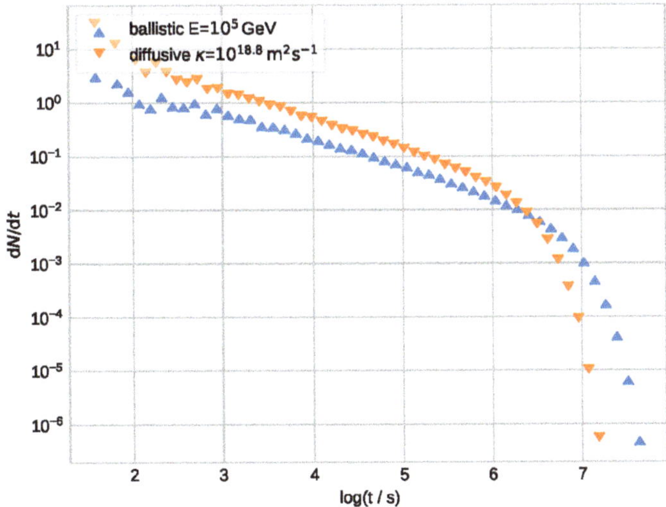

Figure 7. Cosmic-ray flare ($E = 10^5$ GeV) from a blazar in the diffusive propagation model (orange downward triangle) and in the ballistic propagation model (blue upward triangle) as differential particle number per unit time, dN/dt, over time. The total number of particles injected into the simulation is $N_{inj} = 10^5$.

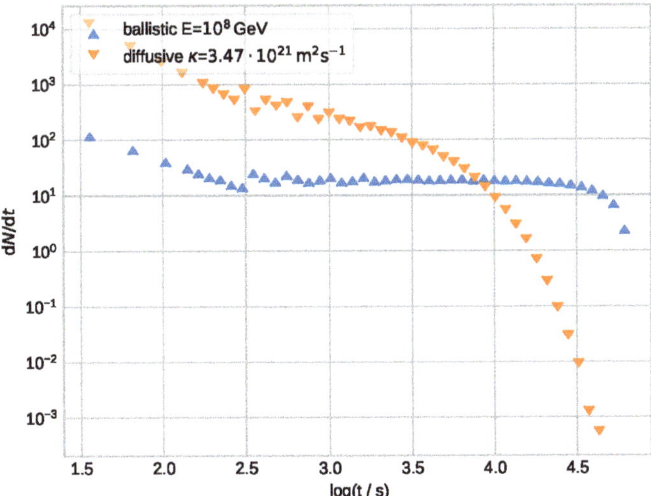

Figure 8. Cosmic-ray flare ($E = 10^8$ GeV) from a blazar in the diffusive propagation model (orange downward triangle) and in the ballistic propagation model (blue upward triangle) as differential particle number per unit time dN/dt over time. The number of injected particles is $N_{inj} = 10^5$.

5. Conclusions

In this paper, the propagation regimes in plasmoids of blazars are investigated as sources of high-energy cosmic rays, which in turn become emitters of high-energy gamma-rays and neutrinos. To explain the spectral energy distributions and lightcurves of this high-energy emission, it is shown that it is necessary to distinguish between the different energy and time regimes of ballistic and diffusive transport. At early times and at high energies, the particles are still in the ballistic regime. At late times or in scenarios, for which the injection of high-energy cosmic rays is significantly longer than the characteristic timescale $\tau \gg R/c$, with R being rthe radius of plasmoid and c the speed of light, the diffusive approach needs to be applied. The details of this transport modeling have an impact on both the spectral energy distribution, and on the temporal evolution of a flare. For the energy behavior, the diffusive part of the spectrum is steepened by the diffusion timescale, which is dominated by the diffusion coefficient. The ballistic part, on the other hand, is connected to an energy-independent escape timescale, thus leading to an emission spectrum close to the acceleration spectrum. When looking at the flaring behavior, it is shown that diffusive approach and transport with the equation-of-motion approach yield about the same result at low energies around 10^5 GeV, where the diffusive approach is accurate at times above $\sim 10^3$ s. That means that if the equation-of-motion approach is performed with the correct parameter setting, the same result is expected at times larger than 10^3 s, which is confirmed within a factor of ~ 2. When approximating this behavior with an escape timescale, the diffusive timescale needs to be applied.

At large energies (10^8 GeV), it can be shown that the diffusive and equation-of-motion approaches lead to quite different flaring behaviors. Here, only the equation-of-motion approach yields a correct result, as it can reproduce the ballistic behavior. It can be approximated in a transport equation approach by applying an energy-independent (ballistic) escape time.

As for the observation of neutrinos, we predict that spectral energy distributions of blazars like TXS0506+056 should have a break in the spectrum at a neutrino energy around $5 \cdot 10^{13}$ eV. If blazars are responsible for the diffuse neutrino emission, a break at the same energy is predicted. For the gamma-ray and neutrino lightcurves of blazars like TXS0506+056, one would expect the behavior of the variability time to be affected by the effect of a ballistic behavior at early times that turns into a diffusive behavior at later times. In particular, when simulating these flares, the shape of the flare will be misinterpreted when using the diffusion approximation at the highest energies, because it breaks down as the particles escape faster than the diffusion time scale.

Author Contributions: Conceptualization, J.B.T.; methodology, J.B.T., M.H., I.J., P.R., W.R., M.S., F.S.; writing—original draft preparation, J.B.T., P.R., I.J.; writing—review and editing, J.B.T., M.H., I.J., P.R., W.R., M.S., F.S.; visualization, P.R., I.J., M.S.; supervision, J.B.T., F.S.; funding acquisition, J.B.T., W.R., F.S. All authors have read and agreed to the published version of the manuscript.

Funding: This research was funded by the German Science Foundation DFG via the Collaborative Research Center SFB1491 "Cosmic Interacting Matters—From Source to Signal". Further funding was received from the DFG via the Grant "Multi-messenger probe of Cosmic Ray Origins (MICRO)", Grant numbers TJ 62/8-1. We would also like to thank the Research Department for Plasmas with Complex Interactions for support.

Data Availability Statement: Data presented in this article can be made available upon request.

Acknowledgments: J.B.T. and W.R. would like to use this opportunity to thank Reinhard Schlickeiser for the long, pleasant journey through physics, administration, and the Ruhr area during the past decades. We hope the journey will continue for long-even if its path might shift in its character. We would like to thank Rainer Grauer for discussions on the plasma physics of blazars, particularly concerning the launching and evolution of plasmoids and their plasma parameters. We also thank Imre Bartos, Peter Biermann, Anna Franckowiak, Francis Halzen, Emma Kun, and Walter Winter for discussions on the modeling of non-thermal blazar emission

Conflicts of Interest: The authors declare no conflict of interest.

References

1. Becker, J.K. High-energy neutrinos in the context of multimessenger astrophysics. *Phys. Rep.* **2008**, *458*, 173–246. [CrossRef]
2. Biermann, P.L.; Becker, J.K.; Caramete, L.; Curuţiu, A.; Engel, R.; Falcke, H.; Gergely, L.Á.; Isar, P.G.; Mariş, I.C.; Meli, A.; et al. Active Galactic Nuclei: Sources for ultra high energy cosmic rays? *Nucl. Phys. Proc. Suppl.* **2009**, *190*, 61–78. [CrossRef]
3. Biermann, P.L.; Strittmatter, P.A. Synchrotron Emission from Shock Waves in Active Galactic Nuclei. *Astrophys. J.* **1987**, *322*, 643. [CrossRef]
4. IceCube Collaboration. Evidence for high-energy extraterrestrial neutrinos at the IceCube detector. *Science* **2013**, *342*, 1242856. [CrossRef]
5. Reynolds, C.; Fabian, A.; Celotti, A.; Rees, M.J. The matter content of the jet in M87: Evidence for an electron—Positronjet. *Mon. Not. R. Astron. Soc.* **1996**, *283*, 873–880. [CrossRef]
6. Wardle, J.; Homan, D.; Ojha, R.; Roberts, D. Electron–positron jets associated with the quasar 3C279. *Nature* **1998**, *395*, 457. [CrossRef]
7. Potter, W.J.; Cotter, G. Synchrotron and inverse-Compton emission from blazar jets–I. A uniform conical jet model. *Mon. Not. R. Astron. Soc.* **2012**, *423*, 756–765. [CrossRef]
8. Böttcher, M.; Reimer, A.; Sweeney, K.; Prakash, A. Leptonic and hadronic modeling of Fermi-detected blazars. *Astrophys. J.* **2013**, *768*, 54. [CrossRef]
9. Becker, J.K.; Biermann, P.L.; Rhode, W. The diffuse neutrino flux from FR-II radio galaxies and blazars: A source property based estimate. *Astropart. Phys.* **2005**, *23*, 355–368. [CrossRef]
10. Becker Tjus, J.; Eichmann, B.; Halzen, F.; Kheirandish, A.; Saba, S.M. High-energy neutrinos from radio galaxies. *Phys. Rev. D* **2014**, *89*, 123005. [CrossRef]
11. Murase, K.; Dermer, C.D.; Takami, H.; Migliori, G. Blazars as ultra-high-energy cosmic-ray sources: Implications for TeV gamma-ray observations. *Astrophys. J.* **2012**, *749*, 63. [CrossRef]
12. Becker Tjus, J.; Merten, L. Closing in on the origin of Galactic cosmic rays using multimessenger information. *Phys. Rep.* **2020**, *872*, 1–98. [CrossRef]
13. Aartsen, M.G. et al. [IceCube Collaboration] Multimessenger observations of a flaring blazar coincident with high-energy neutrino IceCube-170922A. *Science* **2018**, *361*, eaat1378. [CrossRef]
14. Aartsen, M.G. et al. [IceCube Collaboration] Neutrino emission from the direction of the blazar TXS 0506+056 prior to the IceCube-170922A alert. *Science* **2018**, *361*, 147–151. [CrossRef]
15. Halzen, F.; Kheirandish, A.; Weisgarber, T.; Wakely, S.P. On the neutrino flares from the direction of TXS 0506+056. *Astrophys. J. Lett.* **2019**, *874*, L9. [CrossRef]
16. Gao, S.; Fedynitch, A.; Winter, W.; Pohl, M. Modelling the coincident observation of a high-energy neutrino and a bright blazar flare. *Nat. Astron.* **2019**, *3*, 88–92. [CrossRef]
17. Rodrigues, X.; Gao, S.; Fedynitch, A.; Palladino, A.; Winter, W. Leptohadronic Blazar Models Applied to the 2014–2015 Flare of TXS 0506+056. *Astrophys. J. Lett.* **2019**, *874*, L29. [CrossRef]
18. de Bruijn, O.; Bartos, I.; Biermann, P.L.; Becker Tjus, J. Recurrent neutrino emission from supermassive black hole mergers. *Astrophys. J. Lett.* **2020**, *905*, L13. [CrossRef]
19. Hörbe, M.R.; Morris, P.J.; Cotter, G.; Becker Tjus, J. On the relative importance of hadronic emission processes along the jet axis of active galactic nuclei. *Mon. Not. R. Astron. Soc.* **2020**, *496*, 2885–2901. [CrossRef]
20. Kun, E.; Bartos, I.; Tjus, J.B.; Biermann, P.L.; Halzen, F.; Mező, G. Cosmic neutrinos from temporarily gamma-suppressed blazars. *Astrophys. J. Lett.* **2021**, *911*, L18. [CrossRef]
21. Kadler, M.; Krauß, F.; Mannheim, K.; Ojha, R.; Müller, C.; Schulz, R.; Anton, G.; Baumgartner, W.; Beuchert, T.; Buson, S.; et al. Coincidence of a high-fluence blazar outburst with a PeV-energy neutrino event. *Nat. Phys.* **2016**, *12*, 807–814. [CrossRef]
22. Franckowiak, A.; Garrappa, S.; Paliya, V.; Shappee, B.; Stein, R.; Strotjohann, N.L.; Kowalski, M.; Buson, S.; Kiehlmann, S.; Max-Moerbeck, W.; et al. Patterns in the Multiwavelength Behavior of Candidate Neutrino Blazars. *Astrophys. J.* **2020**, *893*, 162. [CrossRef]
23. Giommi, P.; Glauch, T.; Padovani, P.; Resconi, E.; Turcati, A.; Chang, Y.L. Dissecting the regions around IceCube high-energy neutrinos: Growing evidence for the blazar connection. *Mon. Not. R. Astron. Soc.* **2020**, *497*, 865–878. [CrossRef]
24. Murase, K.; Guetta, D.; Ahlers, M. Hidden cosmic-ray accelerators as an origin of TeV-PeV cosmic neutrinos. *Phys. Rev. Lett.* **2016**, *116*, 071101. [CrossRef]
25. Ahlers, M.; Halzen, F. High-energy cosmic neutrino puzzle: A review. *Rep. Prog. Phys.* **2015**, *78*, 126901. [CrossRef] [PubMed]
26. Brodatzki, K.A.; Pardy, D.J.S.; Becker, J.K.; Schlickeiser, R. Internal $\gamma\gamma$ Opacity in active galactic nuclei and the consequences for the TeV observations of M87 and Cen A. *Astrophys. J.* **2011**, *736*, 98. [CrossRef]
27. Eichmann, B.; Schlickeiser, R.; Rhode, W. Differences of leptonic and hadronic radiation production in flaring blazars. *Astrophys. J.* **2012**, *749*, 155. [CrossRef]
28. Giannios, D. Reconnection-driven plasmoids in blazars: Fast flares on a slow envelope. *Mon. Not. R. Astron. Soc.* **2013**, *431*, 355–363. [CrossRef]
29. Morris, P.J.; Potter, W.J.; Cotter, G. The feasibility of magnetic reconnection powered blazar flares from synchrotron self-Compton emission. *Mon. Not. R. Astron. Soc.* **2019**, *486*, 1548–1562. [CrossRef]
30. Rees, M.J. Appearance of relativistically expanding radio sources. *Nature* **1966**, *211*, 468–470. [CrossRef]

31. van der Laan, H. A Model for Variable Extragalactic Radio Sources. *Nature* **1966**, *211*, 1131–1133. [CrossRef]
32. Lyutikov, M.; Komissarov, S.; Sironi, L.; Porth, O. Particle acceleration in explosive relativistic reconnection events and Crab Nebula gamma-ray flares. *J. Plasma Phys.* **2018**, *84*, 635840201. [CrossRef]
33. Sironi, L.; Spitkovsky, A. Relativistic reconnection: An efficient source of non-thermal particles. *Astrophys. J.* **2014**, *783*, L21. [CrossRef]
34. Christie, I.; Petropoulou, M.; Sironi, L.; Giannios, D. Radiative signatures of plasmoid-dominated reconnection in blazar jets. *Mon. Not. R. Astron. Soc.* **2018**, *482*, 65–82. [CrossRef]
35. Keivani, A.; Murase, K.; Petropoulou, M.; Fox, D.B.; Cenko, S.; Chaty, S.; Coleiro, A.; DeLaunay, J.; Dimitrakoudis, S.; Evans, P.; et al. A multimessenger picture of the flaring blazar TXS 0506+ 056: Implications for high-energy neutrino emission and cosmic-ray acceleration. *Astrophys. J.* **2018**, *864*, 84. [CrossRef]
36. Dar, A.; Laor, A. Hadronic production of TeV gamma-ray flares from blazars. *Astrophys. J. Lett* **1997**, *478*, L5. [CrossRef]
37. Araudo, A.T.; Bosch-Ramon, V.; Romero, G.E. Gamma rays from cloud penetration at the base of AGN jets. *Astron. Astrophys.* **2010**, *522*, A97. [CrossRef]
38. Zacharias, M.; Böttcher, M.; Jankowsky, F.; Lenain, J.P.; Wagner, S.; Wierzcholska, A. The extended flare in CTA 102 in 2016 and 2017 within a hadronic model through cloud ablation by the relativistic jet. *Astrophys. J.* **2019**, *871*, 19. [CrossRef]
39. Gergely, L.Á.; Biermann, P.L. The spin-flip phenomenon in supermassive black hole binary mergers. *Astrophys. J.* **2009**, *697*, 1621. [CrossRef]
40. Cerruti, M.; Kreter, M.; Petropoulou, M.; Rudolph, A.; Oikonomou, F.; Böttcher, M.; Dimitrakoudis, S.; Dmytriiev, A.; Gao, S.; Mastichiadis, A.; et al. The blazar hadronic code comparison project. *arXiv* **2021**, arXiv:2107.06377. [CrossRef]
41. Cerruti, M.; Zech, A.; Boisson, C.; Inoue, S. A hadronic origin for ultra-high-frequency-peaked BL Lac objects. *Mon. Not. R. Astron. Soc.* **2015**, *448*, 910–927. [CrossRef]
42. Dimitrakoudis, S.; Mastichiadis, A.; Protheroe, R.J.; Reimer, A. The time-dependent one-zone hadronic model. First principles. *Astron. Astrophys.* **2012**, *546*, A120. [CrossRef]
43. Gao, S.; Pohl, M.; Winter, W. On the Direct correlation between gamma-rays and PeV neutrinos from blazars. *Astrophys. J.* **2017**, *843*, 109. [CrossRef]
44. Rodrigues, X.; Fedynitch, A.; Gao, S.; Boncioli, D.; Winter, W. Neutrinos and ultra-high-energy cosmic-ray nuclei from blazars. *Astrophys. J.* **2018**, *854*, 54. [CrossRef]
45. Merten, L.; Becker Tjus, J.; Fichtner, H.; Eichmann, B.; Sigl, G. CRPropa 3.1—A low energy extension based on stochastic differential equations. *J. Cosmol. Astropart. Phys.* **2017**, *2017*, 046. [CrossRef]
46. Reichherzer, P.; Becker Tjus, J.; Zweibel, E.G.; Merten, L.; Pueschel, M.J. Turbulence-level dependence of cosmic ray parallel diffusion. *Mon. Not. R. Astron. Soc.* **2020**, *498*, 5051–5064. [CrossRef]
47. Reichherzer, P.; Becker Tjus, J.; Hörbe, M.; Jaroschewski, I.; Rhode, W.; Schroller, M.; Schüssler, F. Cosmic-ray transport in blazars: Diffusive or ballistic propagation? *arXiv* **2021**, arXiv:2107.11386. [CrossRef]
48. del Valle, M.V.; de Gouveia Dal Pino, E.M.; Kowal, G. Properties of the first-order Fermi acceleration in fast magnetic reconnection driven by turbulence in collisional magnetohydrodynamical flows. *Mon. Not. R. Astron. Soc.* **2016**, *463*, 4331–4343. [CrossRef]
49. Litvinenko, Y.E.; Schlickeiser, R. The telegraph equation for cosmic-ray transport with weak adiabatic focusing. *Astron. Astrophys* **2013**, *554*, A59. [CrossRef]
50. Litvinenko, Y.E.; Noble, P.L. A Numerical study of diffusive cosmic-ray transport with adiabatic focusing. *Astrophys. J.* **2013**, *765*, 31. [CrossRef]
51. Litvinenko, Y.E.; Effenberger, F.; Schlickeiser, R. The telegraph approximation for focused cosmic-ray transport in the presence of boundaries. *Astrophys. J.* **2015**, *806*, 217. [CrossRef]
52. Tautz, R.C.; Lerche, I. Application of the three-dimensional telegraph equation to cosmic-ray transport. *Res. Astron. Astrophys.* **2016**, *16*, 162. [CrossRef]
53. Shalchi, A.; Skoda, T.; Tautz, R.C.; Schlickeiser, R. Analytical description of nonlinear cosmic ray scattering: Isotropic and quasilinear regimes of pitch-angle diffusion. *Astron. Astrophys.* **2009**, *507*, 589–597. [CrossRef]
54. Reichherzer, P.; Becker Tjus, J.; Zweibel, E.G.; Merten, L.; Pueschel, M.J. Anisotropic cosmic-ray diffusion in isotropic Kolmogorov turbulence. *arXiv* **2021**, arXiv:2112.11827. [CrossRef]
55. Alves Batista, R.; Dundovic, A.; Erdmann, M.; Kampert, K.H.; Kuempel, D.; Müller, G.; Sigl, G.; van Vliet, A.; Walz, D.; Winchen, T. CRPropa 3—A public astrophysical simulation framework for propagating extraterrestrial ultra-high energy particles. *J. Cosmol. Astropart. Phys.* **2016**, *2016*, 38. [CrossRef]
56. Schlegel, L.; Frie, A.; Eichmann, B.; Reichherzer, P.; Becker Tjus, J. Interpolation of turbulent magnetic fields and its consequences on cosmic ray propagation. *Astrophys. J.* **2020**, *889*, 123. [CrossRef]
57. Reichherzer, P.; Merten, L.; Dörner, J.; Becker Tjus, J.; Pueschel, M.J.; Zweibel, E.G. Regimes of cosmic-ray diffusion in Galactic turbulence. *SN Appl. Sci.* **2022**, *4*, 1–14. [CrossRef]

MDPI
St. Alban-Anlage 66
4052 Basel
Switzerland
www.mdpi.com

Physics Editorial Office
E-mail: physics@mdpi.com
www.mdpi.com/journal/physics